Supramolecular Gold Chemistry

Supramolecular Gold Chemistry

From Atomically Precise Thiolate-Protected Gold Nanoclusters to Gold-Thiolate Nanostructures

Special Issue Editor

Rodolphe Antoine

MDPI • Basel • Beijing • Wuhan • Barcelona • Belgrade

Special Issue Editor
Rodolphe Antoine
Institut Lumière Matière,
CNRS and University of Lyon
France

Editorial Office
MDPI
St. Alban-Anlage 66
4052 Basel, Switzerland

This is a reprint of articles from the Special Issue published online in the open access journal *Nanomaterials* (ISSN 2079-4991) from 2018 to 2020 (available at: https://www.mdpi.com/journal/nanomaterials/special_issues/gold_thiolate).

For citation purposes, cite each article independently as indicated on the article page online and as indicated below:

LastName, A.A.; LastName, B.B.; LastName, C.C. Article Title. *Journal Name* **Year**, *Article Number*, Page Range.

ISBN 978-3-3-03928-550-1 (Pbk)
ISBN 978-3-3-03928-551-8 (PDF)

Contents

About the Special Issue Editor

Rodolphe Antoine received his Ph.D. in Molecular Physics from The University of Lyon (Michel Broyer). He was a postdoctoral researcher at the Swiss Federal Institute of Technology Lausanne, EPFL in nonlinear optics at interfaces (Hubert H. Girault). His background spans atomic and molecular physics, laser spectroscopy, and physical chemistry. He has broad, multi-disciplinary interests in both experimental and computational avenues of research related to nanoclusters. He is currently a research group leader focusing on the structure and dynamics of proteins, nanoclusters, and nanoparticles at the Institut Lumière Matière at the University of Lyon and CNRS.

 *nanomaterials*

Editorial

Supramolecular Gold Chemistry: From Atomically Precise Thiolate-Protected Gold Nanoclusters to Gold-Thiolate Nanostructures

Rodolphe Antoine

Institut Lumière Matière UMR 5306, Université Claude Bernard Lyon 1, CNRS, Univ Lyon, F-69100 Villeurbanne, France; rodolphe.antoine@univ-lyon1.fr; Tel.: +33-(0)4-7243-1085

Received: 14 February 2020; Accepted: 18 February 2020; Published: 21 February 2020

Supramolecular chemistry is defined as chemistry beyond the molecule. The supramolecular chemistry of gold and, in particular, of gold metalloligands leads to fascinating structural motifs with enhanced optical properties, as well as innovative catalytic activity. [1] The formation of a gold–sulfur bond is the driving force for the anchoring of thiol ligands on gold surfaces, as exemplified from self-assembled monolayers to nanoclusters (NCs) and nanoparticles (NPs) [2].

The chemistry of the gold–sulfur bond is extremely rich and leads to hybrid materials. Such materials encompass gold thiolate coordination oligomers, for instance $[Au(I)(SR)]_n$, where SR stands for a chemical group containing a sulfur atom, and atomically well-defined clusters $[Au_nSR_m]$, or supramolecular assemblies like $Au(I)(SR)$ coordination polymers. While the majority of gold atoms in the nanoparticles are in the $Au(0)$ state, under strong reducing conditions, gold atoms in supramolecular assemblies, like $Au(I)(SR)$ coordination polymeric NPs, are in the gold(I) state. In atomically well-defined clusters of $[Au_nSR_m]$ stoichiometry, the subtle balance between the $Au(0)$ core and the $Au(I)$–SR shell leads to fascinating material properties and, in particular, to highly tunable optical properties.

The aim of this Special Issue on "Supramolecular Gold Chemistry" was to provide a unique international forum aimed at covering a broad description of results involving the supramolecular chemistry of gold with a special focus on the gold–sulfur interface leading to hybrid materials, ranging from gold–thiolate complexes, [3] to thiolate-protected gold nanoclusters [4–11] and gold–thiolate supramolecular assemblies or nanoparticles. [12–14] The role of thiolates on the structure and optical features of gold nanohybrid systems (ranging from plasmonic gold nanoparticles and fluorescent gold nanoclusters to self-assembled Au-containing thiolated coordination polymers) has been highlighted in the review article by Csapó and coworkers [14].

For gold–thiolate complexes and thiolate-protected gold nanoclusters, the atomically precise nature of their structures enables the elucidation of structure–property relationships, an essential step in their rational design for enhanced performances. From a theoretical point of view, the geometry of the clusters must be determined by quantum chemistry methods, and the optical responses described in terms of molecular transitions whose positions and intensities are predicted by sophisticated calculations of quantum mechanics. Bonačić-Koutecký and coworkers pioneered this concept and reported, in the early 1990s, the absorption spectra obtained with first-principle methods for the most stable structures of small bare metal clusters and nicely illustrated the molecular-like behavior of clusters, leading to an electronic energy quantization and changes in the leading features of the patterns as functions of the cluster sizes [15].

Structural characterization of nanoclusters is an active area of research and X-ray single-crystal diffraction has been the most straightforward and important technique in the structural determination of nanocluster nanomaterials in order to understand their structure–property relationships [16]. Not always applicable for nanoclusters, alternative approaches are to be explored. Separation techniques (liquid chromatography, gas phase ion mobility) can help in discriminating and characterizing structures. In this Special Issue, Antoine and coworkers combine an ion mobility-mass spectrometry approach with density functional theory (DFT) calculations for the determination of the structural and optical properties of gold thiolate oligomers ($Au_{10}(TGA)_{10}$ with TGA: thioglycolic acid) [3]. Whetten and coworkers combine electrospray ionization with high-performance liquid chromatography mass spectrometry (HPLC-MS) to separate and identify 3-MBA (MBA: mercaptobenzoic acid) protected gold nanoclusters, spanning a narrow size range from 13.4 to 18.1 kDa [5]. Theoretical investigations are also useful for structural characterization. Cheng and coworkers theoretically investigate $Au_{70}S_{20}(PPh_3)_{12}$ using density functional theory calculations. The electronic and geometric structure of $Au_{70}S_{20}(PPh_3)_{12}$ is further addressed based on the popular divide and protect concept and the superatom network model [7].

The discrete electronic states of nanoclusters cause molecular-like behavior, leading to fascinating physical–chemical properties, such as luminescence, magnetism, and catalysis, etc. Jin and coworkers highlight this molecular-like behavior by thoroughly exploring the differences in the photophysical properties of small organic molecules, gold–thiolate complexes, nanoclusters, and metallic-state nanoparticles [8]. The luminescence properties of 6-aza-2-thio-thymine stabilized gold nanoclusters [9] and gold thiolate coordination polymers [12] demonstrate the high potential of such nanomaterials for bio-sensing or lighting devices. However, in such nanosystems, the origin of photoluminescence (PL) is still not fully understood. Zhang and coworkers review some general PL mechanisms, from the pure metal-centered quantum confinement mechanism to the ligand-to-metal charge mechanism, as well as introducing a new paradigm, such as the ligand-centered p band intermediate state model [11]. On the other hand, gold nanoclusters have been proposed as a new, promising class of model catalyst [17]. Li and cowokers [6] and Negishi and coworkers [10] nicely review some interesting aspects of nanocluster catalysis for heterogeneous cross-coupling and for energy conversion.

Finally, synthetic routes are at the heart of supramolecular gold chemistry. Innovative strategies include a metal exchange reaction that leads to a new cluster compound in particular alloy nanoclusters. Zhu and coworkers describe a new type of metal exchange: self-alloying induced by intramolecular metal exchange, to produce the $Ag_xAu_{25-x}(SR)_{18}^-$ nanocluster [4]. Moreover, new synthetic routes, beyond wet chemistry using a reducing agent, are being explored. Pulsed laser ablation in liquids is such a new method, in which a solid target immersed in liquid is irradiated with a suitable pulsed laser beam. Kalarikkal and coworkers use this approach for the generation of 2D nanocomposites composed by gold nanoparticles and graphene oxide nanosheets [13]. Such new nanocomposites present remarkable chemical sensing for thiolates.

To conclude this overview on the papers published in the Special Issue "Supramolecular Gold Chemistry: From Atomically Precise Thiolate-Protected Gold Nanoclusters to Gold-Thiolate Nanostructures", I am confident that the readers will enjoy these contributions and may be able to find inspiration for their own research within this Special Issue.

Funding: This research received no external funding.

Acknowledgments: I am grateful to all the authors for submitting their studies to the present Special Issue and for its successful completion. I deeply acknowledge the Nanomaterials reviewers for enhancing the quality and impact of all submitted papers. Finally, I sincerely and warmly thank Melia Wang and the editorial staff of Nanomaterials for their stunning support during the development and publication of the Special Issue. Moreover, the project STIM—REI, Contract Number: KK.01.1.1.01.0003, funded by the European Union through the European Regional Development Fund—the Operational Programme Competitiveness and Cohesion 2014-2020 (KK.01.1.1.01) is gratefully acknowledged.

Conflicts of Interest: The author declares no conflict of interest.

References

1. Gil-Rubio, J.; Vicente, J. The coordination and supramolecular chemistry of gold metalloligands. *Chem. Eur. J.* **2018**, *24*, 32–46. [CrossRef]

2. Bürgi, T. Properties of the gold–sulphur interface: From self-assembled monolayers to clusters. *Nanoscale* **2015**, *7*, 15553–15567. [CrossRef]

3. Comby-Zerbino, C.; Perić, M.; Bertorelle, F.; Chirot, F.; Dugourd, P.; Bonačić-Koutecký, V.; Antoine, R. Catenane Structures of Homoleptic Thioglycolic Acid-Protected Gold Nanoclusters Evidenced by Ion Mobility-Mass Spectrometry and DFT Calculations. *Nanomaterials* **2019**, *9*, 457. [CrossRef] [PubMed]

4. Li, Y.; Chen, M.; Wang, S.; Zhu, M. Intramolecular metal exchange reaction promoted by thiol ligands. *Nanomaterials* **2018**, *8*, 1070. [CrossRef] [PubMed]

5. Black, D.M.; Hoque, M.M.; Plascencia-Villa, G.; Whetten, R.L. New Evidence of the Bidentate Binding Mode in 3-MBA Protected Gold Clusters: Analysis of Aqueous 13–18 kDa Gold-Thiolate Clusters by HPLC-ESI-MS Reveals Special Compositions Aun (3-MBA) p,(n = 48–67, p = 26–30). *Nanomaterials* **2019**, *9*, 1303. [CrossRef] [PubMed]

6. Shi, Q.; Qin, Z.; Xu, H.; Li, G. Heterogeneous Cross-Coupling over Gold Nanoclusters. *Nanomaterials* **2019**, *9*, 838. [CrossRef] [PubMed]

7. Tian, Z.; Xu, Y.; Cheng, L. New Perspectives on the Electronic and Geometric Structure of Au70S20 (PPh3) 12 Cluster: Superatomic-Network Core Protected by Novel Au12 (μ3-S) 10 Staple Motifs. *Nanomaterials* **2019**, *9*, 1132. [CrossRef] [PubMed]

8. Zhou, M.; Zeng, C.; Li, Q.; Higaki, T.; Jin, R. Gold Nanoclusters: Bridging Gold Complexes and Plasmonic Nanoparticles in Photophysical Properties. *Nanomaterials* **2019**, *9*, 933. [CrossRef] [PubMed]

9. Deng, H.-H.; Shi, X.-Q.; Balasubramanian, P.; Huang, K.-Y.; Xu, Y.-Y.; Huang, Z.-N.; Peng, H.-P.; Chen, W. 6-Aza-2-Thio-Thymine Stabilized Gold Nanoclusters as Photoluminescent Probe for Protein Detection. *Nanomaterials* **2020**, *10*, 281. [CrossRef]

10. Kawawaki, T.; Negishi, Y. Gold Nanoclusters as Electrocatalysts for Energy Conversion. *Nanomaterials* **2020**, *10*, 238. [CrossRef]

11. Yang, T.-Q.; Peng, B.; Shan, B.-Q.; Zong, Y.-X.; Jiang, J.-G.; Wu, P.; Zhang, K. Origin of the Photoluminescence of Metal Nanoclusters: From Metal-Centered Emission to Ligand-Centered Emission. *Nanomaterials* **2020**, *10*, 261. [CrossRef] [PubMed]

12. Veselska, O.; Guillou, N.; Ledoux, G.; Huang, C.-C.; Newell, K.D.; Elkaïm, E.; Fateeva, A.; Demessence, A. A New Lamellar Gold Thiolate Coordination Polymer, [Au (m SPhCO2H)] n, for the Formation of Luminescent Polymer Composites. *Nanomaterials* **2019**, *9*, 1408. [CrossRef] [PubMed]

13. Nancy, P.; Nair, A.K.; Antoine, R.; Thomas, S.; Kalarikkal, N. In Situ Decoration of Gold Nanoparticles on Graphene Oxide via Nanosecond Laser Ablation for Remarkable Chemical Sensing and Catalysis. *Nanomaterials* **2019**, *9*, 1201. [CrossRef] [PubMed]

14. Ungor, D.; Dékány, I.; Csapó, E. Reduction of tetrachloroaurate (Iii) ions with bioligands: Role of the thiol and amine functional groups on the structure and optical features of gold nanohybrid systems. *Nanomaterials* **2019**, *9*, 1229. [CrossRef] [PubMed]

15. Blanc, J.; Bonačić-Koutecký, V.; Broyer, M.; Chevaleyre, J.; Dugourd, P.; Koutecký, J.; Scheuch, C.; Wolf, J.P.; Wöste, L. Evolution of the electronic structure of lithium clusters between four and eight atoms. *J. Chem. Phys.* **1992**, *96*, 1793–1809. [CrossRef]

16. Jin, R.; Zeng, C.; Zhou, M.; Chen, Y. Atomically precise colloidal metal nanoclusters and nanoparticles: fundamentals and opportunities. *Chem. Rev.* **2016**, *116*, 10346–10413. [CrossRef] [PubMed]

17. Li, G.; Jin, R. Atomically precise gold nanoclusters as new model catalysts. *Acc. Chem. Res.* **2013**, *46*, 1749–1758. [CrossRef] [PubMed]

 *nanomaterials*

Review

Reduction of Tetrachloroaurate(III) Ions With Bioligands: Role of the Thiol and Amine Functional Groups on the Structure and Optical Features of Gold Nanohybrid Systems

Ditta Ungor [1], **Imre Dékány** [1] and **Edit Csapó** [1,2,*]

[1] Interdisciplinary Excellence Centre, Department of Physical Chemistry and Materials Science, University of Szeged, Rerrich B. square 1, H-6720 Szeged, Hungary
[2] MTA-SZTE Biomimetic Systems Research Group, Department of Medical Chemistry, University of Szeged, Dóm square 8, H-6720 Szeged, Hungary
* Correspondence: juhaszne.csapo.edit@med.u-szeged.hu; Tel.: +36-62-544-476

Received: 23 July 2019; Accepted: 26 August 2019; Published: 29 August 2019

Abstract: In this review, the presentation of the synthetic routes of plasmonic gold nanoparticles (Au NPs), fluorescent gold nanoclusters (Au NCs), as well as self-assembled Au-containing thiolated coordination polymers (Au CPs) was highlighted. We exclusively emphasize the gold products that are synthesized by the spontaneous interaction of tetrachloroaurate(III) ions ($AuCl_4^-$) with bioligands using amine and thiolate derivatives, including mainly amino acids. The dominant role of the nature of the applied reducing molecules as well as the experimental conditions (concentration of the precursor metal ion, molar ratio of the $AuCl_4^-$ ions and biomolecules; pH, temperature, etc.) of the syntheses on the size and structure-dependent optical properties of these gold nanohybrid materials have been summarized. While using the same reducing and stabilizing biomolecules, the main differences on the preparation conditions of Au NPs, Au NCs, and Au CPs have been interpreted and the reducing capabilities of various amino acids and thiolates have been compared. Moreover, various fabrication routes of thiol-stabilized plasmonic Au NPs, as well as fluorescent Au NCs and self-assembled Au CPs have been presented via the formation of $-(Au(I)-SR)_n-$ periodic structures as intermediates.

Keywords: gold nanoparticles; gold nanoclusters; coordination polymer structure; amino acids; template-assisted synthesis; fluorescence; Au(I)-thiolate; gold nanohybrid materials

---

1. Introduction

Nowadays, the development of diverse nanostructured materials have a dominant role in several physical, chemical, medical, etc. fields from the electronics to the food industries [1,2]. The noble metal nanoparticles are extremely investigated nano-objects due to their electric, magnetic and unique morphology, size, and composition-dependent optical features [3,4]. This optical property originates from the so-called localized surface plasmon resonance (LSPR) phenomena, which results in the appearance of a characteristic plasmon band in the 400–800 nm range of the electromagnetic spectra [5,6]. In the last two-three decades, gold nanoparticles (Au NPs) have became increasingly the focus of interests in the material and medical sciences thanks to the advantageous physicochemical properties, such as large specific area, chemical inertness, and tunable optical particularity [7]. Several methods for fabrication of nano-sized Au NPs are known in the literature, including the physical (e.g., physical vapor deposition (PVD), microwave (MW) or ultraviolet (UV) radiation, ball milling or photoreductive routes, etc. [8,9]) and chemical approaches [3,4,10]. In the latter case, depending on the applied reducing and stabilizing agents (e.g., sodium borohydride [11,12], sodium citrate [13–15], surfactants [16,17], various amines [18], peptides [19,20], or biological organisms [21–23]), particles of different shapes and sizes can be produced. In the last decade, the sub-nanometer sized gold nanoclusters (Au NCs) have also became increasingly dominant. Beside the Au NPs, the Au NCs are also in the focus of researches. These ultra-small metal objects consist of only a few of few tens' gold atoms, and generally the oxidation number of the Au is < 1 and Au–Au bonds can be found in the clusters. By the mentioned structure, the Au NCs show unique size-tunable photoluminescence (PL) due to the well-defined molecular structure and discrete electronic transitions [24–26]. The blue-emitting Au NCs usually only contain a few atoms, thus the emission band depends only on the number of atoms in the cluster and the PL lifetime occurs in the nanosecond range. Nevertheless, if the size of the Au NCs achieves the few-nanometer range (d ~1.5–2.0 nm), the characteristic emission band is detected in the orange and in the red visible region. In this case, the surface ligand effect and the oxidation state of the surface metal atoms both influence the location of the emission maximum and the PL lifetime reaches the microsecond range. The larger colloidal Au NPs (d ~2–10 nm) possess weak PL, which is regulated by the surface roughness and the grain size effect [27]. Based on the above-mentioned structure-depending optical features, the sub-nanometer Au NCs can potentially be used as optical probes for biosensing, bio-labelling, and bioimaging applications [24,26,27].

The biomedical applications (cancer therapy, diagnostics, and bioimaging, etc.) of nano-sized functionalized Au particles/clusters require biocompatible preparation routes with mild reaction conditions. Nowadays, the practical one-step "green" preparation protocols of several water-soluble Au NPs/NCs are extremely preferred [21,28–30]. During these processes, mainly the template-assisted preparation approaches are used, where dominant amines, like simple amino acids [31], peptides or proteins [32,33], dendrimers [34,35], and nucleotides [36–39], are applied, which have simultaneously a dual role as reducing and stabilizing ligand. The amines are a crucial class of the possible reducing agents, because they can be found in biological and chemical atmospheres. Main advantages of this relatively simple template-directed reduction technique are that no additional reducing agent is required and based on the well-defined structure of polypeptides and proteins uniform NPs/NCs with tunable optical features can be synthesized. Besides amines, the thiol group-containing molecules (e.g., thiolates) can coordinate and reduce the Au ions at the same time to form periodic –(Au(I)-SR)$_n$– structures/complexes having partially reduced Au(I) ions, which are a well-known intermediates in the fabrication route of thiol-covered gold nanohybrid systems [40–43]. Several researches focus on the better understanding of the unknown structures of so-called atomically precise thiolate-protected Au NCs or the possible utilization of the thiolate-stabilized Au NPs/NCs [43–45]. In addition to the thiol-protected Au NPs/NCs, the study of the formation of Au-thiolate so-called "coordination polymer structure", having Au^0 or mostly Au(I) is in focus of interest. These coordination polymers (CPs) are inorganic-organic hybrid materials, which consist of periodic metal ions/atoms and ligand moieties and possess ordered structure. The self-assembly of this structure results in the formation of lamellar multilayers or helical structures with unique optical properties [41,46,47].

In recent work, we aim to provide an overview that is focused on the summary of the preparation routes, the unique structure, as well as the structure-dependent optical features of Au NPs, Au NCs, and Au CP structures that are synthesized by template-assisted synthesis exclusively using amines (mainly simple amino acids) and thiol-group containing molecules (e.g., thiolates) as possible reducing and stabilizing molecules. We mainly emphasize the formation of Au NPs, Au NCs, and Au CPs, which are fabricated by the direct interaction of tetrachloroaurate(III) ions (AuCl$_4^-$) with amino acids and alkyl- and arylthiolates in the absence of other reducing agents. We clearly summarize the dominant effect of the metal ion concentration, the molar ratio of the precursor aurate ions and reducing bioligands, as well as the experimental conditions (e.g., reaction time, temperature, pH, etc.) on the tunable, structure-dependent optical properties (plasmonic or fluorescence) of the Au nano-objects.

2. Preparation of Amino Acid-Reduced Colloidal Au NPs Having Plasmonic Property

There are several publications all around the world that describe the possible chemical synthesis routes of Au NPs in aqueous or in organic media. The well-known Brust method provides uniform alkyl or arylthiol-protected Au NPs (d = 1–5 nm) reduced by sodium borohydride ($NaBH_4$) in toluene [11], while in aqueous medium the conventional method is the Turkevich process, which results in the formation of water-soluble Au NPs in the range of 5–50 nm reduced and stabilized by sodium citrate [13]. In the last decade, various other reduction and caption possibilities were examined, where bacteria and microorganisms [48,49], plant extracts [50,51], inorganic reagents [52], metal complexes [53,54], organic and physiological molecules [55,56], polymers [57,58], liposomes [59], etc. have been tested. Due to the biocompatible nature, easy accessibility, and remarkable reducing capabilities, the amino acids and their derivatives are used dominantly [60] to produce biocompatible noble metal NPs. As far as we know, to date, all the twenty naturally occurring amino acids were investigated. In 2002, Mandal et al. published firstly the formation of Au NPs having spherical shape and monodisperse size distribution (d = 25 nm) by spontaneous interaction of $AuCl_4^-$ with L-aspartic acid (Asp) under boiling condition while using $AuCl_4^-$:Asp ca. 1:11 molar ratio [61]. Under the same experimental conditions, the synthesis was carried out with L-valine (Val) and L-lysine (Lys), but no reduction of $AuCl_4^-$ was observed and during preparation, the role of the pH was not mentioned. Next year, the reduction capability of Lys was studied again [62], but Au NPs in the range of 6–7 nm could only be prepared at room temperature by the application of extra $NaBH_4$ reductant as well. The hydrogen bonds between the surface-bound Lys molecules of the adjacent Au NPs was confirmed by NMR studies. Through the researches of Mandal, Selvakannan, and Sastry [63], L-tryptophan (Trp)-stabilized gold colloids was also efficiently fabricated. The synthesis was carried out at 50 °C while using $AuCl_4^-$:Trp ca. 1:100 molar ratio. ^{1}H NMR studies clearly indicated the indole-based polymerization of Trp, which contributed to the better understanding of the reduction process of Trp with $AuCl_4^-$ forming Au NPs under mild reaction conditions without application of other harsh reducing agents like $NaBH_4$. In 2005, Bhargava et al. summarized the successful fabrication of Au NPs by spontaneous interaction of potassium tetrabromoaurate(III) precursor ($KAuBr_4$) with L-tyrosine (Tyr) and L-arginine (Arg) at room temperature while using ca. 1:4 metal ion to amino acid molar ratios under alkaline medium [64]. For Tyr-reduced Au NPs having 5–40 nm in size, a slightly polydisperse distribution and coagulations of the NPs were observed. The Arg-produced colloidal NPs have larger size than the average diameter of Tyr-reduced particles, but the size distribution showed much narrower shape. The cyclic voltammetry (CV) studies of Blanchard et al. provided important information regarding the reduction abilities of various amines, including amino acids L-glycine (Gly) and Trp, as well as the proposed reduction mechanism between metal ions and bioligands [65]. Presumably, the reduction of aurate ions occurs thanks to the electron transfer from amines to the metal ions resulting in Au atoms with zero oxidation state and finally the nucleation and growth steps eventuates the formation of NPs. This redox reaction results in the appearance of short chain amine oligomers, which is confirmed by NMR studies. Moreover, the oxidation potential of amines, which are used for the reduction of gold ions, has outstanding impact on the formation of Au NPs considering the reduction potential of $AuCl_4^-$. Amines that have redox potential between the oxidation of Au^0 to gold(I) and the reduction of tetrachloroaurate(III) to Au^0 can be suitable used as reducing agents. L-Glutamic acid (Glu)-reduced Au colloids were also previously fabricated, having a particle size of d = 40 nm, but the synthesis was carried out under refluxing [66]. In 2010, the hydrothermal synthesis of the L-histidine (His)-reduced spherical Au NPs. The average diameter was 11.5 nm reported by Liu et al., where the $AuCl_4^-$:His/1:2.5 molar ratio was used at 150 °C in alkaline (pH 11.50) medium [67]. The structural characterization of His-protected Au NPs supported that the terminal COO^- group of His was not attached of the particle surface, while the imidazole as well as the amino groups were adsorbed on the Au surface. The construction of His-stabilized Au NPs did not occur at room temperature, but the hydrothermal conditions (e.g., high temperature and pressure) facilitate the formation of Au crystals. Besides the above-mentioned amino acids (Asp, Lys, Trp, Tyr, Glu, His), the reduction capabilities of L-aspartate

(Asp), Gly, L-leucine (Leu), Lys, and L-serine (Ser) were also published by the work of Cai et al. in 2014 [68], but they used extra UV irradiation during the synthesis. The different Au NPs have diameters of 15–47 nm and the synthesis was carried out at pH 10.0 while using 1:10/$AuCl_4^-$:amino acid molar ratios. Maruyama et al. studied the spontaneous interaction of each natural amino acids with aurate ions using high bioligand excess (metal ion to ligand ca. 1:100) at 80 °C, and they obtained that L-cysteine (Cys) and L-threonine (Thr) did not provide gold colloids. However, for L-methionine (Met) and L-phenylalanine (Phe), Au NPs were formed, but these colloids were easily precipitated. In 2014, L. Courrol and R. Almeida de Matos summarized their results in a book Chapter [69], where the formation of plasmonic Au colloids was confirmed by spontaneous interaction of aurate ions with Asp, Arg, Thr, Trp and Val using electromagnetic radiation (xenon lamp) at different pH using ca. 1:5 metal ion to amino acid molar ratios. However, the reduction capability of Trp was previously identified [70], but E. Csapó et al. clearly confirmed that the ratio of the precursor $AuCl_4^-$ and the bioligand greatly influences the optical feature of the formed colloids [71]. Using $AuCl_4^-$:Trp/1:0.4 molar ratio in alkaline medium (pH = 12.0), plasmonic Trp-Au NPs (λ_{abs} = 530 nm) were formed (Figure 1B). Based on the best of our belief, this work supported firstly that high ligand excess is no necessary for synthesizing Trp-reduced Au NPs at mild (37 °C) temperature. The presence of stable monodisperse Au NPs was confirmed by DLS (d_{DLS} = 8.8 ± 1.0 nm) and HRTEM (d_{HRTEM} = 7.8 ± 0.3 nm) studies. Moreover, depending on the applied molar ratios of the $AuCl_4^-$:Trp, structure-dependent tunable optical property was also obtained. Namely, at acidic conditions (pH = 1.0), in the case of the mixing of Trp and $AuCl_4^-$ solutions, the intensive yellow color of the solution changed to dark yellow after a few minutes. Below 1:1 ratio, unstable Au colloids was formed, but the application of molar ratio between $AuCl_4^-$:Trp/1:1 and 1:15 resulted in luminescent products. The appearance of the emission peak depends of the ligand excess, namely the maximum value can be detected at λ_{em} = 497 nm ($AuCl_4^-$:Trp/1:1), λ_{em} = 486 nm ($AuCl_4^-$:Trp/1:5), and λ_{em} = 472 nm ($AuCl_4^-$:Trp/1:15). The larger Trp amount causes the decrease of the PL intensities (Figure 1A). This characteristic PL originates from sub-nanometer sized Au nanoclusters (NCs). In the last 8–10 years, the Au NCs, which were synthesized by using template-assisted preparation routes, are in focus of extensive researches. A short summary of only the amino acid-reduced Au NCs is presented in the next chapter.

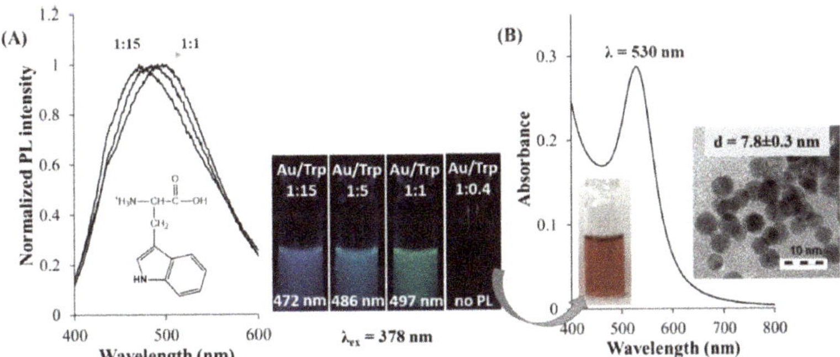

Figure 1. (**A**) The normalized fluorescence spectra (λ_{ex} = 378 nm) of L-tryptophan gold nanoclusters (Trp-Au NCs) with the photos of aqueous dispersions under UV-light. (**B**) Absorbance spectrum of L-tryptophan gold nanoparticles (Trp-Au NPs) with the HRTEM image. $c(AuCl_4^-)$ = 1.0 mM. Reproduced with permission from [71]. Elsevier, 2017.

3. Synthetic Routes of Amino Acid-Reduced Fluorescent Au NCs

Several preparation protocols for Au NCs having sizes less than 2 nm have been established in the last two decades, including both the "top-down" and "bottom-up" approaches, as Figure 2 summarizes [25,72,73].

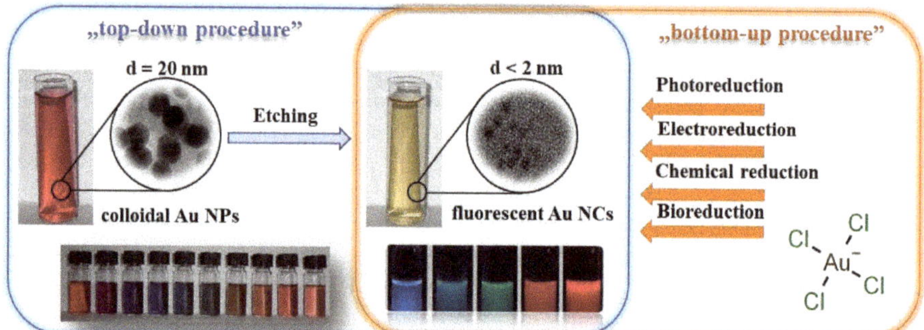

Figure 2. Preparation protocols of Au NCs by "top-down" and "bottom-up" approaches.

For the "top-down" process, the larger colloidal particles undergo so-called "etching" in order to produce smaller clusters, while in case of "bottom-up" methods, the clusters are formed via a reduction of the precursor ions by assembling individual atoms one-by-one [34,74]. The ultra-facile, one-step synthetic processes are in focus of interest, where the execution of the reactions is very convenient, rapid, and mild, exempted from the application of harsh reducing agent, special ambience and media, and high pressure. However, numerous articles were published for the preparation of biocompatible Au NCs that were synthesized by template-assisted preparation protocols while using proteins and peptides [75,76], polymers [77], DNA [78], dendrimers [79], etc., but only a few publications present the possible applicability of simple amino acids as reducing and stabilizing agents.

In this chapter, we clearly focus on the summary of the amino acid-directed fabrication of Au NCs having size-and structure-dependent intense PL features [80,81]. Table 1 clearly summarizes the experimental conditions of amino acid-reduced Au NCs and other Au-based nanohybrid structures. As it can be shown, His, Tyr, Pro, Trp, Cys, and Met amino acids were previously studied. Except for Cys and Met having thiol and thioether side chains, blue-emitting Au_3-Au_{10} NCs can be synthesized by the spontaneous interaction of $AuCl_4^-$ with His, Tyr, Pro, and Trp bioligands, depending on the temperature as well as on the ratio of reactant partners. In case of His, Au_{10} NCs with relatively high QY(%) are formed by using $AuCl_4^-$:amino acid/1:30 molar ratio at room temperature [82]. As Table 1 summarizes, various research groups fabricated His-reduced Au_{10} NCs while using almost the same experimental conditions, where the His-protected Au NCs have been applied for glutathione detection and selective cancer cell imaging [83], while Liu et al. also successfully used the His-Au NCs as ultrasensitive iodide detector system [84]. It can be concluded that, at room temperature, the application of high ligand excess (30-fold excess) results the formation of His-stabilized blue-emitting NCs. Moreover, E. Csapó et al. clearly confirmed that the pH is also a decisive factor during the synthesis in the case of the His/$AuCl_4^-$ system. However, Yang et al. [82] claimed that the emission intensity of the His-stabilized Au_{10} NCs was continually decreased with the increase of pH (from pH = 1.0 to 13.0) and the extreme acidic condition (pH = 1–2) is optimal for these NCs. In contrast with their results, E. Csapó et al. found that (Figure 3A), if the pH is smaller than pH = 5.0 no emission could be detected, but a characteristic emission peak with continually decreasing intensity to pH = 12.0 was evolved at 475 nm at above pH > 6 [71]. The emission maximum values show an interesting correlation with the concentration distribution curves of His. Namely, the emission maximum can be observed in that pH, where the deprotonation of the imidazolium moiety of His eventuates (pK_a = 6.04) [85].

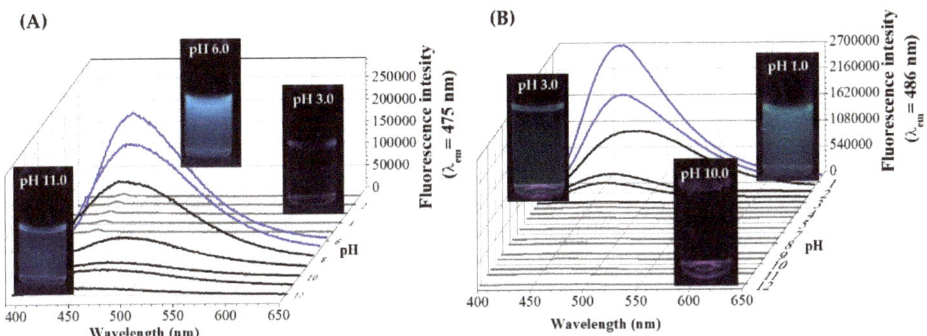

Figure 3. The photoluminescence (c) spectra as a function of the initial pH of the (**A**) $AuCl_4^-$:His/1:30 and (**B**) $AuCl_4^-$:Trp/1:5 systems with representative photos of the samples under UV-light. (λ_{ex} = 378 nm, c_{Au-} = 1.00 mM, T = 37 °C). Published in [71], Elsevier, 2017.

Most probably, the primary coordination of the gold ions to the His occurs via the imidazole-*N* atoms and this aromatic group plays a dominant role in the formation of the fluorescent Au products. Furthermore, it was found that, through the decrease in the concentration of the $AuCl_4^-$ ions from c_{Au} = 2.50 mM to c_{Au} = 1.00 mM, instead of clusters, the presence of blue-emitting polynuclear Au(I) complexes having a well-ordered structure is certifiable by several analytical methods [71].

For Tyr, no high ligand excess is necessary, but at room temperature, the spontaneous interaction of the Tyr with $AuCl_4^-$ ions does not result in the fabrication of Tyr-reduced Au NCs. At higher concentrations (c_{Au} = 2.50 mM), the lower temperature is enough (37 °C), but the boiling condition is essential as the concentration decreases (c_{Au} = 0.07 mM). In the case of Pro, which does not contain an aromatic group in the side chain, the use of extreme high ligand excess (more 100-fold excess) and boiling can result in the production of Au NCs having a few gold atoms. For Trp, the 37 °C and the 100 °C is optimal for the synthesis using from 1:1 to 1:5 $AuCl_4^-$:Trp molar ratio at acidic condition, as in Figure 3B, and the previously mentioned tunable optical feature was found, depending on the reactants ratio, which was summarized in chapter 2 in Figure 1A.

Table 1. Experimental conditions of amino acid-reduced Au NCs and Au nanostructures.

Amino Acid	c_{AuCl4} (mM)	$AuCl_4^-$:Amino Acid Ratio	T (°C)	Product	λ_{ex} (nm)	λ_{em} (nm) QY (%)	Ref.
His	2.50	1:30	25	Au_{10} NCs	386	490 (8.78%)	[82]
His	2.50	1:30	25	Au_{10}–Au_{14} NCs	370	475 (*no inf.*)	[86]
His	2.50	1:30	25	Au NCs *	386	475 (*no inf.*)	[83]
His	2.50	1:30	25	Au NCs *	365	450 (4.60%)	[84]
His	2.50	1:45	25	Au NCs *	386	498 (8.96%)	[87]
His	1.00	1:30	37	Au(I)-His CP	378	475 (3.60%)	[71]
Tyr	2.50	1:1.8	37	Au NCs *	385	470 (2.50%)	[88]
Tyr	0.07	1:0.76	100	Au_{10} NCs	383	498 (1.68%)	[89]
Pro	2.40	1:830	100	Au_7 NCs	365	440 (2.94%)	[90]
Trp	0.43	1:2.7	100	Au_8 NCs	365	450 (*no inf.*)	[91]
	0.50	1:1	37	Au_3–Au_6 NCs	378	497 (1.10%)	[71]
Trp	0.50	1:5	37	Au_3–Au_6 NCs	378	486 (1.30%)	[71]
	0.50	1:15	37	Au_3–Au_6 NCs	378	472 (1.70%)	[71]
Met	4.06	1:20	37	Au NCs *	420	530 (2.80%)	[92]
Cys	1.00	1:10	37	Au(I)-Cys CP	395	620 (*no inf.*)	[93]
Cys	5.00	1:10	25	Au(I)-Cys CP	365	630 (1.10%)	[94]

* no data are available for the number of gold atoms in the clusters.

In case of Met and Cys amino acids, which have thiol and thioether moieties in the side chain, the characteristic PL emission band was detected at higher (in the yellow and orange regions between 520–630 nm) wavelengths. However, for Met, the formation of Au NCs having Au^0 cores was confirmed, but the pH and the temperature were extremely changed during the two-step preparation route. The spontaneous interaction of thiol-group containing Cys with $AuCl_4^-$ does not result in clusters. Instead, a periodic Au(I) CPs was identified at pH = 3.0 by Söptei et al. measurements [94]. This nanohybrid system has a multilayered construction with 1.3 nm of distance and show characteristic fluorescence thanks to the (-S-Au(I)-S-Au(I)-S-)$_n$ cyclical structure, which was verified by previously published similar Au(I)-thiolate systems [95,96]. In conclusion, the application of simple amino acids having aromatic groups (imidazole, indole, benzene) in the side chains dominantly results in the formation of fluorescent Au NCs. In contrast with the larger polypeptides or proteins, which mainly form red-emitting NCs [97], by the utilization of amino acids as reducing agents, only blue-emitting sub-nanometer sized NCs that consist of a few atoms can be synthesized. At lower synthesis temperature (e.g., room temperature), the application of higher ligand excess (ca. 30-fold excess) is advantageous, but, by increasing of the temperature (~40–50 °C), the use of high ligand excess can be reduced. The bioligands like Cys or Cys-containing small peptides, do not produce fluorescent NCs having Au^0, but the formation of partially reduced $-(Au(I)\text{-}SR)_n-$ periodic structures is especially preferred. The preparation possibilities of $-(Au(I)\text{-}SR)_n-$ structures as well as the synthesis routes of thiolate-stabilized Au NPs/NCs and CPs through the $-(Au(I)\text{-}SR)_n-$ are summarized in the next chapter.

4. Fabrication Protocols of Thiolate-Protected Au Nanohybrid Systems

Various publications can be found in the literature, relating to Au nanostructures that are synthesized by the interaction of $AuCl_4^-$ ions with thiolate molecules as Cys amino acid, peptides having Cys residue or alkyl- and arylthiolates. Depending on the applied fabrication parameters (e.g., chemical structure of the reducing ligand, temperature, molar ratio, pH), decisively three different types of gold-thiol nanohybrid systems, such as plasmonic Au NPs or fluorescent Au CPs and Au NCs, as in Figure 4, can be fabricated. Nevertheless, the presence of similar bond (e.g., covalent bond) between the gold and the sulphur atom(s) of the applied bioligands was confirmed for all the nanostructures.

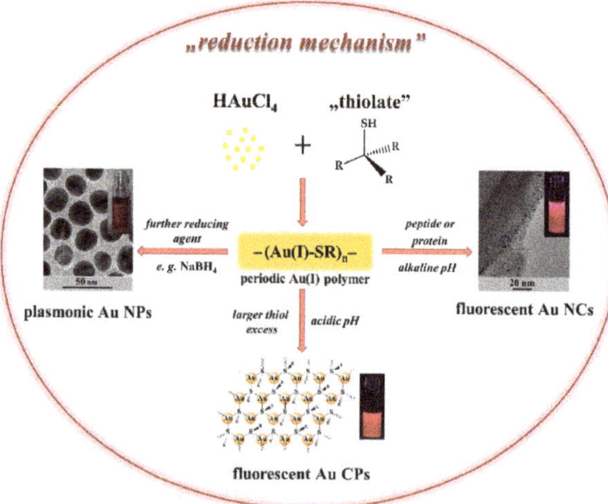

Figure 4. Schematic illustration on the formation mechanisms of different Au nanohybrid systems via interaction of tetrachloroaurate(III) ions with thiolate ligands.

As mentioned in chapter (2.), one of the most commonly used synthesis is the two-phase Brust method for the formation of thiol-protected plasmonic Au NPs [11]. To simplify this method, C. K. Yee et al. developed a protocol, where only tetrahydrofuran was applied as individual solvent [98]. In both methods, several functionalized colloidal particles have been synthesized, which are functionalized by different alkyl- or arylthiols. The size of these Au NPs can be tuned by the molar ratio of the $AuCl_4^-$:thiol-containing molecule, but the one-phase synthesis eventuates larger plasmonic particles [99]. For the exact understanding of these syntheses, Perala and Kumar presented a new synthetic route [100], where the formation of the particle consists of a two-step reduction mechanism, as demonstrated by the Equations (1) and (2).

$$AuCl_4^- + 4RSH \rightarrow \text{-}(Au(I)\text{-}SR)_n\text{-} + RSSR + 4Cl^- + 3H^+ \tag{1}$$

$$\text{-}(Au(I)\text{-}SR)_n\text{-} + BH_4^- + RSH + RSSR \rightarrow Au_x(SR)_y \tag{2}$$

Based on the proposed mechanism, the first two equivalents alkyl- or arylthiol partially reduces the $AuCl_4^-$ ions to Au(I), while next two equivalents involve in the formation of a periodic $-(Au(I)\text{-}SR)_n-$ polymer [101]. The final $Au(I) \rightarrow Au^0$ reduction is carried out by a borohydride salt, which results in the formation of $Au_x(SR)_y$. After reduction, the nucleation, as well as the crystal growth and the particle functionalization, are simultaneously occurred.

As it can be seen, the formation of thiol-protected Au nanohybrid systems occurs through the appearance of a periodic $-(Au(I)\text{-}SR)_n-$ polymer structure. These periodic polymers can simply be further transformed into new gold-containing products having different structure and optical properties (Figure 4). (i) On one hand, the utilization of strong reducing agents (e.g., $NaBH_4$) results in colloidal Au NPs having plasmonic feature; (ii) by the application of a large excess of bioligand having thiol group in the side chain, such as Cys amino acid [40] or glutathione (GSH) tripeptide [102], the formation of Au CPs structures, including self-assembly structure at acidic conditions, is preferred; and, (iii) for the presence of peptide or protein reducing agents excess, fluorescent Au NCs can be synthesized.

These mentioned nanostructures (especially the NCs and CPs) possess intense structure-dependent PL mostly in the orange and red visible or the near infrared (NIR) region. The hybrid electronic states are formed between the sulphur atoms of the ligands and the gold atoms, which results in the emission from the sp to d band transitions [81]. These hybrid bands are below the d band states of Au(I) ions and the excitation wavelength-dependent fluorescence lifetime suggests that the triplet and singlet states are degenerated. In contrast, the hybrid orbitals are above the d band states of gold in case of NIR emission and the microsecond fluorescence lifetime refers to the strong involvement of the Au(I)-S charge transfer in the emission process (Figure 5A,B). In this chapter, the preparation protocols of Au CPs as well as the Au NCs systems were mainly interpreted.

In the case of earlier reports, the pH was not really regulated in the initial stage of the "green" synthesis as well as quite small GSH, Cys, or another thiolates excess was applied. Whereupon, $NaBH_4$ was usually necessary to supplement the reduction process. As a result of the simple reaction of GSH and $HAuCl_4$, T. G. Schaaff and R. L. Whetten identified three different GSH-Au(I) polymers. The $AuCl_4^-$:GSH/1:3 molar ratio, ca. 0.3 mM of $HAuCl_4$ concentration and ten-fold excess of $NaBH_4$ in methanol:water solvent mixture were applied at room temperature, which prevent the polymer from the uncontrolled reduction [103]. The separation of the dark brown products was carried out by polyacrylamide gel filtration (PAGE) and the average sizes of the polymers were 4.3, 5.6, and 8.2 kDa. These nanohybrid systems show strong structure-dependent optical properties in the NIR, visible and UV-region, while the unseparated mixture nor. Y. Negeshi et al. also investigated the effect of the GSH and homo-GSH on the $HAuCl_4$ in two articles. In contrast to the previous result, $AuCl_4^-$:GSH/1:4 molar ratio and 4 mM of tetrachloroaurate(III) concentration were adjusted with a large excess of $NaBH_4$ at 0 °C [104,105]. The identification of the dark-brown powder was accomplished after the PAGE and ultracentrifugation. The nine different Au(I)-polymer structures were recognized by Electrospray Ionization Mass Spectrometry (ESI-MS), optical absorption, and PL spectroscopy (Figure 5C).

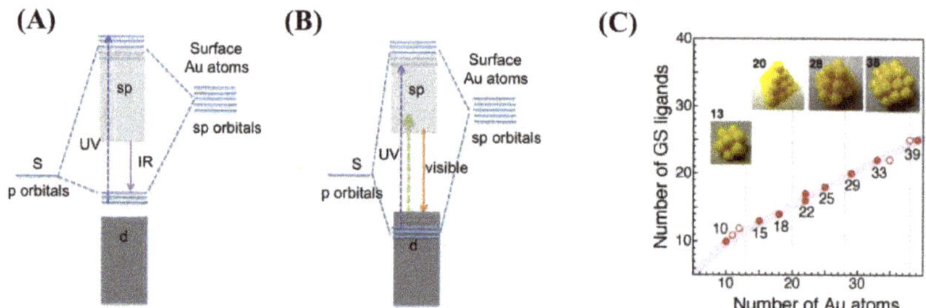

Figure 5. The scheme of the sp and d transitions in case of the (**A**) NIR- and (**B**) visible-emitting thiolate-protected Au nanohybrid systems. Reproduced with permission from [81], RSC, 2012. (**C**) The relationship between the Au atoms and glutathione (GSH) ligands in the most dominant (●) and secondary (○) products. Reproduced with permission from [105], ACS, 2005.

This article presented firstly that, the smaller structures have rather polymeric properties such as the larger emission wavelength and larger binding energy (Au $4f_{7/2}$ ~85 eV), which refers to the decisive presence of Au(I). On the other hand, the systems having larger sizes show cluster-like characteristics with higher emission energy and the binding energy was detected at 84–85 eV. Thereby, the relationship was clearly pointed out between the size, the structure, and the optical behavior of the Au nanohybrid systems.

Neglecting of further reducing agents, R. E. Bachman et al. applied a phenylthiolate to synthesize a fluorescent and self-assembly gold(I) polymeric structure via decomposition of isonitrilegold(I) complex [106]. For the formation of supramolecular system, the dimer units aggregated in an antiparallel fashion at 255 °C, which can be described as a "crinkled tape" motif. It has strong PL in the red region at λ_{em} = 660 nm due to the weak aurophilic interaction in the supramolecular system. I. Odriozola et al. also examined the direct interaction of GSH and $AuCl_4^-$ while using 1:3/gold: ligand molar ratio without the utilization of any further reducing chemicals at room temperature [107]. In their publication, the sol-gel transition was demonstrated, by which the prominent role of the pH on the gold(I)-thiolate structure was discussed. The possible chemical structures of the sol and the gel state were also suggested. H. Nie et al. 3-mercaptopropionic acid, thioglycolic acid, 1-thiogliycerol, and GSH were used to synthesize Au CPs with metal ion: ligand/1:1 stoichiometry [108]. As several CPs have great UV-Vis absorptions that originate from the ligand to metal and the metal-centered charge transfers, thereby the prepared nanohybrids are suitable for the *in-situ* checking the self-assembly of thiol-Au(I) CPs. The synergic effects of the weak interactions were identified with applying different analytical methods (e.g., time-resolved UV-Vis spectrophotometry, HRTEM, X-ray diffraction/XRD, and X-ray photoelectron spectroscopy/XPS). Consequently, it has been proved that the H-bonding, aurophilic and static interactions, and coordination bonding facilitate the evolution of the order structure for Au(I) CPs. C. Lavenn et al. also used phenylthiolate to prepare Au CPs by the development of a hydrothermal method at 120 °C [41]. The formed double helical Au CPs are also stabilized by C-H·π and aurophilic bonds. The product has red emission (λ_{em} = 684 nm) and great quantum yield (~5%). Furthermore, a thermally induced crystallization was presented in solid-state, which rarely occurred in gold(I) polymers. A. T. Royappa et al. applied two different water soluble ethanol-based thiolate molecules to produce of Au(I) CPs while using $AuCl_4^-$:thiol/1:3 molar ratios [109]. The synthesis had a nearly quantitative yield and an amorphous colored gel-like solid was identified as periodic coordination polymer structure, which contains significant aurophilic interactions between the gold atoms.

Besides the previously mentioned, mainly thiolate-based Au CPs, the possible use of biocompatible amino acid Cys is in the focus of interest, especially in the last five years. P. S. Capellari et al. synthesized of ~0.6 nm ultra-small Cys-capped plasmonic Au NPs by precise growth controlling in mild conditions while using pH switching [110]. For understanding the formation mechanism, both acidic and alkaline conditions were examined. The applied molar ratio was ca. $AuCl_4^-$:Cys/1:1 with 5 mM of

HAuCl$_4$ concentration at room temperature. Thanks to their experiments, two very stable polymeric gold(I)-thiolate structure were discerned at the two edges of the pH range and a rather reactive pH interval was identified between 4 < pH < 9. Based on several X-ray analytical methods, the structure of the Cys-Au(I) polymer show strong pH-dependence due to the zwitterionic nature of the Cys. The reactive state was suitable for controlled synthesizing of the plasmonic particle from the stable polymeric structures by pH switching and the adding of NaBH$_4$. For the structural characterization, B. Söptei et al. examined the pale-yellow solid powder by small- and wide-angle X-ray scattering (SWAXS), which was formed by the direct reduction process between the Cys and AuCl$_4^-$. For the preparation, AuCl$_4^-$:Cys/1:10 molar ratio with 5 mM of gold concentration and three different temperature were tested without any regulation of the pH [94]. In their publication, a periodic lamellar structure was presented based on the SWAXS measurements, where the average distance of the lamellas was 1.3 nm. Beside these, the primary coordination bonds were defined by FT-IR spectroscopy. In the IR spectrum of the lamellar structure, the band corresponding to the S-H vibrations was disappeared, while a band was observed at the C = O stretching vibrations. These referred to the Au-S bond in the polymer structure, which were stabilized by strong H-bonds and electrostatic interactions due to the zwitterionic behavior of the Cys amino acid. E. Csapó et al. also examined the spontaneous reaction of the Cys and two cysteine-containing peptides with AuCl$_4^-$ ions while using 1.0 mM of gold concentration at 37 °C in aqueous medium [93]. Depending on the applied pH, the molar ratios and the chemical structure of the Cys and Cys-containing peptides (Cys-Trp, GSH), diverse nanohybrid systems were formed, as in Figure 6. For understanding the ligand-dependent structures of these produced systems, two-dimensional (2D) techniques (surface plasmon resonance and quartz crystal microbalance) were additionally applied. In both cases, orange-emitting products (λ_{em} = 620 and 590 nm) were confirmed while using AuCl$_4^-$:Cys/1:10 and AuCl$_4^-$:GSH/1:15 ratios, respectively. Under acidic conditions (pH 3.0), the coordination polymers were identified and the lamellar architecture with 1.3 nm distance of the Cys-Au(I) CPs is also certified by XRD. Nevertheless, the ordered structure of GSH-Au(I) CPs was not verified, probably for the larger space-filling of the side chain. Under basic conditions, the orange emission was not observed in the GSH-Au system, but a new blue emission band was involved at 445 nm. The XPS studies of this system supposed the formation of ultra-small Au^0 clusters. In contrast of Cys, the redox potential of GSH shows a strong pH-dependent property, thus the tripeptide has stronger reduction capability against the Au(III) ions. Next to the redox feature of the GSH, the hydrolytic process of the aurate(III) ions also influences the structure of final gold products. The presence of AuCl$_4^-$ is dominant between pH = 1–3, but, at basic conditions, the appearance of various hydroxo species (e.g., AuCl(OH)$_3^-$ or Au(OH)$_4^-$) is exclusive.

The amine and thiol-containing dipeptide, named cysteinyl-tryptophan (Cys-Trp), showed mainly amino acid behavior against the AuCl$_4^-$. Depending on the applied ligand amount, the optical properties of the formed gold systems can be tuned. With a small quantity of the Cys-Trp (1:0.5/AuCl$_4^-$:ligand ratio) under basic conditions, plasmonic Au NPs were synthesized with ca. 8–9 nm. In contrast, while using 20-fold dipeptide excess two-coordinated Au(I)-complexes with blue emission (λ_{em} = 470 nm) were identified by the MS techniques. The supramolecular self-assembly of these complexes was not observed, presumably also due to the large size of the ligand. The thioether Met amino acid was used for synthesizing Au NCs by H. H. Deng and co-workers [92]. For the preparation of Met-Au NCs, extreme large Met excess and a two-step thermostated reaction were applied in alkaline medium. The identified cluster shows yellow emission at 530 nm and the quantum yield was 2.9% with two dominant fluorescence lifetimes (181 ns and 1.6 µs). The XPS spectrum suggested that the cluster decisively built up from Au^0. Based on the FT-IR studies, the functional groups of –NH$_2$ and –COOH take part in the formation of the coordinative bonds on the cluster surface, but not on the sulphur atom.

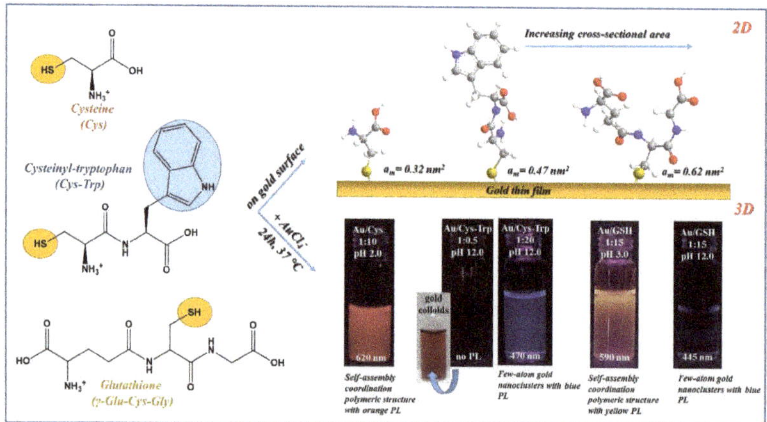

Figure 6. Schematic illustration of the binding of Cys and Cys-containing peptides on gold surface with the corresponding cross-sectional area (above) and the formation of Cys-, Cys-Trp-, and GSH-reduced Au NPs, Au NCs, and Au CPs by spontaneous interaction of the mentioned molecules with AuCl₄⁻ with some representative images. Published in [93], Elsevier, 2016.

As it can be seen, the application of simple (bio)thiolates as simultaneous reducing and stabilizing agent results Au(I)-containing periodic polymer products in most cases. For the synthesis of thiol-reduced Au NCs, either other reducing agents (e.g., borohydride salts) or proteins are usually required. Forasmuch, this article is limited to detailed descriptions of the direct interaction between small amines and thiols, only the brief introduction of the mechanism of the protein-tetrachloroaurate(III) reaction is as follows, because the peptides can be considered as large-sized biocompatible thiolates and amines. Several articles can be found on the syntheses of protein-stabilized Au NCs while using the BSA [111–113], HSA [114,115], LYZ [116–120], trypsin [121], pepsin [122], or immunoglobulin [76]. The typically red-emitting cluster synthesis is carried out under basic conditions (~pH 12) and 10–20-fold protein excess is applied at ca. 40 °C for 24 h. The purification can be done by dialysis or PAGE techniques. The synthesized Au₂₅ NCs contain a core having icosahedral Au₁₃, which are covered by an Au₂₂ shell and they are stabilized by 18 thiolate ligands based on the X-ray crystallographic analysis [123]. Nevertheless, the general accepted mechanism of the cluster formation is the follows. The complete reduction of the Au(III) to Au⁰ also occurred via a precious presented two-steps process. The primary Au(III) → Au(I) progress occurs along the side chain of Trp and Tyr residues. Following a "chain migration", the gold(I) ions are coordinated by the sulphur-containing molecules, where the further reduction is realized by the nearby and suitable amino acids. On one hand, the used extreme basic conditions serve to improve the reduction capability of the Tyr and Trp amino acids. On the other hand, the unfolding of the protein chain is also contributed by applying of alkaline medium, which facilitates easier migration of the partially reduced metal ions along the chain. Based on the above considerations, the presence of the adequate Tyr and Trp beside the thiol-containing amino acids is definitely an important criterion for the success of Au NCs syntheses [124–126]. It can be regarded that the proteins are a great bridge between the biocompatible amine and thiolate ligands.

5. Conclusions

The gold nanoparticles, the ultra-small Au nanoclusters consisting a few or few tens of gold atoms, and the Au-containing self-assembled coordination polymers are in focus of extensive researches thanks to their several excellent properties. Due to the low toxicity as well as their unique, structure-dependent optical feature, they can be used in several fields of medical applications, like as the controlled drug delivery, cancer treatment, fluorescence imaging, diagnostic, and sensing. One of the most important requirements in these medical utilizations is the biocompatibility and the synthesis of these nanostructures under mild reaction conditions in aqueous medium using biocompatible capping agents and avoiding the harsh reducing agents or organic solvents, etc. Based on these expectations, in this review we decisively focused on the short summary of the possible synthetic routes of the formation of colloidal Au NPs, Au NCs, and Au CPs via template-assisted preparation protocol while using amino acids and thiolates as reducing and stabilizing molecules.

For amino acids we can conclude that, almost all amino acids, except Cys, are able to reduce the precursor $AuCl_4^-$ ions at mostly high temperature (T = 50–100 °C), and the formation of stable colloidal Au NPs is preferred. Besides the higher temperature, the high pressure, as well as the extra conditions, like alkaline medium, the high ligand excess or the application of UV light further facilitate the appearance of Au NPs, having sizes larger than 2 nm. In the case of fluorescent amino acids-reduced Au NCs, only the possible utilization of His, Trp, Pro, and Tyr having aromatic residues in the side chain was confirmed to date. At lower synthesis temperature (e.g., room temperature), the application of higher ligand excess (ca. 30-fold excess) is advantageous, but, by increasing of the temperature (~40–50 °C), the use of high ligand excess can be reduced.

The Cys or Cys-containing peptides do not produce fluorescent NCs, but the formation of Au(I)-containing polymers having an ordered structure is especially preferred. The preparation possibilities of these structures through the periodic –(Au(I)-SR)n– as well as the characteristic features of thiolate-stabilized Au NPs/NCs and CPs were also summarized. As presented, the detailed examination of the relationship between the reaction conditions and the optical/structural features of the formed Au-containing nanohybrid systems is extremely important for future applications. Due to the effective PL quenching of Au NCs and Au CPs or the LSPR phenomena of Au NPs, these nanostructures are potential candidates for Photodynamic therapy (PDT), Photothermal therapy (PTT), and X-ray imaging. Moreover, these nanosized noble metal-based nanohybrid structures play a decisive role as possible nanosized controlled drug delivery systems in pharmaceutical applications. Moreover, the sub-nanometer sized fluorescent NCs are excellent nanosensors for rapid and selective detection of essential (Fe(III), Cu(II)) and toxic (Hg(II), Cd(II)) metal ions, anions (e.g., CN^-), or biological molecules (e.g., glucose, folic acid, glutathione, toxins, drugs, etc.)

Author Contributions: Conceptualization, E.C. and D.U.; writing—original draft preparation, D.U. and E.C.; writing—review and editing, E.C. and D.U.; supervision, I.D.

Funding: This research was supported by the National Research, Development and Innovation Office-NKFIH through the project GINOP-2.3.2-15-2016-00038 and FK131446 and FK132067. This paper was supported by the János Bolyai Research Scholarship of the Hungarian Academy of Sciences (E. Csapó) and UNKP-19-4-SZTE-57 New National Excellence Program of the Ministry of Human Capacities (E. Csapó). The Ministry of Human Capacities, Hungary grant TUDFO/47138-1/2019-ITM is also acknowledged.

Conflicts of Interest: The authors declare no conflict of interest.

References

1. Chen, M.; Yin, M. Design and development of fluorescent nanostructures for bioimaging. *Prog. Polym. Sci.* **2014**, *39*, 365–395. [CrossRef]

2. Zaitsev, S.Y.; Solovyeva, D.O. Supramolecular nanostructures based on bacterial reaction center proteins and quantum dots. *Adv. Colloid Interface Sci.* **2015**, *218*, 34–47. [CrossRef]

3. Pérez-Juste, J.; Pastoriza-Santos, I.; Liz-Marzán, L.M.; Mulvaney, P. Gold nanorods: Synthesis, characterization and applications. *Coord. Chem. Rev.* **2005**, *249*, 1870–1901. [CrossRef]

4. Dykman, L.; Khlebtsov, N. Gold nanoparticles in biomedical applications: Recent advances and perspectives. *Chem. Soc. Rev.* **2012**, *41*, 2256–2282. [CrossRef]

5. Faraday, M. The Bakerian Lecture: Experimental Relations of Gold (and Other Metals) to Light. *Philos. Trans. R. Soc. Lond.* **1857**, *147*, 145–181. [CrossRef]

6. Creighton, J.A.; Eadon, D.G. Ultraviolet-visible absorption spectra of the colloidal metallic elements. *J. Chem. Soc. Faraday Trans.* **1991**, *87*, 3881–3891. [CrossRef]

7. Huang, X.; El-Sayed, M.A. Gold nanoparticles: Optical properties and implementations in cancer diagnosis and photothermal therapy. *J. Adv. Res.* **2010**, *1*, 13–28. [CrossRef]

8. Mafuné, F.; Kohno, J.Y.; Takeda, Y.; Kondow, T. Full physical preparation of size-selected gold nanoparticles in solution: Laser ablation and laser-induced size control. *J. Phys. Chem. B* **2002**, *106*, 7575–7577. [CrossRef]

9. Freitas de Freitas, L.; Varca, G.; dos Santos Batista, J.; Benévolo Lugão, A. An Overview of the Synthesis of Gold Nanoparticles Using Radiation Technologies. *Nanomaterials* **2018**, *8*, 939. [CrossRef]

10. Panigrahi, S.; Kundu, S.; Ghosh, S.K.; Nath, S.; Pal, T. General method of synthesis for metal nanoparticles. *J. Nanoparticle Res.* **2004**, *6*, 411–414. [CrossRef]

11. Brust, M.; Walker, M.; Bethell, D.; Schiffrin, D.J.; Whyman, R. Synthesis of thiol-derivatised gold nanoparticles in a two-phase Liquid–Liquid system. *J. Chem. Soc. Chem. Commun.* **1994**, 801–802. [CrossRef]

12. Kuzmann, E.; Csapó, E.; Stichleutner, S.; Garg, V.K.; de Oliveira, A.C.; da Silva, S.W.; Sing, L.H.; Pati, S.S.; Guimaraes, E.M.; Lengyel, A.; et al. Fine structure of gold nanoparticles stabilized by buthyldithiol: Species identified by Mössbauer spectroscopy. *Colloids Surf. A Physicochem. Eng. Asp.* **2016**, *504*, 260–266. [CrossRef]

13. Turkevich, J.; Stevenson, P.C.; Hillier, J. A study of the nucleation and growth processes in the synthesis of colloidal gold. *Discuss. Faraday Soc.* **1951**, *11*, 55. [CrossRef]

14. Majzik, A.; Patakfalvi, R.; Hornok, V.; Dékány, I. Growing and stability of gold nanoparticles and their functionalization by cysteine. *Gold Bull.* **2009**, *42*, 113–123. [CrossRef]

15. Csapó, E.; Oszkó, A.; Varga, E.; Juhász, Á.; Buzás, N.; Kőrösi, L.; Majzik, A.; Dékány, I. Synthesis and characterization of Ag/Au alloy and core(Ag)–shell(Au) nanoparticles. *Colloids Surf. A Physicochem. Eng. Asp.* **2012**, *415*, 281–287. [CrossRef]

16. Xu, F.; Zhang, Q.; Gao, Z. Simple one-step synthesis of gold nanoparticles with controlled size using cationic Gemini surfactants as ligands: Effect of the variations in concentrations and tail lengths. *Colloids Surf. A Physicochem. Eng. Asp.* **2013**, *417*, 201–210. [CrossRef]

17. Bali, K.; Sáfrán, G.; Pécz, B.; Mészáros, R. Preparation of Gold Nanocomposites with Tunable Charge and Hydrophobicity via the Application of Polymer/Surfactant Complexation. *ACS Omega* **2017**, *2*, 8709–8716. [CrossRef]

18. Polavarapu, L.; Xu, Q.H. A single-step synthesis of gold nanochains using an amino acid as a capping agent and characterization of their optical properties. *Nanotechnology* **2008**, *19*, 075601. [CrossRef]

19. Slocik, J.M.; Stone, M.O.; Naik, R.R. Synthesis of gold nanoparticles using multifunctional peptides. *Small* **2005**, *1*, 1048–1052. [CrossRef]

20. Francois, T.; Onani, M.; Madiehe, A.; Meyer, M. Aqueous soluble gold nanoparticle synthesis using polyethyleneimine and reduced glutathione. *Int. J. Mater. Res.* **2014**, *105*, 1025–1037. [CrossRef]

21. Gericke, M.; Pinches, A. Microbial production of gold nanoparticles. *Gold Bull.* **2006**, *39*, 22–28. [CrossRef]

22. Sharma, N.; Pinnaka, A.K.; Raje, M.; FNU, A.; Bhattacharyya, M.S.; Choudhury, A.R. Exploitation of marine bacteria for production of gold nanoparticles. *Microb. Cell Fact.* **2012**, *11*, 1. [CrossRef]

23. Li, J.; Li, Q.; Ma, X.; Tian, B.; Li, T.; Yu, J.; Dai, S.; Weng, Y.; Hua, Y. Biosynthesis of gold nanoparticles by the extreme bacterium Deinococcus radiodurans and an evaluation of their antibacterial properties. *Int. J. Nanomed.* **2016**, *11*, 5931–5944. [CrossRef]

24. Chen, L.Y.Y.; Wang, C.W.W.; Yuan, Z.; Chang, H.T.T. Fluorescent gold nanoclusters: Recent advances in sensing and imaging. *Anal. Chem.* **2015**, *87*, 216–229. [CrossRef]

25. Jin, R.; Zeng, C.; Zhou, M.; Chen, Y. Atomically Precise Colloidal Metal Nanoclusters and Nanoparticles: Fundamentals and Opportunities. *Chem. Rev.* **2016**, *116*, 10346–10413. [CrossRef]

26. Kaur, N.; Aditya, R.N.; Singh, A.; Kuo, T.R. Biomedical Applications for Gold Nanoclusters: Recent Developments and Future Perspectives. *Nanoscale Res. Lett.* **2018**, *13*, 302. [CrossRef]

27. Qu, X.; Li, Y.; Li, L.; Wang, Y.; Liang, J.; Liang, J. Fluorescent Gold Nanoclusters: Synthesis and Recent Biological Application. *J. Nanomater.* **2015**, *2015*, 1–23. [CrossRef]

28. Yin, X.; Chen, S.; Wu, A. Green chemistry synthesis of gold nanoparticles using lactic acid as a reducing agent. *Micro Nano Lett.* **2010**, *5*, 270. [CrossRef]

29. Sharma, R.K.; Gulati, S.; Mehta, S. Preparation of gold nanoparticles using tea: A green chemistry experiment. *J. Chem. Educ.* **2012**, *89*, 1316–1318. [CrossRef]

30. Sujitha, M.V.; Kannan, S. Green synthesis of gold nanoparticles using Citrus fruits (Citrus limon, Citrus reticulata and Citrus sinensis) aqueous extract and its characterization. *Spectrochim. Acta Part A Mol. Biomol. Spectrosc.* **2013**, *102*, 15–23. [CrossRef]

31. Antoine, R.; Bertorelle, F.; Broyer, M.; Compagnon, I.; Dugourd, P.; Kulesza, A.; Mitrič, R.; Bonačić-Koutecký, V. Gas-phase synthesis and intense visible absorption of tryptophangold cations. *Angew. Chem. Int. Ed.* **2009**, *48*, 7829–7832. [CrossRef]

32. Le Guével, X.; Daum, N.; Schneider, M. Synthesis and characterization of human transferrin-stabilized gold nanoclusters. *Nanotechnology* **2011**, *22*, 275103. [CrossRef]

33. Liu, C.L.; Wu, H.T.; Hsiao, Y.H.; Lai, C.W.; Shih, C.W.; Peng, Y.K.; Tang, K.C.; Chang, H.W.; Chien, Y.C.; Hsiao, J.K.; et al. Insulin-directed synthesis of fluorescent gold nanoclusters: Preservation of insulin bioactivity and versatility in cell imaging. *Angew. Chem. Int. Ed.* **2011**, *50*, 7056–7060. [CrossRef]

34. Duan, H.; Nie, S. Etching Colloidal Gold Nanocrystals with Hyperbranched and Multivalent Polymers: A New Route to Fluorescent and Water-Soluble Atomic Clusters. *J. Am. Chem. Soc.* **2007**, *129*, 2412–2413. [CrossRef]

35. Sun, X.; Dong, S.; Wang, E. One-step preparation and characterization of poly(propyleneimine) dendrimer-protected silver nanoclusters. *Macromolecules* **2004**, *37*, 7105–7108. [CrossRef]

36. Zhang, Y.; Jiang, H.; Ge, W.; Li, Q.; Wang, X. Cytidine-directed rapid synthesis of water-soluble and highly yellow fluorescent bimetallic AuAg nanoclusters. *Langmuir* **2014**, *30*, 10910–10917. [CrossRef]

37. Zhang, Y.; Jiang, H.; Wang, X. Cytidine-stabilized gold nanocluster as a fluorescence turn-on and turn-off probe for dual functional detection of Ag^+ and Hg^{2+}. *Anal. Chim. Acta* **2015**, *870*, 1–7. [CrossRef]

38. Ahn, J.K.; Kim, H.Y.; Baek, S.; Park, H.G. A new s-adenosylhomocysteine hydrolase-linked method for adenosine detection based on DNA-templated fluorescent Cu/Ag nanoclusters. *Biosens. Bioelectron.* **2017**, *93*, 330–334. [CrossRef]

39. Ungor, D.; Csapó, E.; Kismárton, B.; Juhász, A.; Dékány, I. Nucleotide-directed syntheses of gold nanohybrid systems with structure-dependent optical features: Selective fluorescence sensing of Fe^{3+} ions. *Colloids Surf. B Biointerfaces* **2017**, *155*, 135–141. [CrossRef]

40. Nafady, A.; Afridi, H.I.; Sara, S.; Shah, A.; Niaz, A. Direct synthesis and stabilization of Bi-sized cysteine-derived gold nanoparticles: Reduction catalyst for methylene blue. *J. Iran. Chem. Soc.* **2011**, *8*, S34–S43. [CrossRef]

41. Lavenn, C.; Okhrimenko, L.; Guillou, N.; Monge, M.; Ledoux, G.; Dujardin, C.; Chiriac, R.; Fateeva, A.; Demessence, A. A luminescent double helical gold(I)–thiophenolate coordination polymer obtained by hydrothermal synthesis or by thermal solid-state amorphous-to-crystalline isomerization. *J. Mater. Chem. C* **2015**, *3*, 4115–4125. [CrossRef]

42. Deák, A.; Jobbágy, C.; Marsi, G.; Molnár, M.; Szakács, Z.; Baranyai, P. Anion-, Solvent-, Temperature-, and Mechano-Responsive Photoluminescence in Gold(I) Diphosphine-Based Dimers. *Chem. A Eur. J.* **2015**, *21*, 11495–11508. [CrossRef]

43. Aljuhani, M.A.; Bootharaju, M.S.; Sinatra, L.; Basset, J.M.; Mohammed, O.F.; Bakr, O.M. Synthesis and Optical Properties of a Dithiolate/Phosphine-Protected Au_{28} Nanocluster. *J. Phys. Chem. C* **2017**, *121*, 10681–10685. [CrossRef]

44. Negishi, Y.; Takasugi, Y.; Sato, S.; Yao, H.; Kimura, K.; Tsukuda, T. Kinetic stabilization of growing gold clusters by passivation with thiolates. *J. Phys. Chem. B* **2006**, *110*, 12218–12221. [CrossRef]

45. Le Guével, X.; Spies, C.; Daum, N.; Jung, G.; Schneider, M. Highly fluorescent silver nanoclusters stabilized by glutathione: A promising fluorescent label for bioimaging. *Nano Res.* **2012**, *5*, 379–387. [CrossRef]

46. Bayse, C.A.; Ming, J.L.; Miller, K.M.; McCollough, S.M.; Pike, R.D. Photoluminescence of silver(I) and gold(I) cyanide 1D coordination polymers. *Inorg. Chim. Acta* **2011**, *375*, 47–52. [CrossRef]

47. Luo, Z.; Yuan, X.; Yu, Y.; Zhang, Q.; Leong, D.T.; Lee, J.Y.; Xie, J. From Aggregation-Induced Emission of Au(I)–Thiolate Complexes to Ultrabright Au(0)@Au(I)–Thiolate Core–Shell Nanoclusters. *J. Am. Chem. Soc.* **2012**, *134*, 16662–16670. [CrossRef]

48. Ahmad, A.; Senapati, S.; Khan, M.I.; Kumar, R.; Ramani, R.; Srinivas, V.; Sastry, M. Intracellular synthesis of gold nanoparticles by a novel alkalotolerant actinomycete, Rhodococcus species. *Nanotechnology* **2003**, *14*, 824–828. [CrossRef]

49. Feng, Y.; Lin, X.; Wang, Y.; Wang, Y.; Hua, J. Diversity of Aurum bioreduction by Rhodobacter capsulatus. *Mater. Lett.* **2008**, *62*, 4299–4302. [CrossRef]

50. Nune, S.K.; Chanda, N.; Shukla, R.; Katti, K.; Kulkarni, R.R.; Thilakavathy, S.; Mekapothula, S.; Kannan, R.; Katti, K.V. Green nanotechnology from tea: Phytochemicals in tea as building blocks for production of biocompatible gold nanoparticles. *J. Mater. Chem.* **2009**, *19*, 2912–2920. [CrossRef]

51. Kasthuri, J.; Veerapandian, S.; Rajendiran, N. Biological synthesis of silver and gold nanoparticles using apiin as reducing agent. *Colloids Surf. B Biointerfaces* **2009**, *68*, 55–60. [CrossRef]

52. Vemula, P.K.; Aslam, U.; Mallia, V.A.; John, G. In situ synthesis of gold nanoparticles using molecular gels and liquid crystals from vitamin-C amphiphiles. *Chem. Mater.* **2007**, *19*, 138–140. [CrossRef]

53. Lee, J.; Ryu, J.; Choi, W. Preparation of Gold and Platinum Nanoparticles Using Visible Light Activated Fe III -complex. *Chem. Lett.* **2007**, *36*, 176–177. [CrossRef]

54. Nasr, G.; Guerlin, A.; Dumur, F.; Baudron, S.A.; Dumas, E.; Miomandre, F.; Clavier, G.; Sliwa, M.; Mayer, C.R. Dithiolate-appended iridium(III) complex with dual functions of reducing and capping agent for the design of small-sized gold nanoparticles. *J. Am. Chem. Soc.* **2011**, *133*, 6501–6504. [CrossRef]

55. Kasthuri, J.; Rajendiran, N. Functionalization of silver and gold nanoparticles using amino acid conjugated bile salts with tunable longitudinal plasmon resonance. *Colloids Surf. B Biointerfaces* **2009**, *73*, 387–393. [CrossRef]

56. Huang, T.; Meng, F.; Qi, L. Controlled synthesis of dendritic gold nanostructures assisted by supramolecular complexes of surfactant with cyclodextrin. *Langmuir* **2010**, *26*, 7582–7589. [CrossRef]

57. Hussain, I.; Brust, M.; Papworth, A.J.; Cooper, A.I. Preparation of Acrylate-Stabilized Gold and Silver Hydrosols and Gold–Polymer Composite Films. *Langmuir* **2003**, *19*, 4831–4835. [CrossRef]

58. Sardar, R.; Park, J.W.; Shumaker-Parry, J.S. Polymer-induced synthesis of stable gold and silver nanoparticles and subsequent ligand exchange in water. *Langmuir* **2007**, *23*, 11883–11889. [CrossRef]

59. Meldrum, F.C.; Heywood, B.R.; Mann, S. Influence of Membrane Composition on the Intravesicular Precipitation of Nanophase Gold Particles. *J. Colloid Interface Sci.* **1993**, *161*, 66–71. [CrossRef]

60. Wangoo, N.; Kaur, S.; Bajaj, M.; Jain, D.V.S.; Sharma, R.K. One pot, rapid and efficient synthesis of water dispersible gold nanoparticles using alpha-amino acids. *Nanotechnology* **2014**, *25*, 435608. [CrossRef]

61. Mandal, S.; Selvakannan, P.R.; Phadtare, S.; Pasricha, R.; Sastry, M. Synthesis of a stable gold hydrosol by the reduction of chloroaurate ions by the amino acid, aspartic acid. *Proc. Indian Acad. Sci. Chem. Sci.* **2002**, *114*, 513–520. [CrossRef]

62. Selvakannan, P.R.; Mandal, S.; Phadtare, S.; Pasricha, R.; Sastry, M. Capping of gold nanoparticles by the amino acid lysine renders them water-dispersible. *Langmuir* **2003**, *19*, 3545–3549. [CrossRef]

63. Selvakannan, P.R.; Mandal, S.; Phadtare, S.; Gole, A.; Pasricha, R.; Adyanthaya, S.D.; Sastry, M. Water-dispersible tryptophan-protected gold nanoparticles prepared by the spontaneous reduction of aqueous chloroaurate ions by the amino acid. *J. Colloid Interface Sci.* **2004**, *269*, 97–102. [CrossRef]

64. Bhargava, S.K.; Booth, J.M.; Agrawal, S.; Coloe, P.; Kar, G. Gold nanoparticle formation during bromoaurate reduction by amino acids. *Langmuir* **2005**, *21*, 5949–5956. [CrossRef]

65. Newman, J.D.S.S.D.S.; Blanchard, G.J.J. Formation of gold nanoparticles using amine reducing agents. *Langmuir* **2006**, *22*, 5882–5887. [CrossRef]

66. Wangoo, N.; Bhasin, K.K.; Mehta, S.K.; Suri, C.R. Synthesis and capping of water-dispersed gold nanoparticles by an amino acid: Bioconjugation and binding studies. *J. Colloid Interface Sci.* **2008**, *323*, 247–254. [CrossRef]

67. Liu, Z.; Zu, Y.; Fu, Y.; Meng, R.; Guo, S.; Xing, Z.; Tan, S. Hydrothermal synthesis of histidine-functionalized single-crystalline gold nanoparticles and their pH-dependent UV absorption characteristic. *Colloids Surf. B Biointerfaces* **2010**, *76*, 311–316. [CrossRef]

68. Cai, H.; Yao, P. Gold nanoparticles with different amino acid surfaces: Serum albumin adsorption, intracellular uptake and cytotoxicity. *Colloids Surf. B Biointerfaces* **2014**, *123*, 900–906. [CrossRef]

69. Courrol, L.C.; de Matos, R.A. Synthesis of Gold Nanoparticles Using Amino Acids by Light Irradiation. In *Catalytic Application of Nano-Gold Catalysts*; InTech: Bergharen, The Netherlands, 2016; Volume i, p. 13.

70. Maruyama, T.; Fujimoto, Y.; Maekawa, T. Synthesis of gold nanoparticles using various amino acids. *J. Colloid Interface Sci.* **2014**, *447*, 254–257. [CrossRef]

71. Csapó, E.; Ungor, D.; Kele, Z.; Baranyai, P.; Deák, A.; Juhász, Á.; Janovák, L.; Dékány, I. Influence of pH and aurate/amino acid ratios on the tuneable optical features of gold nanoparticles and nanoclusters. *Colloids Surf. A Physicochem. Eng. Asp.* **2017**, *532*, 601–608. [CrossRef]

72. Zhang, L.; Wang, E. Metal nanoclusters: New fluorescent probes for sensors and bioimaging. *Nano Today* **2014**, *9*, 132–157. [CrossRef]

73. Khandelwal, P.; Poddar, P. Fluorescent metal quantum clusters: An updated overview of the synthesis, properties, and biological applications. *J. Mater. Chem. B* **2017**, *5*, 9055–9084. [CrossRef]

74. Shang, L.; Dong, S.; Nienhaus, G.U. Ultra-small fluorescent metal nanoclusters: Synthesis and biological applications. *Nano Today* **2011**, *6*, 401–418. [CrossRef]

75. Chevrier, D.M.; Chatt, A.; Zhang, P. Properties and applications of protein-stabilized fluorescent gold nanoclusters: Short review. *J. Nanophotonics* **2012**, *6*, 064504. [CrossRef]

76. Ungor, D.; Horváth, K.; Dékány, I.; Csapó, E. Red-emitting gold nanoclusters for rapid fluorescence sensing of tryptophan metabolites. *Sens. Actuators B Chem.* **2019**, *288*, 728–733. [CrossRef]

77. Tsunoyama, H.; Tsukuda, T. Magic numbers of gold clusters stabilized by PVP. *J. Am. Chem. Soc.* **2009**, *131*, 18216–18217. [CrossRef]

78. Li, Z.; Liu, R.; Xing, G.; Wang, T.; Liu, S. A novel fluorometric and colorimetric sensor for iodide determination using DNA-templated gold/silver nanoclusters. *Biosens. Bioelectron.* **2017**, *96*, 44–48. [CrossRef]

79. Zheng, J.; Petty, J.T.; Dickson, R.M. High Quantum Yield Blue Emission from Water-Soluble Au 8 Nanodots. *J. Am. Chem. Soc.* **2003**, *125*, 7780–7781. [CrossRef]

80. Zheng, J.; Nicovich, P.R.; Dickson, R.M. Highly Fluorescent Noble-Metal Quantum Dots. *Annu. Rev. Phys. Chem.* **2007**, *58*, 409–431. [CrossRef]

81. Zheng, J.; Zhou, C.; Yu, M.; Liu, J. Different sized luminescent gold nanoparticles. *Nanoscale* **2012**, *4*, 4073. [CrossRef]

82. Yang, X.; Shi, M.; Zhou, R.; Chen, X.; Chen, H. Blending of $HAuCl_4$ and histidine in aqueous solution: A simple approach to the Au_{10} cluster. *Nanoscale* **2011**, *3*, 2596–2601. [CrossRef]

83. Zhang, X.; Wu, F.G.; Liu, P.; Gu, N.; Chen, Z. Enhanced fluorescence of gold nanoclusters composed of $HAuCl_4$ and histidine by glutathione: Glutathione detection and selective cancer cell imaging. *Small* **2014**, *10*, 5170–5177. [CrossRef]

84. Liu, X.; Yu, X.; Luo, X. Ultrasensitive iodide detection based on the resonance light scattering of histidine-stabilized gold nanoclusters. *Microchim. Acta* **2014**, *181*, 1379–1384. [CrossRef]

85. Schmidt, L.A.; Kirk, L.; Appleman, W.K. The Appaern Dissociation costants of Arginine and of Lysine and the Apparent Heats of Ionization of Certain Amino Acids. *J. Biol. Chem.* **1930**, *88*, 285–293.

86. Zhang, Y.; Hu, Q.; Paau, M.C.; Xie, S.; Gao, P.; Chan, W.; Choi, M.M.F. Probing histidine-stabilized gold nanoclusters product by high-performance liquid chromatography and mass spectrometry. *J. Phys. Chem. C* **2013**, *117*, 18697–18708. [CrossRef]

87. Guo, Y.; Long, T.; Lin, M.; Liu, Z.; Huang, C.; Zhao, X. Histidine-mediated synthesis of chiral fluorescence gold nanoclusters: Insight into the origin of nanoscale chirality. *RSC Adv.* **2015**, *5*, 61449–61454. [CrossRef]

88. Yang, X.; Luo, Y.; Zhuo, Y.; Feng, Y.; Zhu, S. Novel synthesis of gold nanoclusters templated with *L*-tyrosine for selective analyzing tyrosinase. *Anal. Chim. Acta* **2014**, *840*, 87–92. [CrossRef]

89. Mu, X.; Qi, L.; Qiao, J.; Ma, H. One-pot synthesis of tyrosine-stabilized fluorescent gold nanoclusters and their application as turn-on sensors for Al^{3+} ions and turn-off sensors for Fe^{3+} ions. *Anal. Methods* **2014**, *6*, 6445–6451. [CrossRef]

90. Mu, X.; Qi, L.; Dong, P.; Qiao, J.; Hou, J.; Nie, Z.; Ma, H. Facile one-pot synthesis of *L*-proline-stabilized fluorescent gold nanoclusters and its application as sensing probes for serum iron. *Biosens. Bioelectron.* **2013**, *49*, 249–255. [CrossRef]

91. Zheng, S.; Yin, H.; Li, Y.; Bi, F.; Gan, F. One–step synthesis of *L*-tryptophan-stabilized dual-emission fluorescent gold nanoclusters and its application for Fe^{3+} sensing. *Sens. Actuators B Chem.* **2017**, *242*, 469–475. [CrossRef]

92. Deng, H.H.; Zhang, L.N.; He, S.B.; Liu, A.L.; Li, G.W.; Lin, X.H.; Xia, X.H.; Chen, W. Methionine-directed fabrication of gold nanoclusters with yellow fluorescent emission for Cu^{2+} sensing. *Biosens. Bioelectron.* **2015**, *65*, 397–403. [CrossRef]

93. Csapó, E.; Ungor, D.; Juhász, Á.; Tóth, G.K.; Dékány, I. Gold nanohybrid systems with tunable fluorescent feature: Interaction of cysteine and cysteine-containing peptides with gold in two- and three-dimensional systems. *Colloids Surf. A Physicochem. Eng. Asp.* **2016**, *511*, 264–271. [CrossRef]

94. Söptei, B.; Mihály, J.; Szigyártó, I.C.; Wacha, A.; Németh, C.; Bertóti, I.; May, Z.; Baranyai, P.; Sajó, I.E.; Bóta, A. The supramolecular chemistry of gold and *L*-cysteine: Formation of photoluminescent, orange-emitting assemblies with multilayer structure. *Colloids Surf. A Physicochem. Eng. Asp.* **2015**, *470*, 8–14. [CrossRef]

95. Jin, R. Quantum sized, thiolate-protected gold nanoclusters. *Nanoscale* **2010**, *2*, 343–362. [CrossRef]

96. Nasaruddin, R.R.; Chen, T.; Yan, N.; Xie, J. Roles of thiolate ligands in the synthesis, properties and catalytic application of gold nanoclusters. *Coord. Chem. Rev.* **2018**, *368*, 60–79. [CrossRef]

97. Xavier, P.L.; Chaudhari, K.; Baksi, A.; Pradeep, T. Protein-protected luminescent noble metal quantum clusters: An emerging trend in atomic cluster nanoscience. *Nano Rev.* **2012**, *3*, 14767. [CrossRef]

98. Yee, C.K.; Jordan, R.; Ulman, A.; White, H.; King, A.; Rafailovich, M.; Sokolov, J. Novel One-Phase Synthesis of Thiol-Functionalized Gold, Palladium, and Iridium Nanoparticles Using Superhydride. *Langmuir* **1999**, *15*, 3486–3491. [CrossRef]

99. Frenkel, A.I.; Nemzer, S.; Pister, I.; Soussan, L.; Harris, T.; Sun, Y.; Rafailovich, M.H. Size-controlled synthesis and characterization of thiol-stabilized gold nanoparticles. *J. Chem. Phys.* **2005**, *123*, 184701. [CrossRef]

100. Perala, S.R.K.; Kumar, S. On the Mechanism of Metal Nanoparticle Synthesis in the Brust–Schiffrin Method. *Langmuir* **2013**, *29*, 9863–9873. [CrossRef]

101. Yu, C.; Zhu, L.; Zhang, R.; Wang, X.; Guo, C.; Sun, P.; Xue, G. Investigation on the Mechanism of the Synthesis of Gold(I) Thiolate Complexes by NMR. *J. Phys. Chem. C* **2014**, *118*, 10434–10440. [CrossRef]

102. Briñas, R.P.; Hu, M.; Qian, L.; Lymar, E.S.; Hainfeld, J.F. Gold Nanoparticle Size Controlled by Polymeric Au(I) Thiolate Precursor Size. *J. Am. Chem. Soc.* **2008**, *130*, 975–982. [CrossRef]

103. Schaaff, T.G.; Whetten, R.L. Giant Gold–Glutathione Cluster Compounds: Intense Optical Activity in Metal-Based Transitions. *J. Phys. Chem. B* **2000**, *104*, 2630–2641. [CrossRef]

104. Negishi, Y.; Takasugi, Y.; Sato, S.; Yao, H.; Kimura, K.; Tsukuda, T.; Characterization, S.; Tsukuda, T. Magic-Numbered Au_n Clusters Protected by Glutathione Monolayers (n = 18, 21, 25, 28, 32, 39): Isolation and Spectroscopic Characterization. *J. Am. Chem. Soc.* **2004**, *126*, 6518–6519. [CrossRef]

105. Negishi, Y.; Nobusada, K.; Tsukuda, T. Glutathione-Protected Gold Clusters Revisited: Bridging the Gap between Gold(I)−Thiolate Complexes and Thiolate-Protected Gold Nanocrystals. *J. Am. Chem. Soc.* **2005**, *127*, 5261–5270. [CrossRef]

106. Bachman, R.E.; Bodolosky-Bettis, S.A.; Glennon, S.C.; Sirchio, S.A. Formation of a Novel Luminescent Form of Gold(I) Phenylthiolate via Self-Assembly and Decomposition of Isonitrilegold(I) Phenylthiolate Complexes. *J. Am. Chem. Soc.* **2000**, *122*, 7146–7147. [CrossRef]

107. Odriozola, I.; Loinaz, I.; Pomposo, J.A.; Grande, H.J. Gold–glutathione supramolecular hydrogels. *J. Mater. Chem.* **2007**, *17*, 4843. [CrossRef]

108. Nie, H.; Li, M.; Hao, Y.; Wang, X.; Zhang, S.X.A. Time-resolved monitoring of dynamic self-assembly of Au(I)-thiolate coordination polymers. *Chem. Sci.* **2013**, *4*, 1852. [CrossRef]

109. Royappa, A.T.; Tran, C.M.; Papoular, R.J.; Khan, M.; Marbella, L.E.; Millstone, J.E.; Gembicky, M.; Chen, B.; Shepard, W.; Elkaim, E. Copper(I) and gold(I) thiolate precursors to bimetallic nanoparticles. *Polyhedron* **2018**, *155*, 359–365. [CrossRef]

110. Cappellari, P.S.; Buceta, D.; Morales, G.M.; Barbero, C.A.; Sergio Moreno, M.; Giovanetti, L.J.; Ramallo-López, J.M.; Requejo, F.G.; Craievich, A.F.; Planes, G.A. Synthesis of ultra-small cysteine-capped gold nanoparticles by pH switching of the Au(I)–cysteine polymer. *J. Colloid Interface Sci.* **2015**, *441*, 17–24. [CrossRef]

111. Habeeb Muhammed, M.A.; Verma, P.K.; Pal, S.K.; Retnakumari, A.; Koyakutty, M.; Nair, S.; Pradeep, T. Luminescent Quantum Clusters of Gold in Bulk by Albumin-Induced Core Etching of Nanoparticles: Metal Ion Sensing, Metal-Enhanced Luminescence, and Biolabeling. *Chem. A Eur. J.* **2010**, *16*, 10103–10112. [CrossRef]

112. Hemmateenejad, B.; Shakerizadeh-shirazi, F.; Samari, F. BSA-modified gold nanoclusters for sensing of folic acid. *Sens. Actuators B Chem.* **2014**, *199*, 42–46. [CrossRef]

113. Zhao, S.; Li, Z.; Li, Y.; Yu, J.; Liu, G.; Liu, R.; Yue, Z. BSA-AuNCs based enhanced photoelectrochemical biosensors and its potential use in multichannel detections. *J. Photochem. Photobiol. A Chem.* **2017**, *342*, 15–24. [CrossRef]

114. Gui, R.; Jin, H. Aqueous synthesis of human serum albumin-stabilized fluorescent Au/Ag core/shell nanocrystals for highly sensitive and selective sensing of copper(II). *Analyst* **2013**, *138*, 7197. [CrossRef]

115. Yu, Y.; New, S.Y.; Xie, J.; Su, X.; Tan, Y.N. Protein-based fluorescent metal nanoclusters for small molecular drug screening. *Chem. Commun.* **2014**, *50*, 13805–13808. [CrossRef]

116. Wei, H.; Wang, Z.; Yang, L.; Tian, S.; Hou, C.; Lu, Y. Lysozyme-stabilized gold fluorescent cluster: Synthesis and application as Hg^{2+} sensor. *Analyst* **2010**, *135*, 1406. [CrossRef]

117. Chan, P.H.; Wong, S.Y.; Lin, S.H.; Chen, Y.C. Lysozyme-encapsulated gold nanocluster-based affinity mass spectrometry for pathogenic bacteria. *Rapid Commun. Mass Spectrom.* **2013**, *27*, 2143–2148. [CrossRef]

118. Lu, D.; Liu, L.; Li, F.; Shuang, S.; Li, Y.; Choi, M.M.F.; Dong, C. Lysozyme-stabilized gold nanoclusters as a novel fluorescence probe for cyanide recognition. *Spectrochim. Acta Part A Mol. Biomol. Spectrosc.* **2014**, *121*, 77–80. [CrossRef]

119. Hornok, V.; Csapó, E.; Varga, N.; Ungor, D.; Sebők, D.; Janovák, L.; Laczkó, G.; Dékány, I. Controlled syntheses and structural characterization of plasmonic and red-emitting gold/lysozyme nanohybrid dispersions. *Colloid Polym. Sci.* **2016**, *294*, 49–58. [CrossRef]

120. Russell, B.A.A.; Jachimska, B.; Komorek, P.; Mulheran, P.A.A.; Chen, Y. Lysozyme encapsulated gold nanoclusters: Effects of cluster synthesis on natural protein characteristics. *Phys. Chem. Chem. Phys.* **2017**, *19*, 7228–7235. [CrossRef]

121. Kawasaki, H.; Yoshimura, K.; Hamahuchi, K.; Arakawa, R. Trypsin-Stabilized Fluorescent Gold Nanocluster for Sensitive and Selective Hg^{2+} Detection. *Anal. Sci.* **2011**, *27*, 591. [CrossRef]

122. Kawasaki, H.; Hamaguchi, K.; Osaka, I.; Arakawa, R. pH-Dependent Synthesis of Pepsin-Mediated Gold Nanoclusters with Blue Green and Red Fluorescent Emission. *Adv. Funct. Mater.* **2011**, *21*, 3508–3515. [CrossRef]

123. Zhu, M.; Aikens, C.M.; Hollander, F.J.; Schatz, G.C.; Jin, R. Correlating the Crystal Structure of A Thiol-Protected Au_{25} Cluster and Optical Properties. *J. Am. Chem. Soc.* **2008**, *130*, 5883–5885. [CrossRef]

124. Zhou, R.; Shi, M.; Chen, X.; Wang, M.; Chen, H. Atomically Monodispersed and Fluorescent Sub-Nanometer Gold Clusters Created by Biomolecule-Assisted Etching of Nanometer-Sized Gold Particles and Rods. *Chem. A Eur. J.* **2009**, *15*, 4944–4951. [CrossRef]

125. Xie, J.; Zheng, Y.; Ying, J.Y. Protein-Directed Synthesis of Highly Fluorescent Gold Nanoclusters. *J. Am. Chem. Soc.* **2009**, *131*, 888–889. [CrossRef]

126. Chaudhari, K.; Xavier, P.L.; Pradeep, T. Understanding the Evolution of Luminescent Gold Quantum Clusters in Protein Templates. *ACS Nano* **2011**, *5*, 8816–8827. [CrossRef]

 *nanomaterials*

Article

Catenane Structures of Homoleptic Thioglycolic Acid-Protected Gold Nanoclusters Evidenced by Ion Mobility-Mass Spectrometry and DFT Calculations

Clothilde Comby-Zerbino [1], Martina Perić [2], Franck Bertorelle [1], Fabien Chirot [3], Philippe Dugourd [1], Vlasta Bonačić-Koutecký [2,4] and Rodolphe Antoine [1,*]

[1] Institut Lumière Matière UMR 5306, Université Claude Bernard Lyon 1, CNRS, Univ Lyon, F-69100 Villeurbanne, France; clothilde.zerbino@univ-lyon1.fr (C.C.-Z.); franck.bertorelle@univ-lyon1.fr (F.B.); philippe.dugourd@univ-lyon1.fr (P.D.)

[2] Center of Excellence for Science and Technology-Integration of Mediterranean region (STIM) at Interdisciplinary Center for Advanced Sciences and Technology (ICAST), University of Split, Poljička cesta 35, HR-21000 Split, Croatia; martina@stim.unist.hr (M.P.); vbk@stim.unist.hr (V.B.-K.)

[3] Institut des Sciences Analytiques UMR 5280, Université Claude Bernard Lyon 1, ENS de Lyon, CNRS, Univ Lyon, 5 rue de la Doua, F-69100 Villeurbanne, France; fabien.chirot@univ-lyon1.fr

[4] Department of Chemistry, Humboldt Universitat zu Berlin, Brook-Taylor-Strasse 2, 12489 Berlin, Germany

[*] Correspondence: rodolphe.antoine@univ-lyon1.fr; Tel.: +33-472-43-1085

Received: 21 February 2019; Accepted: 16 March 2019; Published: 19 March 2019

Abstract: Thiolate-protected metal nanoclusters have highly size- and structure-dependent physicochemical properties and are a promising class of nanomaterials. As a consequence, for the rationalization of their synthesis and for the design of new clusters with tailored properties, a precise characterization of their composition and structure at the atomic level is required. We report a combined ion mobility-mass spectrometry approach with density functional theory (DFT) calculations for determination of the structural and optical properties of ultra-small gold nanoclusters protected by thioglycolic acid (TGA) as ligand molecules, $Au_{10}(TGA)_{10}$. Collision cross-section (CCS) measurements are reported for two charge states. DFT optimized geometrical structures are used to compute CCSs. The comparison of the experimentally- and theoretically-determined CCSs allows concluding that such nanoclusters have catenane structures.

Keywords: gold nanoclusters; thiolate; catenane; ion mobility; DFT calculations

---

1. Introduction

Thiolate-protected metal nanoclusters (NCs) are a promising class of nanomaterials due to fascinating molecular-like properties along with well-defined molecular structures [1–3]. However, their physicochemical properties are highly size- and structure-dependent. As a consequence, for the rationalization of their synthesis and for the design of new clusters with tailored properties, a precise characterization of their composition and structure at the atomic level is required.

The structural features of stoichiometric $Au_n(SR)_n$ gold nanoclusters (SR:thiolate ligand) was predicted to change from single rings to interlocked ring motifs (i.e., catenane structures) when $n \geq 10$ [4]. The interlocked ring motif is a unique feature of homoleptic $[Au(I)-SR]_x$ complexes found in $Au_{10}(SR)_{10}$, $Au_{11}(SR)_{11}$, and $Au_{12}(SR)_{12}$ [5–7]. More importantly, the catenane-like staple motifs predicted for $Au_{15}(SR)_{13}$ and $Au_{24}(SR)_{20}$ suggest that, at a Au/SR ratio approaching 1/1, the interlocked staple motifs may become a widespread conformation in thiolate-protected metal nanoclusters [8–10]. Moreover, the $Au_{10}(SR)_{10}$ catenane structure was recently identified as the best structural candidate for the Au local structure in bovine serum albumin protein-stabilized gold nanoclusters [11]. We reported in a recent work, a "one-pot–one-size" synthesis of $Au_{10}(SG)_{10}$

NCs (SG:glutathione:γ-L-glutamyl-L-cysteinylglycine) characterized by electrospray MS. The X-ray diffraction pattern of $Au_{10}(SG)_{10}$ was utilized as fingerprints for homoleptic gold–glutathione catenanes [7]. Regarding optical properties, enhanced second harmonic response and circular dichroism signals in the spectral region of 250–400 nm were observed due to this catenane structure exhibiting a centrosymmetry-broken structure [7]. Recently, Chevrier et al. confirmed the catenane structure by using synchrotron-based X-ray absorption fine structure (XAFS) spectroscopy [11]. As a complement to these powder-based structural characterization techniques requiring X-ray beams or synchrotron facilities, mass spectrometry-based techniques performed on gas phase nanoclusters ions may provide information on 3D molecular structures. In particular, ion mobility spectrometry (IMS) has been used for the characterization of gas-phase ligand-protected metal nanoclusters [12–19]. IMS separation is based on the different velocities adopted by ions travelling in an inert gas under a low electric field. The drift time of the ions through the IMS tube depends on the ratio between their collision cross-section (CCS) with the gas and their charge, thus allowing isomer discrimination. Our groups showed how IMS studies can provide insight into the size of glutathione-protected gold nanoclusters, as well as in the structural determination of inorganic nanoclusters [16,18,19].

In a previous recent work, we reported an ion mobility-mass spectrometry (IM-MS) approach for the analysis of homoleptic $Au_{10\text{-}12}(SG)_{10\text{-}12}$ nanoclusters. CCS measurements were reported for different charge states for $Au_{10}(SG)_{10}$, $Au_{11}(SG)_{11}$, and $Au_{12}(SG)_{12}$ nanoclusters [18]. Strong charge-state effects on experimental CCS values were observed and attributed to charge-induced glutathione unfolding. However, the importance of core structure and the ligand conformations on the total CCS was difficult to disentangle due to conformational effects of such a flexible peptide ligand. The IMS technique was not sufficient to discriminate between different possible structures (in particular catenane structures) for the core.

This discrimination could be easier if smaller and more rigid ligands are used for protection, where charge-induced ligand unfolding effects will be minimized. In this case, the structural characterization of clusters may thus be possible by comparing the arrival time distribution profiles recorded by ion mobility mass spectrometry with theoretical calculations using molecular modelling (density functional theory, DFT) and subsequent collision cross-section calculations using projection approximation. Here, we report a combined ion mobility and spectrometry approach with DFT calculations for the analysis of a stoichiometric gold nanocluster ligated by thioglycolic acid $Au_{10}(TGA)_{10}$ (TGA; see Figure S1 in the Supplementary Materials) as ligand molecules. Collision cross-section (CCS) measurements are reported for two charge states. DFT calculations have been performed to optimize different candidate structures for which CCSs were computed. The comparison of the experimentally- and theoretically-determined CCSs allows concluding about the catenane structures of such nanoclusters.

2. Materials and Methods

Materials and synthesis protocol: All the chemicals were commercially available and were used without purification. $HAuCl_4 \cdot 3H_2O$, trifluoroacetic acid (TFA), and methanol (HPLC grade) were purchased from Carl Roth (Lauterbourg, France). Thioglycolic acid (TGA), NaOH, and NH_4OH were purchased from Sigma-Aldrich (Saint-Quentin Fallavier, France). Milli-Q water with a resistivity of 18.2 MΩ cm^{-1} was used for all experiments. $Au_{10}(TGA)_{10}$ NC was prepared as described in [7] with TGA as the ligand instead of glutathione. Briefly, 70 mg of TGA ($\approx$53 µL) were diluted in 35 mL of methanol and 2 mL of triethylamine. Then, 100 mg of $HAuCl_4 \cdot 3H_2O$ in 15 mL of water were added, and the solution was stirred overnight at ambient temperature. To induce precipitation, 2 mL of 1 M NaOH solution were added, and the solution was centrifuged (10 min at 11,000 rpm).

Ion mobility-mass spectrometry: Ion mobility measurements were performed using an ion mobility spectrometer as described in [20]. Measurements were done using a fresh mixture of $Au_{10}(TGA)_{10}$, prepared in an aqueous solution to a concentration of about 50 µM and directly electrosprayed using a syringe pump. Mobility measurements were done by injecting ion bunches in the drift tube filled with 4.0 Torr helium, in which a constant drift field was maintained through the controlled voltage drop across the tube. The temperature of the whole instrument was kept at 296 K. After their drift, ions were transferred to a reflector time-of-flight mass spectrometer. Mass spectra were finally recorded as a function of the IMS drift time, allowing for extraction of arrival time distributions (ATDs) for ions with any desired mass-to-charge ratio. Collision cross-sections (CCS) were extracted from ATDs as described in [21]. Using this method, the error of the experimental CCS was estimated to be 2%.

Computational: The structural and absorption properties of $Au_{10}(TGA)_{10}$ were determined using the DFT and its time-dependent version TD-DFT approach [22,23]. For gold atoms, a 19-electron relativistic effective core potential (19e-RECP) was employed [24]. The structural and spectroscopic properties of $Au_{10}(TGA)_{10}$ were obtained at the PBE0/Def2-SVP level of theory [25,26].

3. Results and Discussion

3.1. Characterization of $Au_{10}(TGA)_{10}$

The formation of $Au_{10}(TGA)_{10}$ NCs as the product was confirmed by electrospray ionization-mass spectrometry (ESI-MS) in negative mode (see the inset in Figure 1). A charge state distribution of the general formula $[M-nH^+]^{n-}$ ($2 \leq n \leq 4$) was observed for the $Au_{10}(TGA)_{10}$. The additional peaks observed in MS spectra were due to smaller stoichiometric $(AuTGA)_n$ complexes ($n \leq 6$) originating from the "in-source" fragmentation of the $Au_{10}(TGA)_{10}$ clusters (as evidenced by collision-induced dissociation experiments; see Figure S2 in the Supplementary Materials).

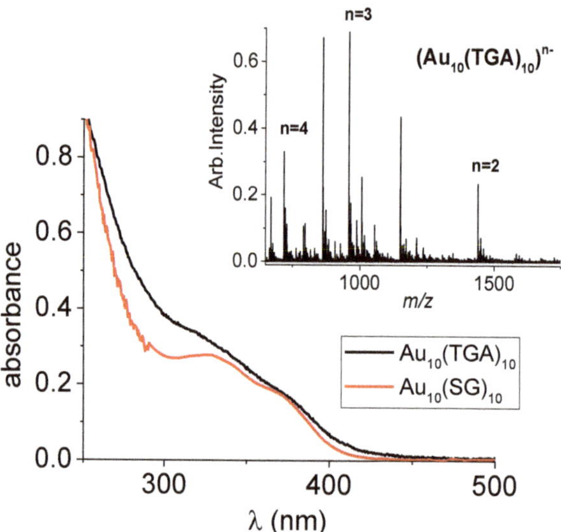

Figure 1. Experimental absorption spectra of $Au_{10}(SR)_{10}$ nanoclusters (NCs) (with SR = thioglycolic acid (TGA) and SG (see [7]). (Inset) Electrospray ionization ESI mass spectrum of the as-synthesized $Au_{10}(TGA)_{10}$ NCs.

Concerning the optical properties, the one-photon absorption spectrum of the as-synthesized $Au_{10}(TGA)_{10}$ NCs showed a monotonic increase of intensity below 390 nm and a shoulder at ~310 nm. There was similarity with the absorption spectrum of the previously-reported $Au_{10}(SG)_{10}$ NCs (see Figure 1) [7].

3.2. Theoretical Investigation of the Structural and Optical Properties of $Au_{10}(TGA)_{10}$

The DFT method has been used to determine the structures of the $Au_{10}(SR)_{10}$ NCs based on the results obtained by a genetic algorithm search method [4]. The [5,5] catenane structure containing two interpenetrating $-AuSR-$ pentagons was found to be the most stable structure (Figure 2a). The [6,4] structure containing four- and six-membered Au rings interpenetrating each other (Figure 2b) and the crown-like structure (Figure 2c) was higher in energy. The structure of these three isomers is shown in Figure 2. Interestingly, the size of TGA ligand along with the size of the crown and the Au-S bond length allowed for a rich hydrogen-bonding network within the TGA ligands, leading to a "ball-like" shape for the crown-like structure.

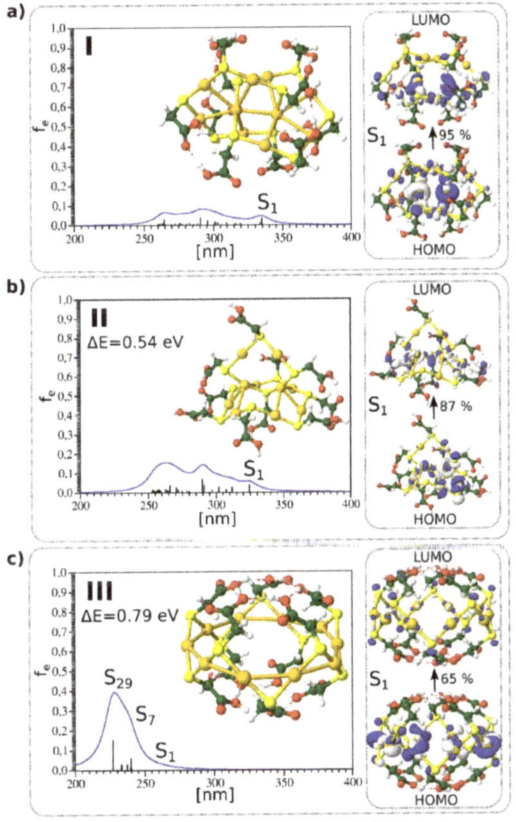

Figure 2. TD-DFT absorption spectrum and structure for three lowest energy isomers of $Au_{10}(TGA)_{10}$ shown in (**a–c**) respectively. Leading excitations responsible for the characteristic features of absorption are illustrated on the right side. HOMO-LUMO for isomers I, II, and III are 4.54, 4.62, and 5.55 eV, respectively.

The absorption spectra calculated using a TD-DFT approach for the three isomers with catenane structures are also shown in Figure 2. For the [5,5] and [6,4] catenane structures, the first excited states were located between 320 and 350 nm. The leading excitations responsible for S_1 and S_2 excited states shown also in Figure 2 involved Au–Au aurophilic subunits bound to neighboring sulfur atoms and arose from the penetration of the two rings into each other. The absorption spectrum for the crown-like structure obtained from the TD-DFT approach differed considerably from those of other two isomers.

3.3. Catenane Structures of Homoleptic Au$_{10}$(TGA)$_{10}$ Evidenced by Ion Mobility-Mass Spectrometry and DFT Calculations

In order to characterize the structural properties of Au$_{10}$(TGA)$_{10}$ NCs, we conducted ion mobility-mass spectrometry (IM–MS) measurements. The extracted arrival time distributions (ATDs) were mainly monomodal for the two- and three-charge states of Au$_{10}$(TGA)$_{10}$, indicating that the corresponding clusters presented essentially a single structural type, and the width of the peaks was compatible with a single structural type being present (see Figures S3 and S4 in the Supplementary Materials). In addition, Figure S4 in the Supplementary Materials shows that the arrival time distributions (ATDs) for [Au$_{10}$(TGA)$_{10}$−2H]$^{2-}$ and [Au$_{10}$(TGA)$_{10}$−3H]$^{3-}$ were very close to the predicted ATDs by the Fick law. The observed ATDs peaks were thus limited by the experimental instrumental resolution. This means that the observed single peaks in ATDs corresponded to single structures, and other possible effects (conformational freedom and especially motion around the Au–S bond in the TGA ligand and possible interconversions between ligand conformations) cannot be resolved.

The experimental CCSs determined for different charge states for Au$_{10}$(TGA)$_{10}$ nanoclusters are given in Table 1. The collision cross-section for the three-charge state was only slightly higher by ~4% than that for the two-charge state. This finding is in contrast with Au$_{10}$(SG)$_{10}$, where a charge-induced unfolding due to Coulomb repulsion between charged moieties was observed, producing more dramatic effects on the CCS [18]. Indeed, for Au$_{10}$(SG)$_{10}$, the increase in the collision cross-section as a function of charge was more important (by ~6.5%). Furthermore, the size of the glutathione ligand was in the same order as the size of the metallic core. This indicates that the charging of the TGA ligand molecule played a minor role in the total collision cross-section of Au$_{10}$(TGA)$_{10}$. This means that the overall structure of the NCs was not significantly modified by the charge, as confirmed by DFT structures obtained for neutral Au$_{10}$(TGA)$_{10}$ and [Au$_{10}$(TGA)$_{10}$−2H]$^{2-}$ (see Figure S5 in the Supplementary Materials). For the two charge state, the CCS value calculated from the [5,5] and [6,4] catenane structures matched the experimental CCS value, confirming that core geometry was consistent with a catenane-like form for Au$_{10}$(TGA)$_{10}$ nanoclusters.

Table 1. Experimental and calculated collision cross-section (CCS) values for three isomers of Au$_{10}$(TGA)$_{10}$ NCs are given. The influence of charge has been experimentally determined (error bars are in brackets). For this purpose, the trajectory method has been used [27]. The DFT structures obtained for [Au$_{10}$(TGA)$_{10}$−2H]$^{2-}$ are given in Figure S5 in the Supplementary Materials.

CCS of Au$_{10}$(TGA)$_{10}$ (Å^2)	Au$_{10}$(TGA)$_{10}$ neutral	[Au$_{10}$(TGA)$_{10}$−2H]$^{2-}$	[Au$_{10}$(TGA)$_{10}$−3H]$^{3-}$
Exp.		225 (5)	235 (5)
[5,5] catenane	212	220	
[6,4] catenane	228	230	
Crown-like	196	196	

4. Conclusions

The chemistry of the sulfur–gold bond is extremely rich and leads to hybrid materials. Such materials encompass gold thiolate coordination oligomers, for instance Au$_n$(SR)$_n$ and atomically well-defined clusters Au$_n$(SR)$_m$, or supramolecular assemblies like −(AuSR)$_\infty$−. The catenane-like structure is a unique feature of Au$_n$(SR)$_n$ complexes, but certainly also in thiolate-protected metal nanoclusters at a low Au/SR ratio limit (i.e., approaching 1:1). Unraveling the total structure of gold nanoclusters is of paramount importance for their characterization. Unfortunately, the use of X-ray crystallography is problematic for homoleptic thiolate-protected metal nanoclusters, because sample crystallization requires extremely high purity and stability. Additional characterization tools able to distinguish structural isomers are thus highly desirable. The DFT approach provides information about catenane-like structures for the two lowest energy isomers. The TDDFT absorption features allows for the structural assignment to experimental data, as well. Ion mobility-mass spectrometry

(IM-MS) has proven to be a useful complement to MS due to its ability to separate ions based on their "shape". In this work, we used this coupling and additionally reported collision cross-sections (CCS) for selected gas phase charge states of $Au_{10}(TGA)_{10}$ cluster ions. Charge effects on the CCS were found negligible for a simple and small thiolated ligand (thioglycolic acid (TGA)). Furthermore, the comparison of CCS values from different structural isomers of $Au_{10}(TGA)_{10}$ obtained at the DFT level of theory has permitted confirming the catenane structure for such nanoclusters.

Supplementary Materials: The following are available online at http://www.mdpi.com/2079-4991/9/3/457/s1: Figure S1: Chemical structure of thioglycolic acid (TGA). Figure S2: Collision-induced dissociation spectra of $(Au_{10}(TGA)_{10})^{3-}$. Figure S3: ATDs recorded for two charge states of $Au_{10}(TGA)_{10}$ in negative mode. Figure S4: ATDs recorded for two charge states of $Au_{10}(TGA)_{10}$ compared to the fick law. Figure S5: DFT structures obtained for $[Au_{10}(TGA)_{10}-2H]^{2-}$.

Author Contributions: R.A. conceived of the initial idea and coordinated the work. F.B. synthesized and prepared the nanoclusters. C.C.-Z. measured CCS and recorded mass spectra. M.P. and V.B.-K. performed and analyzed the theoretical results. C.C.-Z. and F.C. analyzed the results. R.A., P.D. and V.B.-K. supervised and financed the project. R.A. and V.B.-K. wrote the paper. All the authors provided critical feedback and helped to shape the final manuscript.

Funding: This research was partially supported by the project STIM – REI, Contract Number KK.01.1.1.01.0003, funded by the European Union through the European Regional Development Fund—the Operational Programme Competitiveness, and Cohesion 2014–2020 (KK.01.1.1.01). (V.B.-K and M.P.) We would like to acknowledge the financial support from the French-Croatian project "International Laboratory for Nano Clusters and Biological Aging, LIA NCBA".

Acknowledgments: V.B.-K. and M.P acknowledge the Center for Advanced Computing and Modelling (CNRM) for providing computing resources of the supercomputer Bura at the University of Rijeka and SRCE at University of Zagreb, Croatia.

References

1. Jin, R.; Zeng, C.; Zhou, M.; Chen, Y. Atomically Precise Colloidal Metal Nanoclusters and Nanoparticles: Fundamentals and Opportunities. *Chem. Rev.* **2016**, *116*, 10346–10413. [CrossRef] [PubMed]

2. Chakraborty, I.; Pradeep, T. Atomically Precise Clusters of Noble Metals: Emerging Link between Atoms and Nanoparticles. *Chem. Rev.* **2017**, *117*, 8208–8271. [CrossRef] [PubMed]

3. Antoine, R. Atomically precise clusters of gold and silver: A new class of nonlinear optical nanomaterials. *Front. Res. Today* **2018**, *1*, 01001. [CrossRef]

4. Liu, Y.; Tian, Z.; Cheng, L. Size evolution and ligand effects on the structures and stability of (AuL)n (L = Cl, SH, SCH3, PH2, P(CH3)2, n = 1-13) clusters. *RSC Adv.* **2016**, *6*, 4705–4712. [CrossRef]

5. Wiseman, M.R.; Marsh, P.A.; Bishop, P.T.; Brisdon, B.J.; Mahon, M.F. Homoleptic Gold Thiolate Catenanes. *J. Am. Chem. Soc.* **2000**, *122*, 12598–12599. [CrossRef]

6. Chui, S.S.-Y.; Chen, R.; Che, C.-M. A Chiral [2]Catenane Precursor of the Antiarthritic Gold(I) Drug Auranofin. *Angew. Chem. Int. Ed.* **2006**, *45*, 1621–1624. [CrossRef] [PubMed]

7. Bertorelle, F.; Russier-Antoine, I.; Calin, N.; Comby-Zerbino, C.; Bensalah-Ledoux, A.; Guy, S.; Dugourd, P.; Brevet, P.-F.; Sanader, Ž.; Krstić, M.; et al. Au10(SG)10: A Chiral Gold Catenane Nanocluster with Zero Confined Electrons. Optical Properties and First-Principles Theoretical Analysis. *J. Phys. Chem. Lett.* **2017**, *8*, 1979–1985. [CrossRef] [PubMed]

8. Jiang, D.-E.; Overbury, S.H.; Dai, S. Structure of Au15(SR)13 and Its Implication for the Origin of the Nucleus in Thiolated Gold Nanoclusters. *J. Am. Chem. Soc.* **2013**, *135*, 8786–8789. [CrossRef]

9. Pei, Y.; Pal, R.; Liu, C.; Gao, Y.; Zhang, Z.; Zeng, X.C. Interlocked Catenane-Like Structure Predicted in Au24(SR)20: Implication to Structural Evolution of Thiolated Gold Clusters from Homoleptic Gold(I) Thiolates to Core-Stacked Nanoparticles. *J. Am. Chem. Soc.* **2012**, *134*, 3015–3024. [CrossRef] [PubMed]

10. Pei, Y.; Wang, P.; Ma, Z.; Xiong, L. Growth-Rule-Guided Structural Exploration of Thiolate-Protected Gold Nanoclusters. *Accounts Chem. Res.* **2019**, *52*, 23–33. [CrossRef]

11. Chevrier, D.M.; Thanthirige, V.D.; Luo, Z.; Driscoll, S.; Cho, P.; MacDonald, M.A.; Yao, Q.; Guda, R.; Xie, J.; Johnson, E.R.; et al. Structure and formation of highly luminescent protein-stabilized gold clusters. *Chem. Sci.* **2018**, *9*, 2782–2790. [CrossRef]

12. Angel, L.A.; Majors, L.T.; Dharmaratne, A.C.; Dass, A. Ion Mobility Mass Spectrometry of Au25(SCH2CH2Ph)18 Nanoclusters. *ACS Nano* **2010**, *4*, 4691–4700. [CrossRef] [PubMed]

13. Harkness, K.M.; Fenn, L.S.; Cliffel, D.E.; McLean, J.A. Surface Fragmentation of Complexes from Thiolate Protected Gold Nanoparticles by Ion Mobility-Mass Spectrometry. *Anal. Chem.* **2010**, *82*, 3061–3066. [CrossRef] [PubMed]

14. Baksi, A.; Harvey, S.R.; Natarajan, G.; Wysocki, V.H.; Pradeep, T. Possible isomers in ligand protected Ag11 cluster ions identified by ion mobility mass spectrometry and fragmented by surface induced dissociation. *Chem. Commun.* **2016**, *52*, 3805–3808. [CrossRef] [PubMed]

15. Baksi, A.; Ghosh, A.; Mudedla, S.K.; Chakraborty, P.; Bhat, S.; Mondal, B.; Krishnadas, K.R.; Subramanian, V.; Pradeep, T. Isomerism in Monolayer Protected Silver Cluster Ions: An Ion Mobility-Mass Spectrometry Approach. *J. Phys. Chem. C* **2017**, *121*, 13421–13427. [CrossRef]

16. Daly, S.; Choi, C.M.; Zavras, A.; Krstić, M.; Chirot, F.; Connell, T.U.; Williams, S.J.; Donnelly, P.S.; Antoine, R.; Giuliani, A.; et al. Gas-Phase Structural and Optical Properties of Homo- and Heterobimetallic Rhombic Dodecahedral Nanoclusters [Ag14–nCun(C≡CtBu)12X]+ (X = Cl and Br): Ion Mobility, VUV and UV Spectroscopy, and DFT Calculations. *J. Phys. Chem. C* **2017**, *121*, 10719–10727. [CrossRef]

17. Ligare, M.R.; Baker, E.S.; Laskin, J.; Johnson, G.E. Ligand induced structural isomerism in phosphine coordinated gold clusters revealed by ion mobility mass spectrometry. *Chem. Commun.* **2017**, *53*, 7389–7392. [CrossRef] [PubMed]

18. Comby-Zerbino, C.; Bertorelle, F.; Chirot, F.; Dugourd, P.; Antoine, R. Structural insights into glutathione-protected gold Au10−12(SG)10−12 nanoclusters revealed by ion mobility mass spectrometry. *Eur. Phys. J. D* **2018**, *72*, 144. [CrossRef]

19. Soleilhac, A.; Bertorelle, F.; Comby-Zerbino, C.; Chirot, F.; Calin, N.; Dugourd, P.; Antoine, R. Size Characterization of Glutathione-Protected Gold Nanoclusters in the Solid, Liquid and Gas Phases. *J. Phys. Chem. C* **2017**, *121*, 27733–27740. [CrossRef]

20. Simon, A.-L.; Chirot, F.; Choi, C.M.; Clavier, C.; Barbaire, M.; Maurelli, J.; Dagany, X.; MacAleese, L.; Dugourd, P. Tandem ion mobility spectrometry coupled to laser excitation. *Rev. Sci. Instrum.* **2015**, *86*, 094101. [CrossRef]

21. Revercomb, H.E.; Mason, E.A. Theory of plasma chromatography/gaseous electrophoresis. *Anal. Chem.* **1975**, *47*, 970–983. [CrossRef]

22. Bonačić-Koutecký, V.; Kulesza, A.; Gell, L.; Mitrić, R.; Antoine, R.; Bertorelle, F.; Hamouda, R.; Rayane, D.; Broyer, M.; Tabarin, T.; et al. Silver cluster–biomolecule hybrids: From basics towards sensors. *Phys. Chem. Chem. Phys.* **2012**, *14*, 9282–9290. [CrossRef] [PubMed]

23. Bertorelle, F.; Hamouda, R.; Rayane, D.; Broyer, M.; Antoine, R.; Dugourd, P.; Gell, L.; Kulesza, A.; Mitric, R.; Bonacic-Koutecky, V. Synthesis, characterization and optical properties of low nuclearity liganded silver clusters: Ag31(SG)19 and Ag15(SG)11. *Nanoscale* **2013**, *5*, 5637–5643. [CrossRef] [PubMed]

24. Andrae, D.; Häußermann, U.; Dolg, M.; Stoll, H.; Preuß, H. Energy-adjusted ab initio pseudopotentials for the second and third row transition elements. *Theor. Chi. Acta* **1990**, *77*, 123–141. [CrossRef]

25. Adamo, C.; Barone, V. Toward reliable density functional methods without adjustable parameters: The PBE0 model. *J. Chem. Phys.* **1999**, *110*, 6158–6170. [CrossRef]

26. Weigend, F.; Ahlrichs, R. Balanced basis sets of split valence, triple zeta valence and quadruple zeta valence quality for H to Rn: Design and assessment of accuracy. *Phys. Chem. Chem. Phys.* **2005**, *7*, 3297–3305. [CrossRef] [PubMed]

27. Larriba-Andaluz, C.; Hogan, C.J., Jr. Collision cross section calculations for polyatomic ions considering rotating diatomic/linear gas molecules. *J. Chem. Phys.* **2014**, *141*, 194107. [CrossRef] [PubMed]

 *nanomaterials*

Article

New Evidence of the Bidentate Binding Mode in 3-MBA Protected Gold Clusters: Analysis of Aqueous 13–18 kDa Gold-Thiolate Clusters by HPLC-ESI-MS Reveals Special Compositions Au_n(3-MBA)$_p$, (n = 48–67, p = 26–30)

David M. Black [1],[†], M. Mozammel Hoque [1],[*],[†], Germán Plascencia-Villa [1],[2] and Robert L. Whetten [1],[*],[‡]

1   Department of Physics & Astronomy, University of Texas, San Antonio, TX 78249, USA
2   Department of Biology, University of Texas, San Antonio, TX 78249, USA
*   Correspondence: mohammad.hoque64@gmail.com (M.M.H.); whettenz60@gmail.com (R.L.W.)
†   These authors contributed equally to this work.
‡   Present Address: Department of Applied Physics and Material Science, Northern Arizona University, Flagstaff, AZ 86011, USA.

Received: 22 August 2019; Accepted: 9 September 2019; Published: 11 September 2019

Abstract: Gold clusters protected by 3-MBA ligands (MBA = mercaptobenzoic acid, –SPhCO$_2$H) have attracted recent interest due to their unusual structures and their advantageous ligand-exchange and bioconjugation properties. Azubel et al. first determined the core structure of an Au_{68}-complex, which was estimated to have 32 ligands (3-MBA groups). To explain the exceptional structure-composition and reaction properties of this complex, and its larger homologs, Tero et al. proposed a "dynamic stabilization" via carboxyl O–H⸺Au interactions. Herein, we report the first results of an integrated liquid chromatography/mass spectrometer (LC/MS) analysis of unfractionated samples of gold/3-MBA clusters, spanning a narrow size range 13.4 to 18.1 kDa. Using high-throughput procedures adapted from bio-macromolecule analyses, we show that integrated capillary high performance liquid chromatography electrospray ionization mass spectrometer (HPLC-ESI-MS), based on aqueous-methanol mobile phases and ion-pairing reverse-phase chromatography, can separate *several* major components from the nanoclusters mixture that may be difficult to resolve by standard native gel electrophoresis due to their similar size and charge. For each component, one obtains a well-resolved mass spectrum, nearly free of adducts or signs of fragmentation. A consistent set of molecular mass determinations is calculated from detected charge-states tunable from 3– (or lower), to 2+ (or higher). One thus arrives at a series of new compositions (n, p) specific to the Au/3-MBA system. The smallest major component is assigned to the previously unknown (48, 26); the largest one is evidently (67, 30), vs. the anticipated (68, 32). Various explanations for this discrepancy are considered. A prospective is given for the *various* members of this novel series, along with a summary of the advantages and present limitations of the micro-scale integrated LC/MS approach in characterizing such metallic-core macro-molecules, and their derivatives.

Keywords: 3-MBA/Au MPCs; TEA-HFIP; ESI-MS; HPLC-MS; bidentate binding

1. Introduction

This work on the 3-MBA protected gold clusters, or cluster compounds, has, in brief, been motivated by the following circumstances:

(i) A *Science* paper from 2014 identified a medium-sized (68, 32) 3-MBA gold cluster and determined its structure through an HREM-based statistical algorithm [1].

(ii) It has been proposed [2–4] that this ligand has a different 'binding mode' than its sister 4-MBA (pMBA), which is better understood thanks in particular to Vergara et al.'s recent subatomic resolution of the (146, 57) compound [5].

(iii) Previous attempts by the Tsukuda group to analyze these by ESI-MS are limited in scope, as the obtained spectra are "too broad" (unresolved), i.e., inadequate to establish the composition, e.g., is it truly Au_{68} (as the HREM reconstruction indicates)? What is the true ligand count (supplied by computational modeling)?

(iv) According to our best evidence, the main component is (67, 30), rather than the previously published (68, 32), and the smaller main component is (48, 26). These unusual numbers, and indeed the entire graph of the observed composition number (p vs. n), are consistent with the theoretical and experimental (nuclear magnetic resonance, NMR) proposition of a special (bidentate) mode of 3-MBA binding. This trend-line (dependency) is in accord with the idea that as the size increases, the decreasing curvature (of the core's surface) increases the propensity for the bidentate mode. Asymptotically, flat surfaces (self-assembled monolayers, SAMS) may be dominated by this binding mode.

Noble metal clusters, especially of gold and its intermetallic compounds, form highly stable complexes with thiolate and other pseudo-halide ligands. These are often called "monolayer protected clusters" (MPCs), because of their relation to the analogous self-assembled monolayers (SAMs) on planar or extended electrodes [6–8]. They have attracted special attention because of their nobility [9] (tolerance to air, moisture, and light; bio-compatibility, etc.); their facile modification via ligand exchange reactions [10–12]; a high-contrast detection, whether visual/optical or in X-ray and electron scattering [13]; and the fascination and potential utility of their strongly size-dependent optical, electrical and structure-bonding properties [14–16].

By now, much evidence has accumulated to suggest that many of these MPCs may be obtained in high yield as pure macromolecular substances of definite composition [17,18] and structure-bonding characteristics [19–21], as opposed to the more usual metal colloidal or nanoparticle [9] substances that often show heterogeneity. Such a proven structural uniformity of MPCs is essential to precision-intensive applications, as well as to all fundamental physicochemical understanding. The most compelling demonstrations are the cases of a total structure determination by single crystal X-ray [22] or electron diffraction methods [23], which for gold-thiolates have recently been extended to MPCs as large as $Au_{146}(pMBA)_{57}$ (aqueous) [5] and $Au_{279}(TBBT)_{84}$ (nonaqueous) [24].

Azubel et al. [1] determined the core structure of an Au_{68}-complex via cryo-TEM, which was estimated to have 32 ligands (3-MBA groups). Tero et al. [4] proposed a "dynamic stabilization" mechanism via carboxyl O–H----Au interactions to explain its structure, composition and reaction properties, as well as those of its larger homologs [2,3].

Many reports have discussed the challenge of adequately characterizing samples of novel MPCs, particularly in the early stages of identifying the main compounds or components of a mixture, as discussed elsewhere [25,26]. Our approach here has been to adapt a method—electrospray ionization (ESI)-coupled high performance liquid chromatography mass spectrometry (HPLC-MS)—established earlier for bio-macromolecules of a similar size (or mass) and surface chemistry as the MPCs under investigation [27]. Specifically, the larger Au/MBA clusters have many (~24–60) acid-terminated ligands [3], and so are presumed to exist in an aqueous solution at a normal (or higher) pH as poly-anions (plus respective counter-cations). For this case, a long experience with oligonucleotides (DNA or RNA), composed of a similar number, ~24–60 base-sugar-phosphate repeats), seems most instructive.

Our aims in the present work have been (i) to determine whether the unusual solution-phase characteristics of the Au/3-MBA clusters will permit them to yield to be analyzed by the ESI-coupled LC-MS methods that have recently improved the analysis of Au-pMBA clusters ranging from small oligomers and clusters (25, 18) and (36, 24) to the larger species (102, 44), (130, 50) and (144, 60) [27]; (ii) to examine whether ion-pairing agents will work similarly to enable both high-resolution LC separations and reduced-fragmentation ESI-ToF (time-of-flight) mass spectra; (iii) to provide some insight into the powerful selection principles underlying the results in refs. [1–4]; (iv) to search for minor or hidden components (new compositions) as semi-stable or transition MPCs; and (v) to provide additional evidence pertaining to the 'bidentate' or dynamical carboxyl-gold interactions described in reference [4].

Herein, we report the first results of an ESI-coupled LC-MS analysis of unfractionated samples of Au/3-MBA clusters that span a narrow mass range, 13.4–18.2 kDa. Using procedures adapted from oligonucleotide analyses, we show that integrated capillary HPLC-ESI-MS, based on aqueous-methanol mobile phases and ion-pairing reverse-phase chromatography, can separate *at least two* major components (and several minor ones) that are present in all sources. For each component, a well-resolved mass spectrum, nearly free of adducts or signs of fragmentation, allows the determination of a consistent assignment of molecular masses, as calculated from detected charge-states tunable from 3– (or lower), to 2+ (or higher). One thus arrives at a set of proposed compositions (*n*, *p*), as characteristic of the Au/3-MBA system. The smaller major component is assigned to the previously unknown (48, 26); the larger one is assigned to (67, 30), vs. the anticipated (68, 32).

2. Materials and Methods

2.1. Synthesis

The size-uniform samples prepared at the University of Texas at San Antonio by Germán Placencia-Villa (*GPV*) for this work are synthesized according to a modified Brust-Schiffrin approach described elsewhere [1]. In brief, a stirred solution of 3:1 molar ratio of 3-MBA—HAuCl$_4$ (Sigma Aldrich, Saint Louis, MO, USA) was allowed to equilibrate for 16 h under basic conditions in 30% methanol (Fisher Scientific, Hampton, NH, USA) solution prior to the cluster-forming reduction reaction initiated by the addition of sodium borohydride (Sigma Aldrich, Saint Louis, MO, USA). This altered method has been shown to produce uniformly sized clusters, as opposed to the production of many discretely sized particles.

2.2. 3-MBA/Au System Characterization

The characterization of molecular nanoparticle preparations is carried out by a variety of methods for the characterization of the system of interest. Size-exclusion [28], gel-permeation [29], thin-layer chromatography [30], gel-electrophoresis [31], reversed-phase [32], and hydrophobic interaction [33] liquid chromatography has all been extensively used as essential analytical tools for the characterization of such nanoclusters. These methods separate the various components of a mixture according to one or more physical and/or chemical attributes, including differences in size, polarity, hydrophobic character, and electrophoretic mobility (related the size-to-charge ratio). Ion-pairing can be combined with reversed-phase LC for the analysis of acidic and basic clusters [34]. Separation methods may be used alone for sample fractionation or in conjunction with various detectors.

The analysis of nanoclusters through these methods is only possible for those samples exhibiting a certain degree of modal- or multi-modal distribution—with each mode showing a minimal variance. Samples that exhibit a continual distribution, as is the case of nanoparticles exceeding an approximately 3-nm core diameter, are not amenable to LC or MS analysis. 'Magic-number' nanocluster preparations are good candidates for a characterization by liquid chromatography and mass spectrometry because these clusters form in a multi-modal fashion, with only a few compositions exhibiting a high degree of stability. The LC-MS data acquired from these samples may be used to assign a specific cluster identity as well as for the semi-quantitative determination of each of the components present in a sample. Aqueous nanoparticles, like those investigated here, are of interest because of their potential application in medical and life sciences [2,3].

2.2.1. Coupled Chromatography—ESI-MS

In the present work, efforts were focused to determine whether the 3-MBA/Au systems were amenable to analysis by HPLC-ESI-MS in the same way as previously observed for the analogous

4-MBA/Au systems (aka p-MBA/Au). Specifically of interest was the possibility of ion-pairing with triethylammonium cations (TEAH+) [27] for the retention and separation of these poly-acid clusters via reversed phase chromatography. Additionally, of interest was an understanding on the effectiveness of this ion-pairing strategy for electrospray ionization (ESI) and if the necessary conditions could be implemented for a supported determination of cluster compositions with some degree of clarity by minimizing fragmentation. The successful implementation of HPLC-MS to these systems may help reveal 'hidden components' [35]—not otherwise known or detectable by native PAGE gel-electrophoresis. Any evidence to support or refute the proposed 'bidentate' bonding (H-bonding of carboxyl to Au) is also of interest in these studies.

Although gel-electrophoresis is a standard technique for the analysis of nanoparticle preparations, it is a relatively coarse size separation method. An exact determination of size and uniformity requires confirmation by a secondary analytical technique, since it is possible for the components having different sizes, shapes, or charges to share the same, or similar, electrophoretic mobilities.

2.2.2. HP-LC–ESI-MS Sample Preparation

The obtained Au/3-MBA samples were either re-dispersed or diluted—if a solid or solution, respectively—approximately 10× in the appropriate solution. LC separation was performed with coupled electrospray time-of-flight mass spectrometry detection (ToF-MS). The separations were carried out on a C_{18} stationary phase using gradient methods, whereby the initial mobile phase composition was replaced with a higher organic concentration in a linear fashion over a period of twenty minutes. The instrumental procedures, i.e., mass spectrometer, HPLC columns used in this work, are described elsewhere [36] in the HPLC-MS and UV−Vis Method Conditions section. The mobile phases were prepared containing 400 mM hexfluoroisopropanol (HFIP)-15 mM triethylamine (TEA), TEA-HFIP *or* 10 mM triethylammonium acetate (TEAA) in ddH_2O (mobile phase A) and methanol (mobile phase B). The separation behavior of the nanoclusters predominantly depends on the selected combination of the stationary phase, mobile phase, gradient, and mobile phase modifier. Starting from the conditions used to obtain a satisfactory separation and ionization of our previous report of larger p-MBA/Au MPCs [27], the gradient and modifier selection were varied to find the conditions for the satisfactory separations for this current 3-MBA (aka m-MBA)/Au MPCs work. In this work, 10–40% MeOH (mobile phase B) gradient over 20 min at ambient temperature were used for the efficient separation of the clusters, though the selection of the modifier (ion-paring agent) plays the vital rules as demonstrated in our results. The near-baseline separation of the various components is crucial for providing a differentiation and correlation between the various MS signals observed, which aids the MS interpretation and reduces the possibilities for ion-suppression artifacts. The gradient method can be adapted to produce a greater separation between components, and the mobile phase modifier is essential for a good chromatographic performance compatible with acceptable electrospray ionization. Solution phase ion-pairing effectively neutralizes the MBA's carboxylate ($-COO^-$) group by association with $TEAH^+$, enhancing the interaction of the mercaptobenzoic acid ligands with the C_{18} stationary phase.

3. Results

Recent reports have demonstrated the possibility of producing uniformly-sized batches of 3-mercaptobenzoic acid (MBA) protected nanoparticles [1–4]. Smaller nanoparticles, or nanoclusters, are noteworthy for their interesting properties, and because certain stoichiometries (i.e., gold-to-ligand ratios) form in abundance due to their relatively higher thermodynamic stability [4,37]. This phenomenon makes it possible to produce specific nanomolecular particles in abundance. However, because there exist various "magic-number" sizes (e.g., Au_{25}, Au_{38}, Au_{68}, Au_{102}, Au_{144}, etc.), nanocluster preparations may still exhibit heterogeneity or mixtures varying from one batch to the next. These improved synthetic procedures make it possible to produce size-focused preparations of nanoparticles, thus enabling the production of a higher-quality product.

Synthetic procedures such as these, in tandem with analytical methods that can be used to characterize these preparations, may provide the needed capabilities for the development of various nanoparticle applications. The 3-MBA/Au systems demonstrate 'certain advantages' over other thiolates (organic or hydrophobic) for the purposes of ligand exchange and conjugation, as well as for bio-applications.

Figure 1 shows negative-ionization (-ESI) LC-MS mode data acquired following the sample (prepared by the size-uniform synthesis procedure) provided by M. Azubel, as prepared at Stanford University. Two dominant components are readily identified: (67, 30; 17.8 kDa) and (48, 26; 13.4 kDa). These appear at the short (long) retention times and high (low) mass ends of the spectrum.

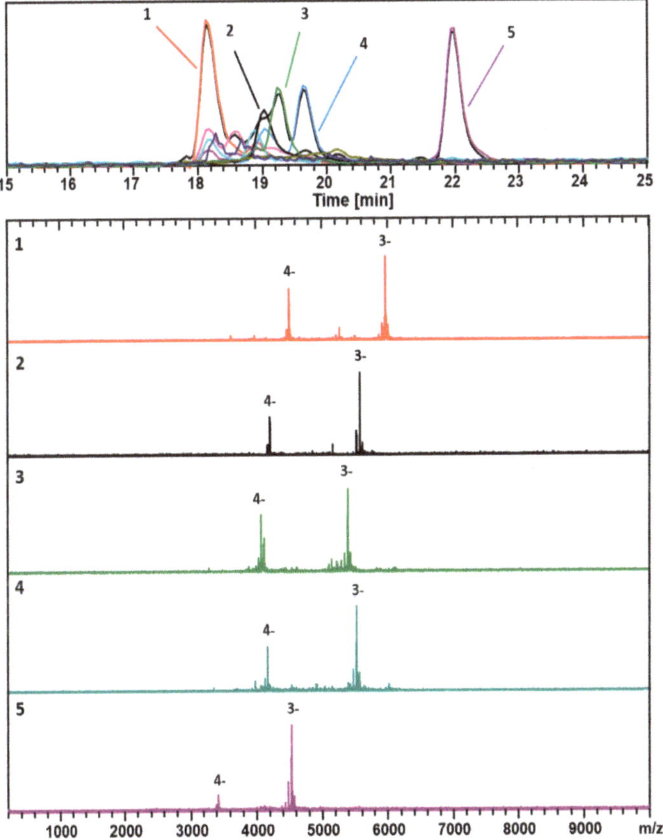

Figure 1. ESI-coupled LC-MS analysis of Au/3-MBA clusters from the Azubel-preparation. Detection is set for negative ions, under conditions that generate mainly 3− and 4− charge states. The top frame shows the chromatograms, i.e., the base peak chromatogram (*m/z* 100–10,000), and an extracted-ion chromatogram (EIC) for each identified component. The color-coded EIC chromatographic peaks track with the coded and numbered mass spectra listed herein with compositions assigned as follows: (1, Red) (67, 30), 17.8 kDa; (2, Black) (60, 31), 16.6 kDa; (3, Green) (58, 30), 16.0 kDa; (4, Blue) (60, 30), 16.4 kDa; and (5, Purple) (48, 26), 13.4 kDa. The fine-structure of the $(67, 30)^{3-}$ complexes is presented in Figure S1.

Figure 2 shows the results from the analysis of the same sample, obtained in the positive-ionization (+ESI) LC-MS mode. The mass spectra are shown for each of the 5 major components: (67, 30; 18.1 Da), (60, 31; 16.9 kDa), (58, 30; 16.2 kDa), (60, 30; 16.6 kDa), and (48, 26; 13.6 kDa). These are also considered when assigning the compositions listed in Figure 1. In both cases, 10 mM TEAA was used as the ion-pairing agent to facilitate the ionization process.

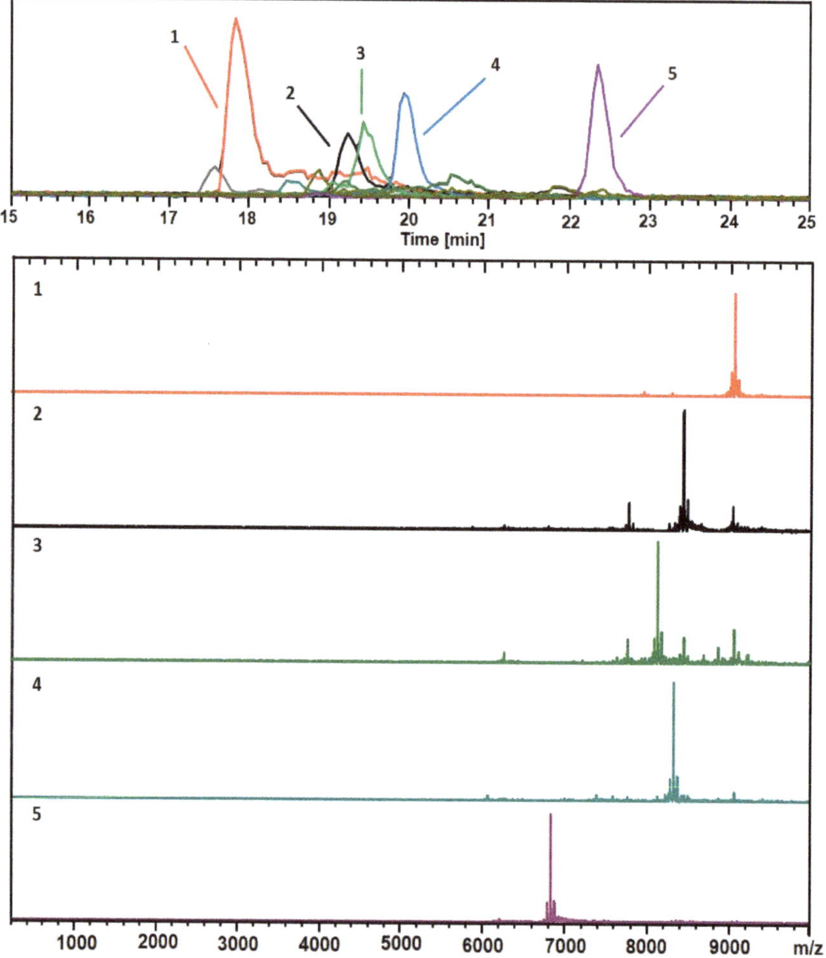

Figure 2. As in Figure 1, but with a positive (ESI+) mode for detection. This analysis shows mainly 2+ charge-states. The black trace corresponds to the base peak chromatogram (*m/z* 100–10,000). The color-coded EIC chromatographic peaks track with the coded and numbered mass spectra listed herein, with compositions assigned as follows: (1, Red) (67, 30), 18.1 kDa; (2, Black) (60, 31), 16.9 kDa; (3, Green) (58, 30), 16.2 kDa; (4, Blue) (60, 30), 16.6 kDa; and (5, Purple) (48, 26), 13.6 kDa. The fine structures of the $(67, 30)^{2+}$ complexes are presented in Figure S2.

Figures 3 and 4 show two analyses with two different ion-pairing agents TEAA and TEA-HFIP, respectively of a separate preparation (*GPV*) of 3-MBA clusters.

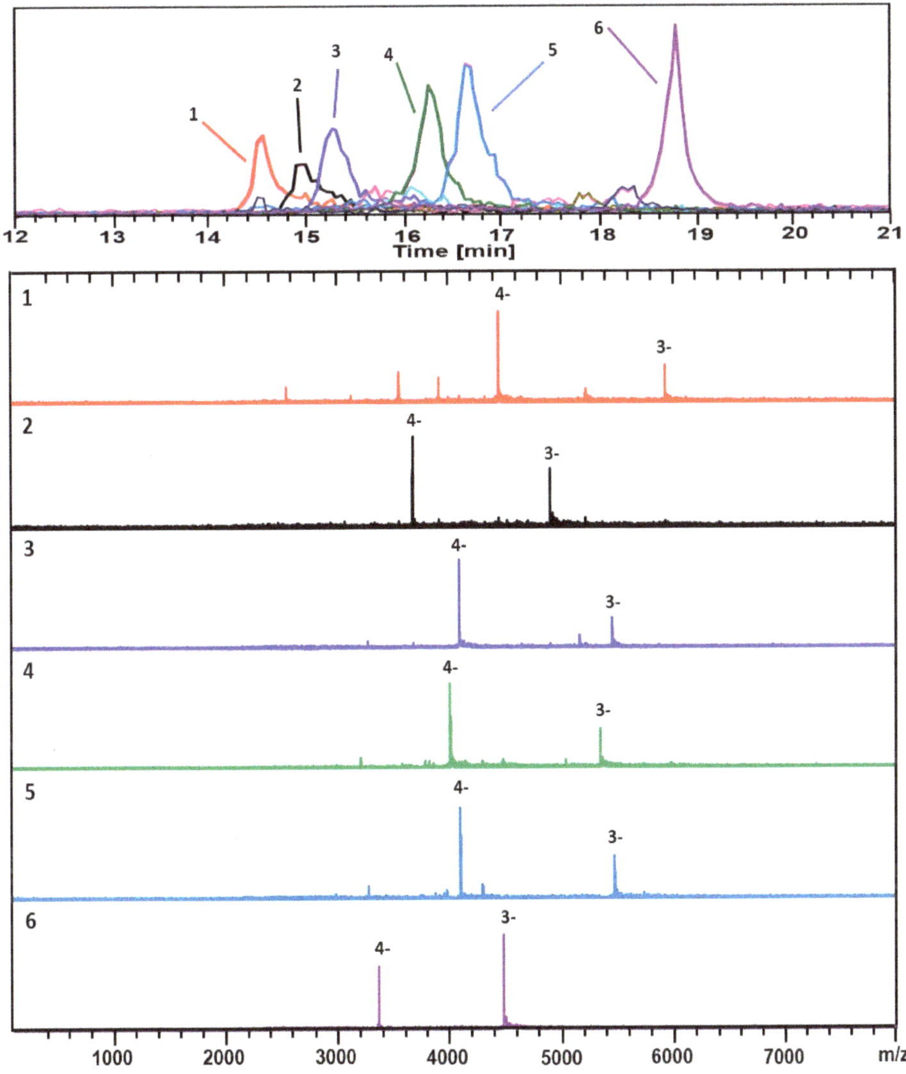

Figure 3. The analysis of a second preparation (*GPV*) of Au/3-MBA clusters. Negative ionization mode (-ESI) detection shows mainly 3− & 4− charge-states. The black trace corresponds to the base peak chromatogram (*m/z* 100–8,000). The color-coded EIC chromatographic peaks track with the coded and numbered mass spectra listed herein, with compositions assigned as follows: (1, Red) (67, 30), 17.8 kDa; (2, Black) (53, 28), 14.7 kDa; (3, Blue) (59, 31), 16.3 kDa; (4, Green) (58, 30), 16.0 kDa; (5, Light Blue) (60, 30), 16.4 kDa; and (6, Purple) (48, 26), 13.4 kDa. For the singly charged (z = 1−) of the same sample, see Figures S3 and S4. The polyacrylamide gel-electrophoresis (PAGE) analysis and corresponding HPLC-ESI-MS chromatogram are presented in Figure S4.

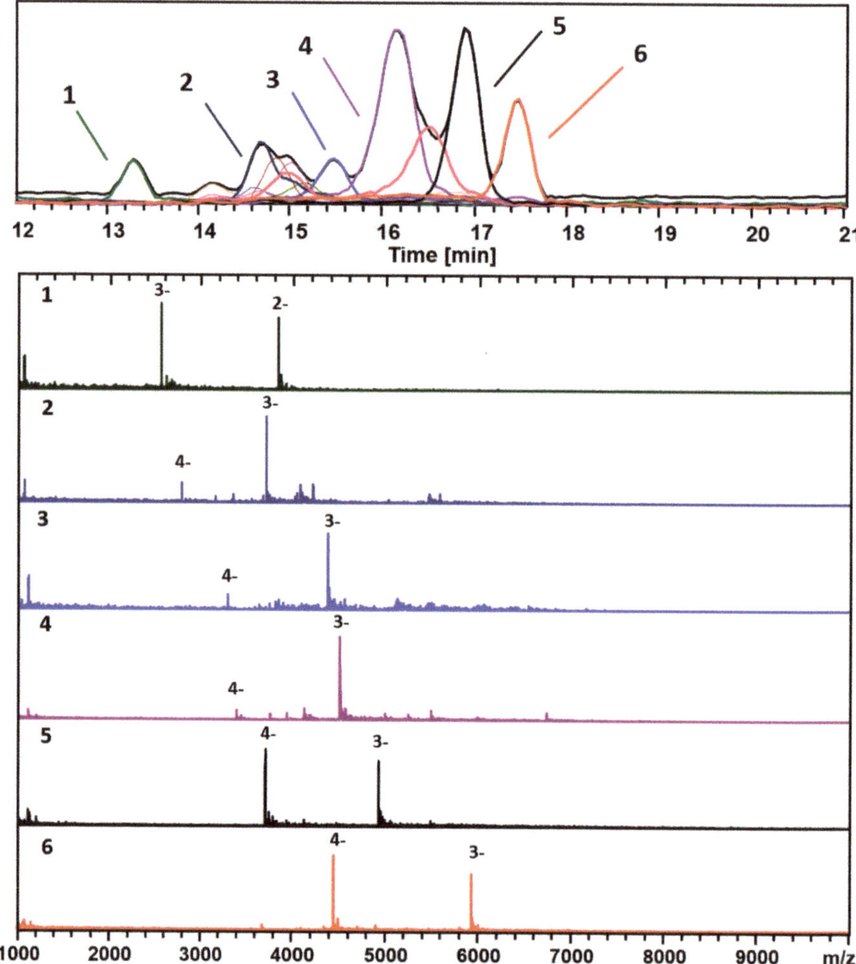

Figure 4. The analysis of the second preparation of 3-MBA clusters using the TEA–HFIP mobile phase buffer composition. The negative ionization mode was used for the analysis, mainly 3– and 4– charge-states. The black trace corresponds to the base peak chromatogram (*m/z* 1000–10,000). The color-coded EIC chromatographic peaks track with coded and numbered mass spectra listed, and the compositions are assigned as follows: (1, Dark green) (25, 18), 7.7 kDa; (2, Dark Blue) (38, 24), 11.1 kDa; (3, Blue) (46, 26), 13.0 kDa; (4, Purple) (48, 26), 13.4 kDa; (5, Black) (53, 28), 14.7 kDa; and (6, Red) (67, 30), 17.8 kDa.

Besides the main components, (67, 30; 17.8 kDa) and (48, 26; 13.4 kDa), identified in Figures 1 and 2 (*Azubel*'s sample), several other minor ones, (25, 18; 7.7 kDa), (38, 24; 11.1 kDa), are identified with our ESI-MS method, especially at a smaller mass. Figure 4 shows results for the same sample analyzed using a combination of a more volatile weak acid HFIP than acetic, and TEA. Interestingly, while the components observed in each analysis are essentially identical, the order of elution is significantly altered for the two modifiers. The TEAA modifier produces a chromatography whereby the larger clusters generally elute first, followed by smaller ones. The TEA-HFIP reverses this general trend, so that smaller clusters elute first, followed by larger ones.

Figure 5 contains a comparison among the mass spectra above (Figures 1–4), as they pertain to the putative "Au$_{68}$(3-MBA)$_{32}$" compound (calculated mass of 18.3 kDa), and also to an extract from the mass spectrum provided in reference [1]. In negative-ion detection, as appropriate to polyacids, the evidence all points toward 17.8 kDa, the mass of (67, 30). In positive-ion detection, where TEAH+ adducts provide the charge, the mass of 18.1 kDa also agrees with (67,30), assuming triple-adduction, i.e., 3 TEAH+, in which case the (67, 30) complex carries an intrinsic (core) charge of 1−.

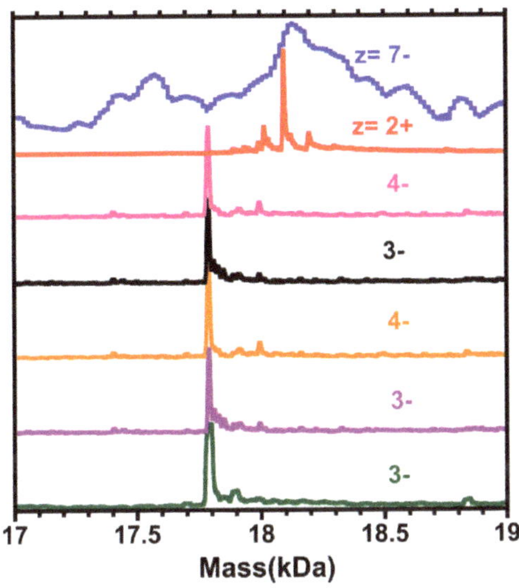

Figure 5. A comparison of the deconvoluted mass spectra in the region of the 17.8 kDa compound, putatively "Au$_{68}$(3-MBA)$_{32}$", vs. the ESI-MS of reference [1] (blue curve at top). Selected portions of the ESI mass spectra of gold cluster samples are depicted, in which the independent variable has been converted from the (*m/z*) scale to the total mass (kDa), using the charge (*z*) assignments indicated in Figure 2 (red), Figure 4 (pink and purple), Figure 1 (orange and black), and Figure 3 (dark green). Note that in the case of the positive ion mode (*z* = 2+), the peak is shifted higher by ~ +0.3 kDa, consistent with three (3) TEAH$^+$ adducts, to the (67,30)$^{1-}$ complex. [Mass of TEAH$^+$ = 101 Da.]

Figure 6 shows plots of the various chemical compositions observed for each of the different samples analyzed here. Although a number of different compositions were observed for each sample, a clear difference between the size-uniformity of the two samples can be observed. When a long equilibration prior to reduction is carried out, a much narrower range of cluster sizes is formed. If the procedure is varied—even slightly—to reduce this time period, a wider range of cluster sizes is formed, and this is supported by the recent reported "captamino", a base side, thiolated gold clusters in the size range of 25–144 numbers of Au, or even larger ones [17,18]. In Figure 3, cluster compositions ranging from (47, 26) to (71, 34) are observed; whereas in Figure 4, compositions ranging from (25, 18) to (67, 30) are observed.

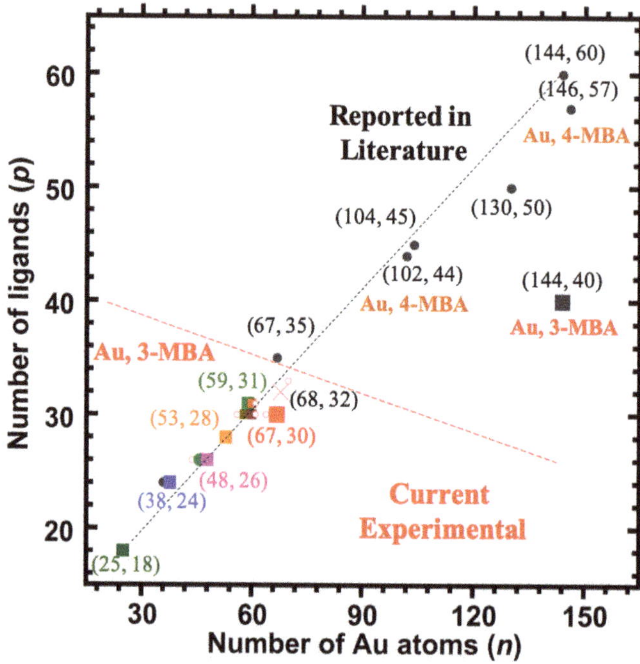

Figure 6. The number of ligands (p) vs. the number of Au atoms (n) found in this work on the Au_n(3-MBA)$_p$, (n = 48–67, p = 26–30) and reported in the literature are shown to see the trends. Legend: symbols designate whether the component identified was detected prominently (colored large square) and for both samples under all ESI-coupled LC/MS conditions, or not (small hollow red circle); a red cross represents the expected (68, 32) clusters, a grey large square represents 3-MBA thiolated larger (144, 40) clusters, whereas small grey circles (filled) represent reported different compositions of other thiolated clusters, for example 4-MBA (a sister molecule of 3-MBA) thiolated Au_{146}(4-MBA)$_{57}$.

4. Discussion

4.1. General Remarks

As mentioned in the Introduction, our general objective in this research has been to advance the analytical chemistry of thiolate-protected gold clusters. Specifically, we have aimed to adapt the standard HPLC-ESI-MS methodology, as applied for example to oligonucleotides, which are acidic (poly-anionic), developing optimized strategies for gold clusters protected by a monolayer of thiolate ligands with terminal (solution-exposed) acidic groups. The recent progress reported by Black et al. includes the HPLC-ESI-MS identification of a long series of Au-pMBA clusters as large as (146, 57) [5], or as small as (25, 18) and (36, 24). Through the use of a suitable ion-pairing agent (TEA+), well resolved mass spectra could be obtained under gentler conditions (to reduce poly-anion fragmentation in electrospray ionization) and nearly free from alkali-ion and solvent adducts [27]. In a related work, silver-lipoate clusters (29, 12)$^{3-}$ have been effectively resolved, where lipoate acts as a bidentate (di-thiolate) ligand with a terminal carboxylate ("thioctic acid") [36,38].

4.2. Contrasting 3-MBA (or Meta-MBA) and 4-MBA (Aka Para-MBA)

In turning from the 4-MBA (pMBA) to 3-MBA (or "mMBA") ligands, one faces a more challenging analytical situation, as described most recently in a 2017 *ACS Nano* report by Tero et al. [4], as well as in the earlier reports of Azubel et al. [1–3], dating to the 2014 *Science* article:

- There has been no total-structure determination of any 3-MBA protected gold clusters.
- There has been no adequately resolved ESI-MS identification of any of these: no composition-determination by any standard analytical method.
- Electron microscopy (or diffraction) provides the gold structure and atom count, in both (2) reported cases. (Ligands/S-atoms are not located by this method). Models are then constructed, which include the ligands, and these are tested (refined) by DFT computations.
- The compositions arrived at by these procedures, (68, 32) and (144, ~40), are respectively distinctly and strikingly different from those determined previously for aliphatic ligands, i.e., (67, 35) and (144, 60), or from the more directly relevant water-soluble aromatic pMBA ligand (146, 57). [Figure 6 presents these compositions in a graphical format.]

In practice, (via the same HPLC-ESI-MS optimized procedures) we were able to readily obtain clear results on the samples believed to be dominated by the (68, 32), but not on the samples labeled as larger compounds (144, ~40). This is not particularly surprising, for large polyanionic assemblies have a reputation for presenting a difficult ESI-MS analysis. For the same reason, the presence of readily detectable smaller components, such as the major one assigned to (48, 26), or even the minor one (25, 18) in one instance, is unsurprising, as their signal levels may be disproportionate to their concentration in solution.

Perhaps the major positive result of our work is the greatly improved (vs. 2014 Science report[1]) quality of ESI mass spectra (Figure 5) that led us to identify (67, 30) as the composition of the compound previously assigned to (68, 32). This is only a minor difference, amounting to a single gold atom and two (2) 3-MBA ligands, but could suggest a reinterpretation of its structure and bonding.

However, one should note that this suggested revision (reducing the ligand-count to 30 from 32) only serves to increase its distinctiveness, as compared to the reference (aliphatic) case, i.e., (67, 35) vs. (67, 30). Now, the 'ligand deficiency' (below) is five (5) rather than two (2), as indicated in Figure 6. The other components, and specifically the one identified as (48, 26), may also be interpreted within this same context of 'ligand deficiency'. Figure 6 shows that, for gold clusters protected by thiophenol-class thiolates, the important compositions (36, 24) and (44, 28) lie on a distinct 'curve'. Yet the smallest (minor) compound identified here as (25, 18) is the same regardless of the thiolate.

Below, we suggest how the 'dynamical stabilization' model—a form of bidentate ligation—of Tero et al. [4] can be generalized, from the two cases {(68, 32), (144, ~40)} investigated by them, to account for the entire range of compositions identified and presented in Figure 6. The dynamic-stabilization model (DSM) [4] was reported to account for the optical (FTIR) spectra as well as other analytical observations, in a way that also explains the ligand-count deficiency and the lability of these two compounds when exposed to other (non-3-MBA) thiolates in solution. In particular, the basis for this model is described as follows:

- In the carbonyl (C=O) stretching region, the vibrational FTIR spectra show "distinct peak[s] around 1730 cm^{-1}, observable only in 3-MBA-passivated clusters, and interpreted as the signal of the O=C−OH···Au interaction." [4].
- Molecular dynamic (MD) simulations were based on structure models for each cluster. "Visual inspection of MD trajectories revealed several weak interactions in the ligand layer and at the ligand−gold interface, such as formation of inter-ligand hydrogen bonds, inter-ligand π stacking (aromatic contacts), π−Au interaction where the aromatic ring lies "flat" on the gold core, and hydrogen bonding-like O=C−OH···Au interaction when the hydroxyl group is rotated toward the gold core." "*We thus assigned the highest frequency observed for both Au$_{144}$(3-MBA)~$_{40}$ and Au$_{68}$(3-MBA)$_{32}$ to the O=C−OH···Au interaction visualized* ... This interaction at the ligand−metal interface has not been reported before for any thiolate protected gold nanocluster." [4].

In the report, we have referred to any such interaction involving a second functional group (other than thiolate sulfur) as a "bidentate" bonding mode, whether the carboxyl group is protonated or de-protonated (as is more typical in solution-phase conditions, pH neutral or > 7).

First, the ligand-deficiency count, which ranges from 15–20 in the case of (144, ~40), amounts to five (~5) for (67, 30), to perhaps a couple (2) in the case of (48, 26), and finally to zero (0) in (25, 18), is taken to represent *the number of ligands bound in a bidentate fashion*. It is represented as a fraction of the whole: In the extreme case of (144, ~40), more than one-third of the ligands are bound in the bidentate mode. In the special case of (67, 30), one-sixth (5/30) are bidentate, and for (48, 26) only one-twelfth (2/26) are so indicated.

Second, as usual the key step is to relate these fractions to the estimated curvature (1/R) of the structure as measured at its surface, where R is the radial distance at which the Au–S or Au–X bonds lie. A high curvature removes the driving force for bidentate coordination, because the steric constraints are greatly reduced (the position meta to sulfur should be well exposed to the solvent and counter-cation).

For now, we leave this as a semiquantitative argument suitable for guiding further work on both the larger, or previously identified compounds, as well as the ones newly identified in this report. The need for this was well predicted in the closing remarks of reference [4]: "Several currently unknown compositions and sizes of 3-MBA-protected gold nanoclusters will undoubtedly be found by variations of the known syntheses, which will open unexplored possibilities for applications of these materials in biolabeling, catalyzing biochemical reactions, imaging, detection, and theranostics."

Supplementary Materials: The following are available online at http://www.mdpi.com/2079-4991/9/9/1303/s1; Figures S1–S4, fine-structure in the electrospray negative ionization mode mass spectrometric analysis of the (67, 30), complex (67, 30)$^{z-}$ in solution, electrospray positive ionization (ESI+) mass spectrometric analysis of the component identified as (67, 30), by HPLC-ESI-MS as in Figure 2, ESI-MS Analysis of *GPV* sample preparation, under conditions wherein mainly the singly charged (z = 1−) ions are detected, and the polyacrylamide gel electrophoresis (PAGE) and HPLC analyses of *GPV* sample preparation.

Author Contributions: The manuscript was written through contributions of all authors. All authors have given approval to the final version of the manuscript. D.M.B. and M.M.H. contributed equally.

Funding: The Welch Foundation, via Grant AX-1857, provided financial support for this work.

Acknowledgments: The co-authors thank Maia Azubel for providing the Stanford University samples used in this research, and Azubel and Roger Kornberg for their advice and encouragement.

Conflicts of Interest: The authors declare no conflict of interest.

References

1. Azubel, M.; Koivisto, J.; Malola, S.; Bushnell, D.; Hura, G.L.; Koh, A.L.; Tsunoyama, H.; Tsukuda, T.; Pettersson, M.; Häkkinen, H.; et al. Electron microscopy of gold nanoparticles at atomic resolution. *Science* **2014**, *345*, 909–912. [CrossRef] [PubMed]

2. Azubel, M.; Kornberg, R.D. Synthesis of water-soluble, thiolate-protected gold nanoparticles uniform in size. *Nano. Lett.* **2016**, *16*, 3348–3351. [CrossRef] [PubMed]

3. Azubel, M.; Koh, A.L.; Koyasu, K.; Tsukuda, T.; Kornberg, R.D. Structure determination of a water-soluble 144-gold atom particle at atomic resolution by aberration-corrected electron microscopy. *ACS Nano* **2017**, *11*, 11866–11871. [CrossRef] [PubMed]

4. Tero, T.-R.; Malola, S.; Koncz, B.; Pohjolainen, E.; Lautala, S.; Mustalahti, S.; Permi, P.; Groenhof, G.; Pettersson, M.; Häkkinen, H. Dynamic stabilization of the ligand–metal interface in atomically precise gold nanoclusters Au_{68} and Au_{144} protected by meta-mercaptobenzoic acid. *ACS Nano* **2017**, *11*, 11872–11879. [CrossRef] [PubMed]

5. Vergara, S.; Lukes, D.A.; Martynowycz, M.W.; Santiago, U.; Plascencia-Villa, G.; Weiss, S.C.; de la Cruz, M.J.; Black, D.M.; Alvarez, M.M.; López-Lozano, X.; et al. MicroED Structure of $Au_{146}(p\text{-MBA})_{57}$ at Subatomic Resolution Reveals a Twinned FCC Cluster. *J. Phys. Chem. Lett.* **2017**, *8*, 5523–5530. [CrossRef] [PubMed]

6. Kumara, C.; Hoque, M.M.; Zuo, X.; Cullen, D.A.; Whetten, R.L.; Dass, A. Isolation of a 300-kDa, $Au_{\sim1400}$ Gold Compound, the Standard 3.6-nm Capstone to a Series of Plasmonic Nanocrystals Protected by Aliphatic-Like-Thiolates. *J. Phys. Chem. Lett.* **2018**, *9*, 6825–6832. [CrossRef] [PubMed]

7. Aikens, C.M. Electronic and Geometric Structure, Optical Properties, and Excited State Behavior in Atomically Precise Thiolate-Stabilized Noble Metal Nanoclusters. *Acc. Chem. Res.* **2018**, *51*, 3065–3073. [CrossRef]

8. Templeton, A.C.; Wuelfing, W.P.; Murray, R.W. Monolayer-Protected Cluster Molecules. *Acc. Chem. Res.* **2000**, *33*, 27–36. [CrossRef]

9. Hoque, M.M.; Mayer, K.M.; Ponce, A.; Alvarez, M.M.; Whetten, R.L. Toward Smaller Aqueous-Phase Plasmonic Gold Nanoparticles: High-Stability Thiolate-Protected ~4.5 nm Cores. *Langmuir* **2019**, *35*, 10610–10617. [CrossRef]

10. Huang, Z.; Ishida, Y.; Narita, K.; Yonezawa, T. Kinetics of Cationic-Ligand-Exchange Reactions in Au_{25} Nanoclusters. *J. Phys. Chem. C* **2018**, *122*, 18142–18150. [CrossRef]

11. Ishida, Y.; Narita, K.; Yonezawa, T.; Whetten, R.L. Fully Cationized Gold Clusters: Synthesis of $Au_{25}(SR^+)_{18}$. *J. Phys. Chem. Lett.* **2016**, *7*, 3718–3722. [CrossRef] [PubMed]

12. Ishida, Y.; Suzuki, J.; Akita, I.; Yonezawa, T. Ultrarapid Cationization of Gold Nanoparticles via a Single-Step Ligand Exchange Reaction. *Langmuir* **2018**, *34*, 10668–10672. [CrossRef] [PubMed]

13. Gutierrez, E.; Powell, R.; Furuya, F.; Hainfeld, J.; Schaaff, T.; Shafigullin, M.; Stephens, P.; Whetten, R. Greengold, a giant cluster compound of unusual electronic structure. *Eur. Phys. J. D* **1999**, *9*, 647–651. [CrossRef]

14. Nieto-Ortega, B.; Bürgi, T. Vibrational Properties of Thiolate-Protected Gold Nanoclusters. *Acc. Chem. Res.* **2018**, *51*, 2811–2819. [CrossRef] [PubMed]

15. Hesari, M.; Ding, Z. A grand avenue to Au nanocluster electrochemiluminescence. *Acc. Chem. Res.* **2017**, *50*, 218–230. [CrossRef] [PubMed]

16. Bauld, R.; Hesari, M.; Workentin, M.S.; Fanchini, G. Tessellated gold nanostructures from $Au_{144}(SCH_2CH_2Ph)_{60}$ molecular precursors and their use in organic solar cell enhancement. *Nanoscale* **2014**, *6*, 7570–7575. [CrossRef] [PubMed]

17. Hoque, M.M.; Dass, A.; Mayer, K.M.; Whetten, R.L. Protein-Like Large Gold Clusters Based on the ω-Aminothiolate DMAET: Precision Thermal and Reaction Control Leading to Selective Formation of Cationic Gold Clusters in the Critical Size Range, n = 130–144 Gold Atoms. *J. Phys. Chem. C* **2019**, *123*, 14871–14879. [CrossRef]

18. Hoque, M.M.; Black, D.M.; Mayer, K.M.; Dass, A.; Whetten, R.L. Base Side of Noble Metal Clusters: Efficient Route to Captamino-Gold, $Au_n(-S(CH_2)_2N(CH_3)_2)p$, n = 25–144. *J. Phys. Chem. Lett.* **2019**, *10*, 3307–3311. [CrossRef] [PubMed]

19. Whetten, R.L.; Weissker, H.-C.; Pelayo, J.J.; Mullins, S.M.; López-Lozano, X.; Garzón, I.L. Chiral-Icosahedral (I) Symmetry in Ubiquitous Metallic Cluster Compounds (145A,60X): Structure and Bonding Principles. *Acc. Chem. Res.* **2019**, *52*, 34–43. [CrossRef]

20. Sakthivel, N.A.; Dass, A. Aromatic Thiolate-Protected Series of Gold Nanomolecules and a Contrary Structural Trend in Size Evolution. *Acc. Chem. Res.* **2018**, *51*, 1774–1783. [CrossRef]

21. Rambukwella, M.; Sakthivel, N.A.; Delcamp, J.H.; Sementa, L.; Fortunelli, A.; Dass, A. Ligand Structure Determines Nanoparticles' Atomic Structure, Metal-Ligand Interface and Properties. *Front. Chem.* **2018**, *6*. [CrossRef] [PubMed]

22. Yan, N.; Xia, N.; Liao, L.; Zhu, M.; Jin, F.; Jin, R.; Wu, Z. Unraveling the long-pursued Au_{144} structure by x-ray crystallography. *Sci. Adv.* **2018**, *4*, eaat7259. [CrossRef] [PubMed]

23. Hoque, M.M.; Vergara, S.; Das, P.P.; Ugarte, D.; Santiago, U.; Kumara, C.; Whetten, R.L.; Dass, A.; Ponce, A. Structural Analysis of Ligand-Protected Smaller Metallic Nanocrystals by Atomic Pair Distribution Function under Precession Electron Diffraction. *J. Phys. Chem. C* **2019**, *123*, 19894–19902. [CrossRef]

24. Sakthivel, N.A.; Theivendran, S.; Ganeshraj, V.; Oliver, A.G.; Dass, A. Crystal Structure of Faradaurate-279: $Au_{279}(SPh-tBu)_{84}$ Plasmonic Nanocrystal Molecules. *J. Am. Chem. Soc.* **2017**, *139*, 15450–15459. [CrossRef] [PubMed]

25. Chakraborty, I.; Pradeep, T. Atomically Precise Clusters of Noble Metals: Emerging Link between Atoms and Nanoparticles. *Chem. Rev.* **2017**, *117*, 8208–8271. [CrossRef] [PubMed]

26. Jin, R.; Zeng, C.; Zhou, M.; Chen, Y. Atomically Precise Colloidal Metal Nanoclusters and Nanoparticles: Fundamentals and Opportunities. *Chem. Rev.* **2016**, *116*, 10346–10413. [CrossRef]

27. Black, D.M.; Alvarez, M.M.; Yan, F.; Griffith, W.P.; Plascencia-Villa, G.; Bach, S.B.; Whetten, R.L. Triethylamine Solution for the Intractability of Aqueous Gold–Thiolate Cluster Anions: How Ion Pairing Enhances ESI-MS and HPLC of aq-$Au_n(pMBA)_p$. *J. Phys. Chem. C* **2016**, *121*, 10851–10857. [CrossRef]

28. Knoppe, S.; Boudon, J.; Dolamic, I.; Dass, A.; Bürgi, T. Size exclusion chromatography for semipreparative scale separation of $Au_{38}(SR)_{24}$ and $Au_{40}(SR)_{24}$ and larger clusters. *Anal. Chem.* **2011**, *83*, 5056–5061. [CrossRef]

29. Gies, A.P.; Hercules, D.M.; Gerdon, A.E.; Cliffel, D.E. Electrospray Mass Spectrometry Study of Tiopronin Monolayer-Protected Gold Nanoclusters. *J. Am. Chem. Soc.* **2007**, *129*, 1095–1104. [CrossRef]

30. Ghosh, A.; Hassinen, J.; Pulkkinen, P.; Tenhu, H.; Ras, R.H.; Pradeep, T. Simple and efficient separation of atomically precise noble metal clusters. *Anal. Chem.* **2014**, *86*, 12185–12190. [CrossRef]

31. Schaaff, T.G.; Knight, G.; Shafigullin, M.N.; Borkman, R.F.; Whetten, R.L. Isolation and selected properties of a 10.4 kDa gold: Glutathione cluster compound. *J. Phys. Chem. B* **1998**, *102*, 10643–10646. [CrossRef]

32. Choi, M.M.; Douglas, A.D.; Murray, R.W. Ion-pair chromatographic separation of water-soluble gold monolayer-protected clusters. *Anal. Chem.* **2006**, *78*, 2779–2785. [CrossRef] [PubMed]

33. Niihori, Y.; Shima, D.; Yoshida, K.; Hamada, K.; Nair, L.V.; Hossain, S.; Kurashige, W.; Negishi, Y. High-performance liquid chromatography mass spectrometry of gold and alloy clusters protected by hydrophilic thiolates. *Nanoscale* **2018**, *10*, 1641–1649. [CrossRef] [PubMed]

34. Niihori, Y.; Kikuchi, Y.; Shima, D.; Uchida, C.; Sharma, S.; Hossain, S.; Kurashige, W.; Negishi, Y. Separation of Glutathionate-Protected Gold Clusters by Reversed-Phase Ion-Pair High-Performance Liquid Chromatography. *Ind. Eng. Chem. Res.* **2017**, *56*, 1029–1035. [CrossRef]

35. Alvarez, M.M.; Chen, J.; Plascencia-Villa, G.; Black, D.M.; Griffith, W.P.; Garzón, I.L.; José-Yacamán, M.; Demeler, B.; Whetten, R.L. Hidden Components in Aqueous "Gold-144" Fractionated by PAGE: High-Resolution Orbitrap ESI-MS Identifies the Gold-102 and Higher All-Aromatic Au-pMBA Cluster Compounds. *J. Phys. Chem. B* **2016**, *120*, 6430–6438. [CrossRef] [PubMed]

36. Black, D.M.; Robles, G.; Lopez, P.; Bach, S.B.H.; Alvarez, M.; Whetten, R.L. Liquid Chromatography Separation and Mass Spectrometry Detection of Silver-Lipoate $Ag_{29}(LA)_{12}$ Nanoclusters: Evidence of Isomerism in the Solution Phase. *Anal. Chem.* **2018**, *90*, 2010–2017. [CrossRef] [PubMed]

37. Goswami, N.; Yao, Q.; Chen, T.; Xie, J. Mechanistic exploration and controlled synthesis of precise thiolate-gold nanoclusters. *Coord. Chem. Rev.* **2016**, *329*, 1–15. [CrossRef]

38. Lopez, P.; Lara, H.H.; Mullins, M.S.; Black, M.D.; Ramsower, H.M.; Alvarez, M.M.; Williams, T.L.; Lopez-Lozano, X.; Weissker, H.C.; García, A.P.; et al. Tetrahedral (T) Closed-Shell Cluster of 29 Silver Atoms & 12 Lipoate Ligands, $[Ag_{29}(R\text{-}\alpha\text{-}LA)_{12}]^{(3-)}$: Antibacterial and Antifungal Activity. *ACS Appl. Nano. Mater.* **2018**, *1*, 1595–1602.

 *nanomaterials*

Article

New Perspectives on the Electronic and Geometric Structure of $Au_{70}S_{20}(PPh_3)_{12}$ Cluster: Superatomic-Network Core Protected by Novel $Au_{12}(\mu_3\text{-}S)_{10}$ Staple Motifs

Zhimei Tian [1,2], **Yangyang Xu** [3] **and Longjiu Cheng** [1,4,*]

1   Department of Chemistry, Anhui University, Hefei 230601, China
2   School of Chemistry and Materials Engineering, Fuyang Normal University, Fuyang 236037, China
3   School of Social and Public Administration, East China University of Science and Technology, Shanghai 200237, China
4   Anhui Province Key Laboratory of Chemistry for Inorganic/Organic Hybrid Functionalized Materials, Anhui University, Hefei 230601, China
*   Correspondence: clj@ustc.edu; Tel.: +86-0551-63861279

Received: 5 July 2019; Accepted: 30 July 2019; Published: 6 August 2019

Abstract: In order to increase the understanding of the recently synthesized $Au_{70}S_{20}(PPh_3)_{12}$ cluster, we used the divide and protect concept and superatom network model (SAN) to study the electronic and geometric of the cluster. According to the experimental coordinates of the cluster, the study of $Au_{70}S_{20}(PPh_3)_{12}$ cluster was carried out using density functional theory calculations. Based on the superatom complex (SAC) model, the number of the valence electrons of the cluster is 30. It is not the number of valence electrons satisfied for a magic cluster. According to the concept of divide and protect, $Au_{70}S_{20}(PPh_3)_{12}$ cluster can be viewed as Au-core protected by various staple motifs. On the basis of SAN model, the Au-core is composed of a union of 2e-superatoms, and 2e-superatoms can be Au_3, Au_4, Au_5, or Au_6. $Au_{70}S_{20}(PPh_3)_{12}$ cluster should contain fifteen 2e-superatoms on the basis of SAN model. On analyzing the chemical bonding features of $Au_{70}S_{20}(PPh_3)_{12}$, we showed that the electronic structure of it has a network of fifteen 2e-superatoms, abbreviated as $15 \times 2e$ SAN. On the basis of the divide and protect concept, $Au_{70}S_{20}(PPh_3)_{12}$ cluster can be viewed as $Au_{46}^{16+}[Au_{12}(\mu_3\text{-}S)_{10}^{8-}]_2[PPh_3]_{12}$. The Au_{46}^{16+} core is composed of one Au_{22}^{12+} innermost core and ten surrounding 2e-Au_4 superatoms. The Au_{22}^{12+} innermost core can either be viewed as a network of five 2e-Au_6 superatoms, or be considered as a 10e-superatomic molecule. This new segmentation method can properly explain the structure and stability of $Au_{70}S_{20}(PPh_3)_{12}$ cluster. A novel extended staple motif $[Au_{12}(\mu_3\text{-}S)_{10}]^{8-}$ was discovered, which is a half-cage with ten μ_3-S units and six teeth. The six teeth staple motif enriches the family of staple motifs in ligand-protected Au clusters. $Au_{70}S_{20}(PPh_3)_{12}$ cluster derives its stability from SAN model and aurophilic interactions. Inspired by the half-cage motif, we design three core-in-cage clusters with cage staple motifs, $Cu_6@Au_{12}(\mu_3\text{-}S)_8$, $Ag_6@Au_{12}(\mu_3\text{-}S)_8$ and $Au_6@Au_{12}(\mu_3\text{-}S)_8$, which exhibit high thermostability and may be synthesized in future.

Keywords: $Au_{70}S_{20}(PPh_3)_{12}$ cluster; superatom network model; electronic structure; geometric structure

---

1. Introduction

Due to the applications in catalysis, optoelectronics, and photoluminescence, ligand-protected gold (Au-L) nanoclusters have drew much attention in both experiment [1–6] and theory [7–11]. The synthesis of Au-L clusters contributes much to many areas of science and technology because they have interesting structures [12,13]. In the past few years, the metalloid thiolate-protected Au nanoclusters with μ_3-S atoms have extended the family and potential applications of Au-L clusters. The experimentally determined metalloid Au-L clusters containing one or two μ_3-S include $Au_{21}S(SR)_{15}$ [14], $Au_{30}S(SR)_{18}$ [15,16], $Au_{38}S_2(SR)_{20}$ [17], and $Au_{103}S_2(SR)_{41}$ clusters [18], while $Au_{30}S_2(SR)_{18}$ cluster is a structure from theoretical prediction [19]. A large metalloid $Au_{108}S_{24}(PPh_3)_{16}$ cluster with 24 μ_3-S has been revealed, which consists of an octahedral Au_{44} core, an $Au_{48}S_{24}$ shell and 16 $Au(PPh_3)$ elements [20]. Very recently, Kenzler et al. has synthesized an intermediate size metalloid gold cluster $Au_{70}S_{20}(PPh_3)_{12}$, revealing an Au_{22} core surrounded by the $Au_{48}S_{20}(PPh_3)_{12}$ shell [21]. According to their report, Au_4S_4 unit is a central structural motif in the shell and they suggest that they could not elucidate a definite superatom character or distinct shell structure in the cluster. Thus, it is necessary to give a detailed study for the cluster, which may help to deeply understand the stability and structural nature of the cluster.

Häkkinen et al. proposed the divide-and-protect concept [22], and Au-L clusters are composed of Au-core and staple motifs; Au-core is protected by staple motifs. The concept has been widely used to predict and analyze the structures of Au-L clusters [23–32]. The idea of staple motif has been introduced since the synthesis of $Au_{102}(SR)_{44}$ cluster, and Jadzinsky et al. termed it [1]. To date, various forms of staple motifs ($-SR-(AuSR)_{x-}$) present in experimentally determined and theoretically predicted Au-L clusters. Monomer and dimer staple motifs present in $Au_{102}(SR)_{44}$ cluster [1]. Dimer staple motif also presents in $Au_{36}(SR)_{24}$ cluster [33]. Bridging -SR ligand and trimer staple motif exist in $Au_{23}(SR)_{16-}$ cluster [4]. In addition, gold-thiolate rings present in $Au_{20}(SR)_{16}$ and $Au_{22}(SR)_{18}$ clusters [8,34]. The protecting motifs include Au and SR, or only SR; moreover, they have two legs. We have predicted a tridentate staple motif with three S legs in the synthesized $Au_{30}S(SR)_{18}$ cluster before [19]. According to the superatom complex (SAC) concept proposed by Häkkinen et al [35], the number of valence electrons (V) for $Au_mS_n(SR)_p^q$ cluster is computed as bellow: $V = m - 2n - p - q$, in which m, n and p are the numbers of Au, S and SR, respectively, whereas q is the charge of the cluster. The super shells for spherical Au clusters is $|1S^2|1P^6|1D^{10}|2S^21F^{14}|2P^61G^{18}|\ldots$ (S–P–D–F–G–H– denote angular-momentum characters), corresponded to magic numbers 2, 8, 18, 34, 58, ... According to SAC model, clusters with valence electrons 2, 8, 18, 34, 58, … present special stability and they are magic number clusters. The theoretically predicted $Au_{12}(SR)_9^+$ and $Au_8(SR)_6$ are 2e magic clusters. $Au_{25}(SR)_{18}^-$, $Au_{44}(SR)_{28}^{2-}$ and $Au_{102}(SR)_{44}$ are 8e, 18e and 58e magic number clusters, respectively. Cheng et al. introduced the superatom-network (SAN) model, which has been used to explore the stability of $Au_{18}(SR)_{14}$, $Au_{20}(SR)_{16}$, $Au_{24}(SR)_{20}$, $Au_{44}(SR)_{28}$ and $Au_{22}(SR)_{18}$ clusters [8,36,37]. Based on the concept of SAN model, the Au-core of Au-L cluster can be viewed as a network of 2e Au_n (n = 3, 4, 5 or 6) superatoms. The interactions between the superatoms are main non-bond interactions.

Here, we investigate the electronic and geometric structure of $Au_{70}S_{20}(PPh_3)_{12}$ to obtain deep understanding of it. Based on the superatom complex (SAC) model, this cluster is a 30e compound [35]. The number of valence electrons for $Au_{70}S_{20}(PCH_3)_{12}$ cluster does not satisfy the magic number electrons of SAC model. Kenzler et al. reported that the Au core of $Au_{70}S_{20}(PPh_3)_{12}$ cluster is Au_{22}, and the protecting tetrahedral shell is composed of four Au_4S_4 units, four S atoms and 32 gold atoms, and no staple motif presents [21]. We are interested in the synthesized $Au_{70}S_{20}(PPh_3)_{12}$ cluster, which has 20 μ_3-S atoms. Now that the number of the valence electrons does not satisfy the SAC model, why it is stable? How do the 20 μ_3-S atoms protect the Au-core? What are the protecting motifs of the cluster? With these questions in mind, we tried to analyze the electronic and geometric structure of the cluster using existing theories and models. This work attempts to explain the structure and properties from a new perspective.

2. Materials and Methods

We start from the experimental structure of $Au_{70}S_{20}(PPh_3)_{12}$ determined as reported by Kenzler et al [21] and the total charge is set to zero. Considering the calculation amount, we used CH_3 instead of all the Ph ligands, and the structure was then relaxed using the Gaussian 09 software (Revision B 01;

Gaussian, Inc., Wallingford, CT, USA) [38]. Density-functional theory (DFT) calculations were employed to optimize the geometric structure using Perdew–Burke–Ernzerhof (PBE0) functional [39]. The basis set of Au element is Lanl2dz, while 6-31G * is used for S, P, C, H elements. The molecular orbital (MO) and natural bond orbital (NBO) calculations of Au cores were also carried out at the same level, whereas the basis set of Au element was Lanl2mb. The adaptive natural density partitioning (AdNDP) method was used to analyze the chemical bonding patterns [40]. MOLEKEL software (version 5.4.0.8, Swiss National Supercomputing Centre, Manno, Switzerland) [41] was used to view the chemical bonding patterns. The superatom-network (SAN) model was taken to analyze the chemical bonds in $Au_{70}S_{20}(PPh_3)_{12}$ cluster [36].

3. Results and Discussion

3.1. Geometric Structure

The structure of the relaxed $Au_{70}S_{20}(PCH_3)_{12}$ cluster is given in Figure 1b, which is in D_2 symmetry. The structural parameters computed here reproduce well with the experimental results.

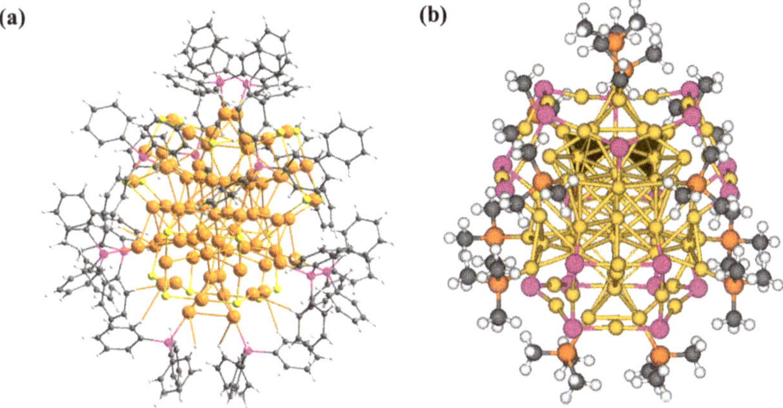

Figure 1. (**a**) Single crystal XRD structure of $Au_{70}S_{20}(PPh_3)_{12}$ from [21], reproduced with permission, Royal Society of Chemistry, 2017; (**b**) The optimized structure of $Au_{70}S_{20}(PCH_3)_{12}$ cluster. The cluster is obtained at the PBE0/LanL2dz(Au) and 6-31G *(S, C, P, H) level of theory. Au, yellow, S, purple; P, orange; C, gray, H, white.

Based on the divide-and-protect concept [22], different building blocks were tried to find the proper segmentation mode. The cluster can be viewed as Au-core and protecting motifs. Through analysis on the structure, the protecting motifs include twelve separate PCH_3 and $(Au-S)_n$ motifs. According to the segmentation analysis in Supplementary Materials, $Au_{70}S_{20}(PCH_3)_{12}$ cluster is divided into three parts as Figures 2 and 3 show. $Au_{70}S_{20}(PCH_3)_{12}$ cluster can be written as $Au_{70}S_{20}(PCH_3)_{12}$ = $[Au_{46}{}^{16+}][Au_{12}(\mu_3\text{-}S)_{10}{}^{8-}]_2[PCH_3]_{12}$. The core of the cluster is $Au_{46}{}^{16+}$ with two new $[Au_{12}(\mu_3\text{-}S)_{10}]^{8-}$ staple motifs and 12 PCH_3 protecting it. The $[Au_{12}(\mu_3\text{-}S)_{10}]^{8-}$ staple motif containing ten μ_3-S atoms is observed for the first time in Au-L clusters. As shown in Figures 2c and 3c, $[Au_{12}(\mu_3\text{-}S)_{10}]^{8-}$ motif can be easily identified from the cluster. Worth noting is that $[Au_{12}(\mu_3\text{-}S)_{10}]^{8-}$ motif has six branches, which is obviously different from common staple motif and it is unprecedented in $Au_m(SR)_n$ clusters. Each S atom is triply coordinated to the neighboring Au atoms in a μ_3 bridging form. $[Au_{12}(\mu_3\text{-}S)_{10}]^{8-}$ motif has six S legs, thus we term it six-tooth staple motif. According to the theoretical studies by Jiang et al., other motifs than common staple motifs may exist [42]. Moreover, Au_xS_y unit is theoretically predicted existing in core-shell structures of Au_mS_n clusters [43,44]. $[Au_{12}S_8]^{4-}$ anion presented in the synthesized crystal thioaurate $[Ph_4As]_4[Au_{12}S_8]$. The framework of $[Au_{12}S_8]^{4-}$ anion is a distorted cube, moreover, sulfur, and gold atoms locate at the corners and edge midpoints of the cubic structure, respectively [45].

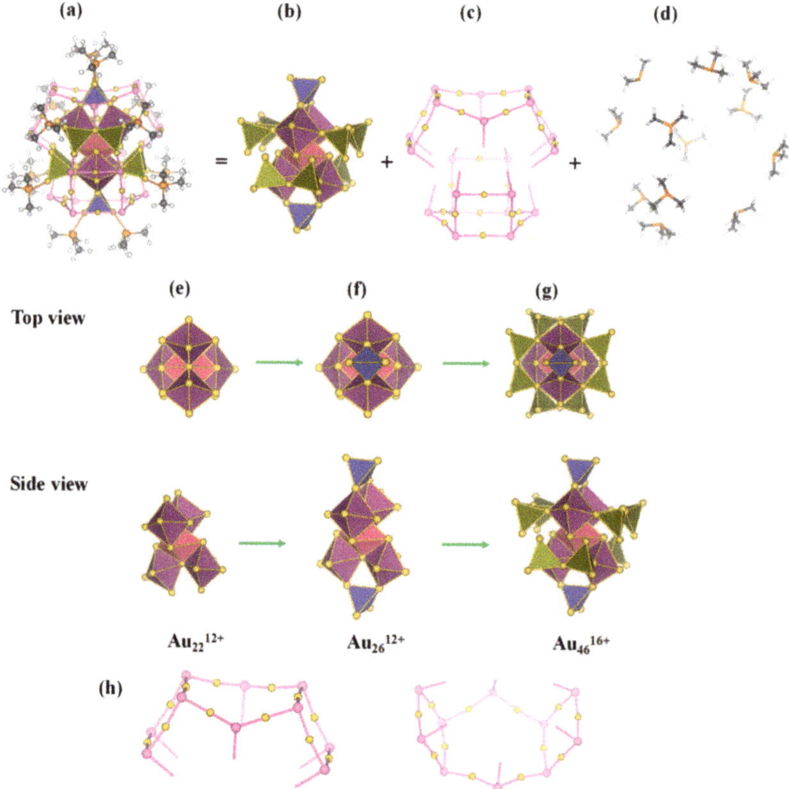

Figure 2. (**a**) Structural model of $Au_{70}S_{20}(PCH_3)_{12}$. The $[Au_{12}(\mu_3\text{-}S)_{10}]^{8-}$ and PCH_3 protecting motifs are given as ball-and-stick models (Au, yellow; P, orange; S, pink; C, gray; H, white). The Au cores are shown as polyhedra. (**b**) Model of Au_{46}^{16+} core, (**c**) Two $[Au_{12}(\mu_3\text{-}S)_{10}]^{8-}$ six-tooth staple motifs, (**d**) Model of twelve PCH_3 protecting motif, (**e**) $6 \times 2e$ SAN of Au_{22}^{12+} core, (**f**) Au_{26}^{12+} core, (**g**) Au_{46}^{16+} core, (**h**) Two views of $[Au_{12}(\mu_3\text{-}S)_{10}]^{8-}$ staple motif.

The $Au_3(\mu_3\text{-}S)$ unit has been proposed as an elementary block and used to design a group of quasi-fullerence hollow-cage $[Au_{3n}(\mu_3\text{-}S)_{2n}]^{n-}$ clusters with high stability [46]. $[Au_{12}(\mu_3\text{-}S)_{10}]^{8-}$ motif can be viewed as a part of $[Au_{15}(\mu_3\text{-}S)_{10}]^{5-}$ cluster, which is a half cage. Here, $[Au_{12}(\mu_3\text{-}S)_{10}]^{8-}$ six-tooth staple motif as a whole protects Au_{46}^{16+} core. The configuration of the vertex-sharing Au_7 core in Au_{46}^{16+} core resembles those in $Au_{28}(SR)_{20}$ and $Au_{20}(SR)_{16}$ clusters [34,47]. From Figure 2, Au_{46}^{16+} core is composed of five edge-sharing Au_6, four vertex-sharing Au_7 and two Au_4 superatoms. The five Au_6 superatoms compose an Au_{22}^{12+} kernel. The valence electrons of Au_{22}^{12+} core is 10e, which is also a 10e superatomic molecule (Figures 1c and 3).

From Figure 2, we can see that the 12 terminal S legs in two $[Au_{12}(\mu_3\text{-}S)_{10}]^{8-}$ staple motifs connect to the neighboring Au_7 cores. The two $[Au_{12}(\mu_3\text{-}S)_{10}]^{8-}$ motifs protect Au_{46}^{16+} core from both top and bottom sides stabilizing the cluster. The average bond length of Au-S in $[Au_{12}(\mu_3\text{-}S)_{10}]^{8-}$ is 2.39 Å suggesting a covalent single bond. The average bond angle of $\angle$Au-S-Au is 94.7° and thus deviate only slightly from the ideal 90° expected for bonding involving the sulfur 3p orbitals. Gold attempts to maintain linearity with average bond angle of $\angle$S-Au-S being 171.4°.

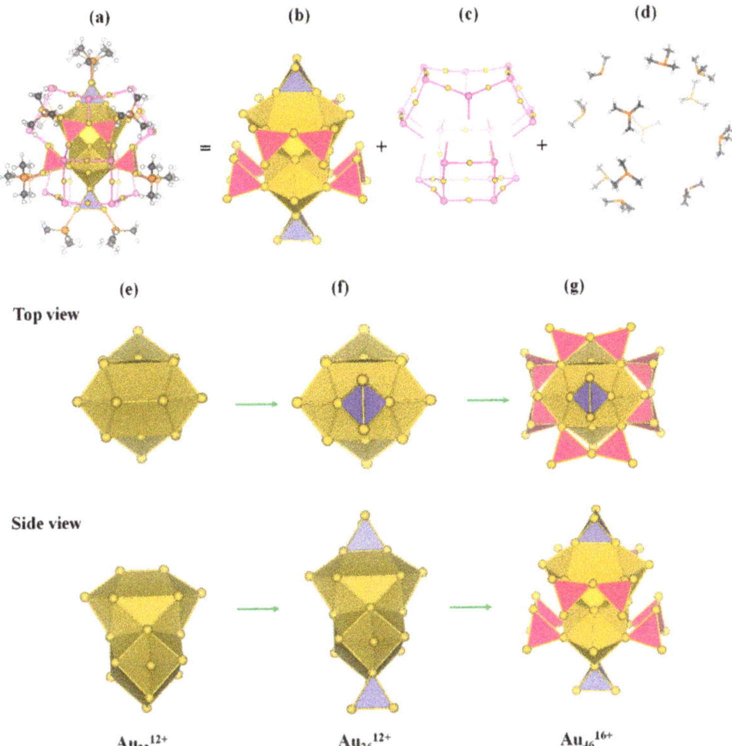

Figure 3. (**a**) Structural model of $Au_{70}S_{20}(PCH_3)_{12}$. The $[Au_{12}(\mu_3\text{-}S)_{10}]^{8-}$ and PCH_3 protecting motifs are given as ball-and-stick models (Au, yellow; P, orange; S, pink; C, gray; H, white). The Au cores are shown as polyhedra. (**b**) Model of Au_{46}^{16+} core, (**c**) Two $[Au_{12}(\mu_3\text{-}S)_{10}]^{8-}$ staple motifs, (**d**) Model of twelve PCH_3 protecting motif, (**e**) Au_{22}^{12+} superatomic molecule, (**f**) Au_{26}^{12+} core, (**g**) Au_{46}^{16+} core.

Figure S1 gives Au–Au contacts in the optimized structure of $Au_{70}S_{20}(PCH_3)_{12}$ cluster: (a) Au_{22} innermost core is $5 \times 2e$ SAN, (b) Au_{22} innermost core is a 10e-superatomic molecule. Also given are the aurophilic contacts between motifs and superatoms and the aurophilic contacts between superatoms. Noticeable gold–gold interactions (baby blue and black lines in Figure S1a,b, Supporting Materials) between the Au atoms in $[Au_{12}(\mu_3\text{-}S)_{10}]^{8-}$ and neighboring gold cores are present. The Au–Au aurophilic distances range from 2.82–3.01 Å, with the average Au–Au distance being 2.91 Å smaller than the Au–Au van der Waals radii (3.32 Å) [48,49]. The blue lines in Figure S1a label the aurophilic interactions between Au_6 and Au_4 cores, and the interactions between Au_4 cores. The green lines in Figure S1b label the aurophilic interactions between Au_{22} and neighboring Au_4 cores. The Au–Au distances range from 2.86–3.01 Å, and the average Au–Au distance is 2.93 Å. The short bond distance between Au and Au indicates strong aurophilic interactions. Thus, the interaction mode between six-tooth staple motifs and Au cores includes clamping and aurophilic interactions, which stabilize the $Au_{70}S_{20}(PCH_3)_{12}$ cluster. Here, the staple motif can extend to six-tooth mode. The staple motif only includes Au and S elements, which is obviously different from previous staple motifs. From above analysis, we can see that both the position of the six-tooth staple motifs and Au–Au contacts in the cluster dedicate to the stability of $Au_{70}S_{20}(PCH_3)_{12}$ cluster.

3.2. Chemical Bonding Analysis

In order to verify the electronic structure of $Au_{70}S_{20}(PCH_3)_{12}$ cluster, we carried out chemical bonding analysis. The electronic structure of the cluster followed the SAN model, that is, it had a network of fifteen 2e-superatoms, abbreviated as 15 × 2e SAN, which contained five 2e-Au_6 and ten 2e-Au_4 superatoms. We took the Au_{46}^{16+} core out of the cluster separately while keeping the structure identical to that in $Au_{70}S_{20}(PCH_3)_{12}$ cluster to analyze the chemical bonds. As expected, AdNDP analysis in Figure 4 indicated that there are 10 four-center-two-electron (4c–2e) bonds with occupancy numbers (ON) = 1.54–1.56 |e|, five 6c–2e bonds with ONs = 1.63–1.68 |e|. Vertex-sharing Au_4 superatoms were present in the experimentally determined $Au_{20}(SR)_{16}$ and $Au_{36}(SR)_{24}$ clusters [33,34].

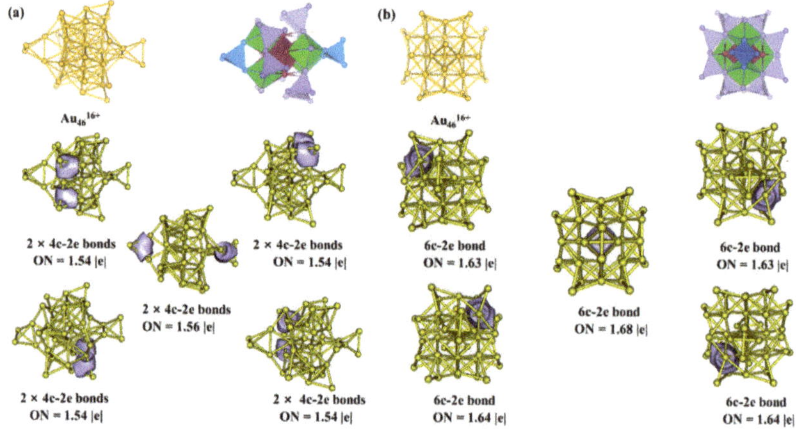

Figure 4. Structures, superatom-network models and adaptive natural density partitioning (AdNDP) localized natural bonding orbitals of (**a**) 4c–2e bonds (side view), (**b**) 6c–2e bonds (top view) in Au_{46}^{16+} core of $Au_{70}S_{20}(PCH_3)_{12}$ cluster.

For purposes of confirming the segmentation scheme, the difference of Au–Au distances inside the Au_{46} core and those between Au_{46} core and two six-tooth staple motifs were recorded. Figure S2 (Supporting Materials) displays all the Au–Au distances, which include the distances between Au_{46} core and two six-tooth staple motifs (black dots), the Au–Au distances in Au_{22} core (red dots), in two Au_4 superatoms on top and bottom of the cluster (blue dots), in the four pairs of vertex-sharing Au_4 superatoms (purple dots). The average Au–Au distances of the above four groups were 2.90, 2.91, 2.82, and 2.86 Å, respectively. From the figure, we can see that, the Au–Au distances between Au_{46} core and two six-tooth staple motifs and distances in the Au_{22} core were relatively bigger than other two groups. The Au_{22} core was consistent with the former report [21]. The reason for the Au–Au distances in Au_{22} core being relatively bigger are probably that the repulsive interactions of Au atoms can be reduced in this way. Lower repulsion is helpful to form a Au_{22} core. The Au–Au distances in the ten Au_4 superatoms were shorter than those between the Au-core and staple motifs, which follow the concept of SAN model. The shorter Au–Au distances were helpful to the formation of Au_4 superatoms. In short, the existence of ten Au_4 superatoms were reasonable, which has been supported from the viewpoint of Au–Au distances.

Further analysis of the innermost Au_{22}^{12+} core was performed and the structure of Au_{22}^{12+} core stayed the same as that in Au_{46}^{16+} core. The results are given in Figure 5. From Figure 5a, we can see that Au_{22}^{12+} core can be viewed as five edge-sharing Au_6 superatoms. AdNDP analysis confirms that there are five 6c–2e bonds. ON is 1.83 |e| for the middle 6c–2e bond, while ONs are all 1.77 |e| for the marginal 6c–2e bonds. The Au_9 kernel in $Au_{18}(SR)_{14}$ cluster consists of two Au_6 superatoms [50,51]. Au_{22}^{12+} core has 10 valence electrons which is identical to a N_2 molecule, and it can be viewed as a super-N_2 molecule. From Figure 5b, AdNDP analysis demonstrates that Au_{22}^{12+} has two 11c–2e super 1S lone pairs with ONs being 1.91 |e|, one 22c–2e super-σ bond and two 22c–2e super π bonds with ONs being 2.00 |e|.

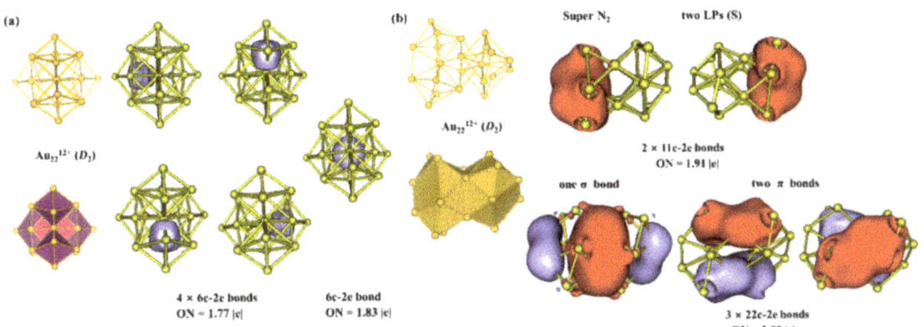

Figure 5. (**a**) Structure, superatom-network model and AdNDP localized natural bonding orbitals of Au_{22}^{12+} core. (**b**) Structure, superatomic molecular model and AdNDP localized natural bonding orbitals of Au_{22}^{12+} superatomic molecule.

3.3. Aromatic Analysis

NICS-scan method is proposed by Stanger, which is similar to the screen method of aromatic center and has been used to predict the aromatic properties of molecules and clusters [52–54]. Here, we use NICS-scan method to further verify the existence of Au_4 superatoms and we have demonstrated the existence of Au_4 superatoms in $Au_{20}(SR)_{16}$, $Au_{28}(SR)_{20}$ and $Au_{30}S_2(SR)_{18}$ clusters in our former work [19,36]. Figure 6 is the NICS-scan curve of Au_{46}^{16+} core along the centers of two neighboring Au_4 superatoms in the range of −6.0–6.0 Å. The position of NICS(0) is set at the midpoint of the geometric centers of two Au_4 superatoms. Two views of the scan in Au_{46}^{16+} have been given in the figure. Considering the symmetry of Au_{46}^{16+}, we only give one scan curve of the cluster. It is obvious that there are two dotted ovals in the figure, indicating two non-conjugate Au_4 superatoms, which further support the SAN model. The NICS(0) values of the two Au_4 superatoms are both −32.2 ppm much smaller than benzene molecule (−9.7 ppm), indicating strong aromaticity. The NICS-scan method is applied to verify Au_4 superatoms in Au_{46}^{16+} core, thus the Au_4 superatoms are further verified from the aromatic view.

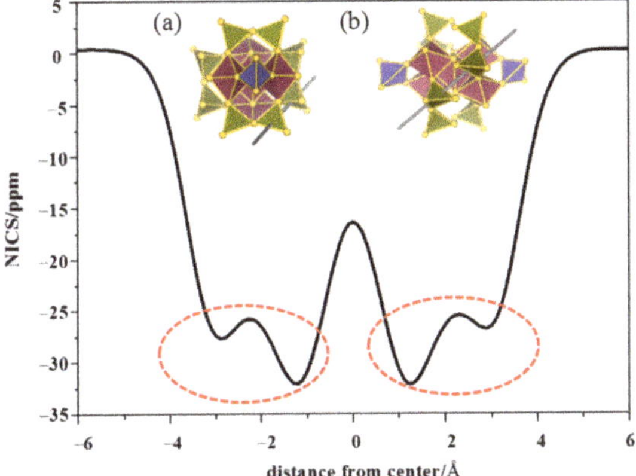

Figure 6. NICScc-scan curve of the Au_{46}^{16+} core, which is the scan along the centers of the neighboring Au_4 superatoms in the range of –6.0–6.0 Å. The red dotted ovals in the figure signal the presence of Au_4 superatoms. The structures labeled in (**a**) and (**b**) indicate two views of the scan.

3.4. $Cu_6@Au_{12}(\mu_3\text{-}S)_8$, $Ag_6@Au_{12}(\mu_3\text{-}S)_8$, and $Au_6@Au_{12}(\mu_3\text{-}S)_8$ Clusters

Worth noting is that $[Au_{12}(\mu_3\text{-}S)_8]^{4-}$ was experimentally crystallized earlier [45]. Meanwhile, the structure of $[Au_{12}(\mu_3\text{-}S)_8]^{4-}$ was theoretically studied [46]. We obtained the optimized structure and harmonic frequencies of $[Au_{12}(\mu_3\text{-}S)_8]^{4-}$ cluster at the level of PBE0/Lanl2dz(Au), 6-31G *(S). The optimized structure presented a cubic structure in O_h symmetry and the Au-S bond length is 2.38 Å. It was found that the harmonic vibrational frequencies of $[Au_{12}(\mu_3\text{-}S)_8]^{4-}$ were all positive. The HOMO-LUMO gap was 2.90 eV, further indicating its high stability.

Jiang et al. have predicted several core-in-cage gold sulfide $Au_xS_y^-$ clusters observed in MALDI fragmentation of $Au_{25}(SR)_{18}^-$ cluster theoretically [42]. They stated that the Au core in the core-in-cage cluster may catalyze reactions. Inspired by the half-cage $[Au_{12}(\mu_3\text{-}S)_{10}]^{8-}$ staple motif, the cubic $[Au_{12}(\mu_3\text{-}S)_8]^{4-}$ cluster can be regarded as a cage staple motif. Thus, we designed three core-in-cage clusters, $Cu_6@Au_{12}(\mu_3\text{-}S)_8$, $Ag_6@Au_{12}(\mu_3\text{-}S)_8$, and $Au_6@Au_{12}(\mu_3\text{-}S)_8$. The structures, models and AdNDP analysis of the three designed clusters are collected in Figure 7. The core-in-cage clusters can keep O_h symmetry after relaxation. The harmonic vibrational frequencies of the three clusters are all positive, indicating they are real local minima on potential energy surfaces. The infrared spectrograms (IR) of them are given in Figure S3. The HOMO-LUMO gaps of $Cu_6@Au_{12}(\mu_3\text{-}S)_8$, $Ag_6@Au_{12}(\mu_3\text{-}S)_8$, and $Au_6@Au_{12}(\mu_3\text{-}S)_8$ clusters are 3.59, 2.97, and 2.87 eV, suggesting their high stability. $Cu_6@Au_{12}(\mu_3\text{-}S)_8$ is more stable than other two clusters because Ag_6 and Au_6 are too large. The Cu–Au, Ag–Au and Au–Au distances between the atoms in core and cage of the three clusters are 2.63, 2.74 and 2.74 Å (Figure 7), respectively. All of them are smaller than the sum of their van der Waals radii (3.12, 3.38, and 3.32 Å) [55], demonstrating that Cu–Au, Ag–Au, and Au–Au interactions play a dominant role in stabilizing the clusters. The cores of the designed clusters are all-metal, which are reminiscent of all-metal aromatic. Thus it is necessary to calculate the NICS(0) values to evaluate the stabilities. The NICS(0) values of $Cu_6@Au_{12}(\mu_3\text{-}S)_8$, $Ag_6@Au_{12}(\mu_3\text{-}S)_8$, and $Au_6@Au_{12}(\mu_3\text{-}S)_8$ are −19.6, −17.0, and −17.9 ppm, respectively. The largely negative NICS(0) values of the cores exhibit that they are aromatic and stable. The aromaticity of the centers contributes to the stabilities of the clusters.

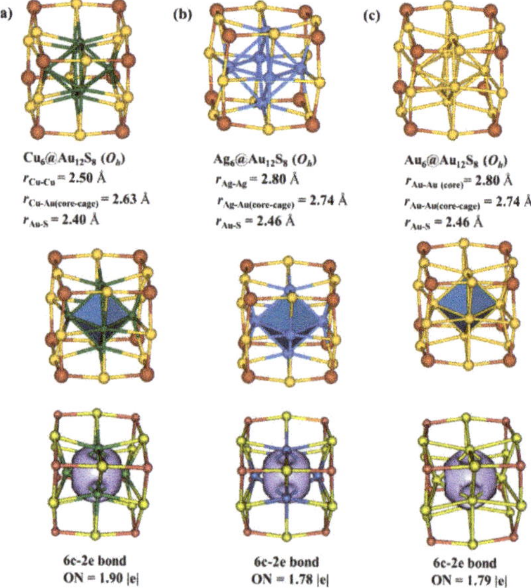

Figure 7. Structures, superatom models and AdNDP localized natural bonding orbitals of 6c-2e bonds in (**a**) $Cu_6@Au_{12}(\mu_3\text{-}S)_8$, (**b**) $Ag_6@Au_{12}(\mu_3\text{-}S)_8$, and (**c**) $Au_6@Au_{12}(\mu_3\text{-}S)_8$ clusters. Cu, green; Ag, blue; Au, yellow; S, brown.

In order to study the thermodynamic stability of the $Cu_6@Au_{12}(\mu_3\text{-}S)_8$, $Ag_6@Au_{12}(\mu_3\text{-}S)_8$ and $Au_6@Au_{12}(\mu_3\text{-}S)_8$ clusters, $Cu_6@Au_{12}(\mu_3\text{-}S)_8$ cluster is taken as a test case. The thermodynamic stabilities of $Cu_6@Au_{12}(\mu_3\text{-}S)_8$ cluster is further confirmed by ab initio molecular dynamics (AIMD) simulations. The AIMD studies of the cluster is carried out using Vienna ab initio simulation package (VASP) with PBE0 method [39,56]. Four different temperatures at 300, 500, 700, and 1000 K with a simulation time of 8ps have been performed. The AIMD simulations of $Cu_6@Au_{12}(\mu_3\text{-}S)_8$ cluster are plotted in Figure S4. From the figure, it is obvious that the structure of $Cu_6@Au_{12}(\mu_3\text{-}S)_8$ cluster can keep after simulation in the temperature range of 300–1000 K, indicating its high thermostability.

The chemical bonding patterns of the three clusters have been analyzed. According to the results of AdNDP analysis (see Figure S5), each $M_6@Au_{12}(\mu_3\text{-}S)_8$ (M = Cu, Ag, and Au) cluster has 24 2c-2e Au-S σ bonds with ONs being 1.85, 1.83 and 1.82 |e|, respectively. From Figure 7, each cluster has one 6c–2e bond, and occupancy numbers of the three 6c–2e bonds in $Cu_6@Au_{12}(\mu_3\text{-}S)_8$, $Ag_6@Au_{12}(\mu_3\text{-}S)_8$ and $Au_6@Au_{12}(\mu_3\text{-}S)_8$ are 1.90, 1.78, and 1.79 |e|, respectively.

4. Conclusions

In conclusion, we have explored the electronic and geometric structure of the recently determined $Au_{70}S_{20}(PPh_3)_{12}$ cluster on the basis of the divide-and-protect concept and SAN model. $Au_{70}S_{20}(PPh_3)_{12}$ cluster is a 30e-compound, which does not satisfy the magic number of SAC concept. Based on SAN model, the cluster has fifteen 2e-superatoms. The Au_{46}^{16+} core is composed of one Au_{22}^{12+} innermost core and ten surrounding 2e-Au_4 superatoms. The Au_{22}^{12+} innermost core can either be viewed as a network of five 2e-Au_6 superatoms, or be considered as a 10e-superatomic molecule. When Au_{22}^{12+} innermost core is viewed as a network of five 2e-Au_6 superatoms, the Au_{46}^{16+} core can be described as a 15 × 2e SAN consisting of 10 × 2e Au_4 and 5 × 2e Au_6 superatoms. The vertex-sharing Au_7 core exists in the experimentally determined $Au_{20}(SR)_{16}$ and $Au_{36}(SR)_{24}$ clusters. A new branching staple motif, six-tooth staple motif, $[Au_{12}(\mu_3\text{-}S)_{10}]^{8-}$, is discovered in Au-L clusters for the first time. The six-tooth staple motif is obviously different from common staple motifs, which have six S legs. Here the newly discovered staple motif enriches the staple motif family. The NICS-san method has been used to confirm the presence of Au_4 superatoms. The new segmentation method here can properly explain the structure and stability of $Au_{70}S_{20}(PPh_3)_{12}$ cluster. The reason for the stability and the nature of bonds have been given. Concretely, the six-tooth staple motifs, the superatom network, the aromatic of the superatoms and Au–Au interactions contribute to the stability of the cluster. We have designed three core-in-cage $Cu_6@Au_{12}(\mu_3\text{-}S)_8$, $Ag_6@Au_{12}(\mu_3\text{-}S)_8$, and $Au_6@Au_{12}(\mu_3\text{-}S)_8$ clusters based on $[Au_{12}(\mu_3\text{-}S)_8]^{4-}$. The three clusters are stable in O_h symmetry. Each of them has one 6c-2e bond in the core. Aromatic analysis reveals that they are aromatic molecules. The $[Au_{12}(\mu_3\text{-}S)_8]^{4-}$ cluster has been experimentally synthesized, and the three constructed clusters are stable based on our computation, thus the three designed clusters may be synthesized in future. Our work will provide some new perspectives to the electronic structure and stability of $Au_{70}S_{20}(PPh_3)_{12}$ cluster. The concept of half-cage and cage staple motif could offer some reference to future synthesis of Au-L clusters.

Supplementary Materials: The following are available online at http://www.mdpi.com/2079-4991/9/8/1132/s1, The segmentation method of $Au_{70}S_{20}(PCH_3)_{12}$ cluster. Figure S1: The Au–Au contacts in the optimized structure of $Au_{70}S_{20}(PCH_3)_{12}$ cluster. (a) Au_{22} innermost core is 5 × 2e SAN, (b) Au_{22} innermost core is a 10e-superatomic molecule. The baby blue and black lines in the structure indicate aurophilic contacts between motifs and superatoms, while blue and green lines show aurophilic contacts between superatoms. Figure S2: (a) The Au–Au bond distances between Au_{46} core and staple motifs, (b) the Au–Au distances in Au_{22} core, (c) the Au–Au distances in two Au_4 superatoms on top and bottom of the cluster, (d) the Au–Au distances in the four pairs of vertex-sharing Au_4 superatoms. Figure S3: IR spectra of $Cu_6@Au_{12}(\mu_3\text{-}S)_8$, $Ag_6@Au_{12}(\mu_3\text{-}S)_8$ and $Au_6@Au_{12}(\mu_3\text{-}S)_8$ clusters. Figure S4: Geometric configuration of $Cu_6@Au_{12}(\mu_3\text{-}S)_8$ at after 8 ps AIMD simulations at (a) 300 K, (b) 500 K, (c) 700 K and (d) 1000 K, respectively. Cu, green; Au, yellow; S, brown. Figure S5: Geometries (Cu, green; Ag, blue; Au, yellow) and AdNDP localized natural bonding orbitals of Au-S σ-bonds in (a) $Cu_6@Au_{12}(\mu_3\text{-}S)_8$, (b) $Ag_6@Au_{12}(\mu_3\text{-}S)_8$ and (c) $Au_6@Au_{12}(\mu_3\text{-}S)_8$ clusters.

Author Contributions: Density functional theory (DFT) computations, Z.M.T.; data analysis, Z.M.T. and L.J.C., writing—original draft preparation, Z.M.T; writing—review and editing, Z.M.T. and L.J.C.; supervision, L.J.C.

Funding: This research was funded by the National Natural Science Foundation of China (21873001), the Foundation of Distinguished Young Scientists of Anhui Province and the financial support of the Fuyang Normal University (2017FSKJ01ZD).

Acknowledgments: The authors acknowledge the high-performance computing center of Anhui university.

Conflicts of Interest: The authors declare no conflict of interest.

References

1. Jadzinsky, P.D.; Calero, G.; Ackerson, C.J.; Bushnell, D.A.; Kornberg, R.D. Structure of a thiol monolayer-protected gold nanoparticle at 1.1 Å resolution. *Science* **2007**, *318*, 430–433. [CrossRef]
2. Zhu, M.Z.; Aikens, C.M.; Hollander, F.J.; Schatz, G.C.; Jin, R.C. Correlating the crystal structure of a thiol-protected Au_{25} cluster and optical properties. *J. Am. Chem. Soc.* **2008**, *130*, 5883–5885. [CrossRef] [PubMed]
3. Yuan, S.F.; Li, P.; Tang, Q.; Wan, X.K.; Nan, Z.A.; Jiang, D.E.; Wang, Q.M. Alkynyl-protected silver nanoclusters featuring an anticuboctahedral kernel. *Nanoscale* **2017**, *9*, 11405–11409. [CrossRef] [PubMed]
4. Das, A.; Li, T.; Nobusada, K.; Zeng, C.; Rosi, N.L.; Jin, R.C. Nonsuperatomic $[Au_{23}(SC_6H_{11})_{16}]^-$ nanocluster featuring bipyramidal Au_{15} kernel and trimeric $Au_3(SR)_4$ motif. *J. Am. Chem. Soc.* **2013**, *135*, 18264–18267. [CrossRef] [PubMed]
5. Dass, A.; Theivendran, S.; Nimmala, P.R.; Kumara, C.; Jupally, V.R.; Fortunelli, A.; Sementa, L.; Barcaro, G.; Zuo, X.B.; Noll, B.C. $Au_{133}(SPh\text{-}tBu)_{52}$ nanomolecules: X-ray crystallography, optical, electrochemical, and theoretical Analysis. *J. Am. Chem. Soc.* **2015**, *137*, 4610–4613. [CrossRef] [PubMed]
6. Li, Y.F.; Chen, M.; Wang, S.X.; Zhu, M.Z. Intramolecular metal exchange reaction promoted by thiol ligands. *Nanomaterials* **2018**, *8*, 1070. [CrossRef] [PubMed]
7. Pei, Y.; Gao, Y.; Shao, N.; Zeng, X.C. Thiolate-protected $Au_{20}(SR)_{16}$ cluster: Prolate Au_8 core with new $[Au_3(SR)_4]$ staple motif. *J. Am. Chem. Soc.* **2009**, *131*, 13619–13621. [CrossRef]
8. Pei, Y.; Tang, J.; Tang, X.Q.; Huang, Y.Q.; Zeng, X.C. New structure model of $Au_{22}(SR)_{18}$: Bitetrahederon golden kernel enclosed by $[Au_6(SR)_6]$ Au(I) complex. *J. Phys. Chem. Lett.* **2015**, *6*, 1390–1395. [CrossRef]
9. Jiang, D.E.; Walter, M.; Akola, J. On the structure of a thiolated gold cluster: $Au_{44}(SR)_{28}{}^{2-}$. *J. Phys. Chem. C* **2010**, *114*, 15883–15889. [CrossRef]
10. Malola, S.; Lehtovaara, L.; Knoppe, S.; Hu, K.J.; Palmer, R.E.; Bürgi, T.; Häkkinen, H. $Au_{40}(SR)_{24}$ cluster as a chiral dimer of 8-electron superatoms: Structure and optical properties. *J. Am. Chem. Soc.* **2012**, *134*, 19560–19563. [CrossRef]
11. Tlahuice-Flores, A. Ligand effects on the optical and chiroptical properties of the thiolated Au_{18} cluster. *Phys. Chem. Chem. Phys.* **2016**, *18*, 27738–27744. [CrossRef] [PubMed]
12. Muñoz-Castro, A.; Maturana, R.G. Understanding planar ligand-supported MAu_5 and MAu_6 cores. Theoretical survey of $[MAu_5(Mes)_5]$ and $[MAu_6(Mes)_6]$ (M = Cu, Ag, Au; Mes = 2, 4, 6-$Me_3C_6H_2$) under the planar superatom model. *J. Phys. Chem. C* **2014**, *118*, 21185–21191. [CrossRef]
13. Jin, R.C.; Zeng, C.J.; Zhou, M.Z.; Chen, Y.X. Atomically precise colloidal metal nanoclusters and nanoparticles: Fundamentals and opportunities. *Chem. Rev.* **2016**, *116*, 10346–10413. [CrossRef] [PubMed]
14. Jones, T.C.; Sementa, L.; Stener, M.; Gagnon, K.J.; Thanthirige, V.D.; Ramakrishna, G.; Fortunelli, A.; Dass, A. $Au_{21}S(SAdm)_{15}$: Crystal structure, mass spectrometry, optical spectroscopy, and first-principles theoretical analysis. *J. Phys. Chem. C* **2017**, *121*, 10865–10869. [CrossRef]
15. Yang, H.Y.; Wang, Y.; Edwards, A.J.; Yan, J.Z.; Zheng, N.F. High-yield synthesis and crystal structure of a green Au_{30} cluster co-capped by thiolate and sulfide. *Chem. Commun.* **2014**, *50*, 14325–14327. [CrossRef] [PubMed]
16. Crasto, D.; Malola, S.; Brosofsky, G.; Dass, A.; Häkkinen, H. Single crystal XRD structure and theoretical analysis of the chiral $Au_{30}S(S\text{-}t\text{-}Bu)_{18}$ cluster. *J. Am. Chem. Soc.* **2014**, *136*, 5000–5005. [CrossRef] [PubMed]
17. Liu, C.; Li, T.; Li, G.; Nobusada, K.; Zeng, C.J.; Pang, G.; Rosi, N.L.; Jin, R.C. Observation of body-centered cubic gold nanocluster. *Angew. Chem. Int. Ed.* **2015**, *54*, 9826–9829. [CrossRef]
18. Higaki, T.; Liu, C.; Zhou, M.; Luo, T.Y.; Rosi, N.L.; Jin, R.C. Tailoring the structure of 58-electron gold nanoclusters: $Au_{103}S_2(S\text{-}Nap)_{41}$ and its implications. *J. Am. Chem. Soc.* **2017**, *139*, 9994–10001. [CrossRef] [PubMed]

19. Tian, Z.M.; Cheng, L.J. Electronic and geometric structures of Au_{30} clusters: A network of 2e-superatom Au cores protected by tridentate protecting motifs with μ_3-S. *Nanoscale* **2016**, *8*, 826–834. [CrossRef] [PubMed]

20. Kenzler, S.; Schrenk, C.; Schnepf, A. $Au_{108}S_{24}(PPh_3)_{16}$: A highly symmetric nanoscale gold cluster confirms the general concept of metalloid clusters. *Angew. Chem. Int. Ed.* **2017**, *56*, 393–396. [CrossRef]

21. Kenzler, S.; Schrenk, C.; Frojd, A.R.; Hakkinen, H.; Clayborne, A.Z.; Schnepf, A. $Au_{70}S_{20}(PPh_3)_{12}$: An intermediate sized metalloid gold cluster stabilized by the Au_4S_4 ring motif and Au-PPh_3 groups. *Chem. Commun.* **2018**, *54*, 248–251. [CrossRef] [PubMed]

22. Häkkinen, H.; Walter, M.; Grönbeck, H. Divide and protect: Capping gold nanoclusters with molecular gold-thiolate rings. *J. Phys. Chem. B* **2006**, *110*, 9927–9931. [CrossRef] [PubMed]

23. Heaven, M.W.; Dass, A.; White, P.S.; Holt, K.M.; Murray, R.W. Crystal structure of the gold nanoparticle $[N(C_8H_{17})_4]$ $[Au_{25}(SCH_2CH_2Ph)_{18}]$. *J. Am. Chem. Soc.* **2008**, *130*, 3754–3755. [CrossRef] [PubMed]

24. Pei, Y.; Gao, Y.; Zeng, X.C. Structural prediction of thiolate-protected Au_{38}: A face-fused bi-icosahedral Au core. *J. Am. Chem. Soc.* **2008**, *130*, 7830–7832. [CrossRef] [PubMed]

25. Lopez-Acevedo, O.; Akola, J.; Whetten, R.L.; Gronbeck, H.; Häkkinen, H. Structure and bonding in the ubiquitous icosahedral metallic gold cluster $Au_{144}(SR)_{60}$. *J. Phys. Chem. C* **2009**, *113*, 5035–5038. [CrossRef]

26. Knoppe, S.; Wong, O.A.; Malola, S.; Häkkinen, H.; Bürgi, T.; Verbiest, T.; Ackerson, C.J. Chiral phase transfer and enantioenrichment of thiolate-protected Au_{102} clusters. *J. Am. Chem. Soc.* **2014**, *136*, 4129–4132. [CrossRef]

27. Xu, W.W.; Gao, Y.; Zeng, X.C. Unraveling structures of protection ligands on gold nanoparticle $Au_{68}(SH)_{32}$. *Sci. Adv.* **2015**, *1*, e1400211. [CrossRef] [PubMed]

28. Xu, W.W.; Li, Y.; Gao, Y.; Zeng, X.C. Medium-sized $Au_{40}(SR)_{24}$ and $Au_{52}(SR)_{32}$ nanoclusters with distinct gold-kernel structures and spectroscopic features. *Nanoscale* **2016**, *8*, 1299–1304. [CrossRef]

29. Xiong, L.; Peng, B.; Ma, Z.; Wang, P.; Pei, Y. A ten-electron (10e) thiolate-protected $Au_{29}(SR)_{19}$ cluster: Structure prediction and a 'gold-atom insertion, thiolate-group elimination' mechanism. *Nanoscale* **2017**, *9*, 2895–2902. [CrossRef]

30. Xu, W.W.; Zhu, B.; Zeng, X.C.; Gao, Y. A grand unified model for liganded gold clusters. *Nat. Commun.* **2016**, *7*, 13574. [CrossRef]

31. Xu, W.W.; Zeng, X.C.; Gao, Y. The structural isomerism in gold nanoclusters. *Nanoscale* **2018**, *10*, 9476–9483. [CrossRef] [PubMed]

32. Ma, Z.Y.; Wang, P.; Xiong, L.; Pei, Y. Thiolate-protected gold nanoclusters: Structural prediction and the understandings of electronic stability from first principles simulations. *WIREs Comput. Mol. Sci.* **2017**, *7*, e1315. [CrossRef]

33. Nimmala, P.R.; Knoppe, S.; Jupally, V.R.; Delcamp, J.H.; Aikens, C.M.; Dass, A. $Au_{36}(SPh)_{24}$ nanomolecules: X-ray crystal structure, optical spectroscopy, electrochemistry, and theoretical analysis. *J. Phys. Chem. B* **2014**, *118*, 14157–14167. [CrossRef] [PubMed]

34. Zeng, C.J.; Liu, C.; Chen, Y.X.; Rosi, N.L.; Jin, R.C. Gold-thiolate ring as a protecting motif in the $Au_{20}(SR)_{16}$ nanocluster and implications. *J. Am. Chem. Soc.* **2014**, *136*, 11922–11925. [CrossRef] [PubMed]

35. Walter, M.; Akola, J.; Lopez-Acevedo, O.; Jadzinsky, P.D.; Calero, G.; Ackerson, C.J.; Whetten, R.L.; Grönbeck, H.; Häkkinen, H. A unified view of ligand-protected gold clusters as superatom complexes. *Proc. Natl. Acad. Sci. USA* **2008**, *105*, 9157–9162. [CrossRef] [PubMed]

36. Cheng, L.J.; Yuan, Y.; Zhang, X.Z.; Yang, J.L. Superatom networks in thiolate-protected gold nanoparticles. *Angew. Chem. Int. Ed.* **2013**, *52*, 9035–9039. [CrossRef] [PubMed]

37. Pei, Y.; Lin, S.S.; Su, J.C.; Liu, C.Y. Structure prediction of $Au_{44}(SR)_{28}$: A chiral superatom cluster. *J. Am. Chem. Soc.* **2013**, *135*, 19060–19063. [CrossRef] [PubMed]

38. Frisch, M.J.; Schlegel, H.B.; Scuseria, G.E.; Robb, M.A.; Cheeseman, J.R.; Scalmani, G.; Barone, V.; Mennucci, B.; Petersson, G.A.; Nakatsuji, H.; et al. *Gaussian 09, Revision B 01*; Gaussian, Inc.: Wallingford, CT, USA, 2009.

39. Perdew, J.P.; Burke, K.; Ernzerhof, M. Generalized gradient approximation made simple. *Phys. Rev. Lett.* **1996**, *77*, 3865–3868. [CrossRef]

40. Zubarev, D.Y.; Boldyrev, A.I. Developing paradigms of chemical bonding: Adaptive natural density partitioning. *Phys. Chem. Chem. Phys.* **2008**, *10*, 5207–5217. [CrossRef]

41. Varetto, U. *MOLEKEL*, version 5.4.0.8; Swiss National Supercomputing Centre: Manno, Switzerland, 2009. Available online: http://ugovaretto.github.io/molekel/wiki/pmwiki.php/Main/HomePage.html (accessed on 1 August 2019).

42. Jiang, D.E.; Walter, M.; Dai, S. Gold sulfide nanoclusters: A unique core-in-cage structure. *Chem. Eur. J.* **2010**, *16*, 4999–5003. [CrossRef]

43. Pei, Y.; Shao, N.; Li, H.; Jiang, D.E.; Zeng, X.C. Hollow polyhedral structures in small gold-sulfide clusters. *ACS Nano* **2011**, *5*, 1441–1449. [CrossRef] [PubMed]

44. Feng, Y.Q.; Cheng, L.J. Structural evolution of $(Au_2S)_n$ (n = 1-8) clusters from first principles global optimization. *RSC Adv.* **2015**, *5*, 62543–62550. [CrossRef]

45. Gerolf, M.; Joachim, S. Synthesis and crystal Structure of $[Ph_4As]_4[Au_{12}S_8]$, a distorted cubane-like thioaurate(I). *Angew. Chem. Int. Ed. Engl.* **1984**, *23*, 715–716. [CrossRef]

46. Xu, W.W.; Zeng, X.C.; Gao, Y. $(Au_3(\mu_3\text{-}S)$ (0e) elementary block: New insights into ligated gold clusters with μ_3-sulfido motifs. *Nanoscale* **2017**, *9*, 8990–8996. [CrossRef] [PubMed]

47. Knoppe, S.; Malola, S.; Lehtovaara, L.; Bürgi, T.; Häkkinen, H. Electronic structure and optical properties of the thiolate-protected $Au_{28}(SMe)_{20}$ Cluster. *J. Phys. Chem. A* **2013**, *117*, 10526–10533. [CrossRef]

48. Pyykkö, P.; Mendizabal, F. Theory of d^{10}-d^{10} closed-shell attraction. III. rings. *Inorg. Chem.* **1998**, *37*, 3018–3025. [CrossRef]

49. Pekka, P.; Nino, R.; Fernando, M. Theory of the d^{10}-d^{10} closed-shell attraction: 1. dimers near equilibrium. *Chem. Eur. J.* **1997**, *3*, 1451–1457. [CrossRef]

50. Chen, S.; Wang, S.X.; Zhong, J.; Song, Y.B.; Zhang, J.; Sheng, H.T.; Pei, Y.; Zhu, M.Z. The structure and optical properties of the $[Au_{18}(SR)_{14}]$ nanocluster. *Angew. Chem. Int. Ed.* **2015**, *54*, 3145–3149. [CrossRef]

51. Das, A.; Liu, C.; Byun, H.Y.; Nobusada, K.; Zhao, S.; Rosi, N.; Jin, R.C. Structure determination of $[Au_{18}(SR)_{14}]$. *Angew. Chem.* **2015**, *127*, 3183–3187. [CrossRef]

52. Stanger, A. Nucleus-independent chemical shifts (NICS): Distance dependence and revised criteria for aromaticity and antiaromaticity. *J. Org. Chem.* **2006**, *71*, 883–893. [CrossRef]

53. Tian, Z.M.; Cheng, L.J. First principles study on the structural evolution and properties of $(MCl)_n$ (n = 1–12, M = Cu, Ag) clusters. *RSC Adv.* **2016**, *6*, 30311–30319. [CrossRef]

54. Yuan, Y.; Cheng, L. B_{14}^{2+}: A magic number double-ring cluster. *J. Chem. Phys.* **2012**, *137*, 044308. [CrossRef] [PubMed]

55. Bondi, A. Van der Waals volumes and radii. *J. Phys. Chem.* **1964**, *68*, 441–451. [CrossRef]

56. Kresse, G.; Furthmüller, J. Efficient iterative schemes for ab initio total-energy calculations using a plane-wave basis set. *Phys. Rev. B* **1996**, *54*, 11169–11186. [CrossRef] [PubMed]

 *nanomaterials*

Article

Gold Nanoclusters: Bridging Gold Complexes and Plasmonic Nanoparticles in Photophysical Properties

Meng Zhou, Chenjie Zeng, Qi Li, Tatsuya Higaki and Rongchao Jin *

Department of Chemistry, Carnegie Mellon University, Pittsburgh, PA 15213, USA
* Correspondence: rongchao@andrew.cmu.edu

Received: 29 May 2019; Accepted: 25 June 2019; Published: 28 June 2019

Abstract: Recent advances in the determination of crystal structures and studies of optical properties of gold nanoclusters in the size range from tens to hundreds of gold atoms have started to reveal the grand evolution from gold complexes to nanoclusters and further to plasmonic nanoparticles. However, a detailed comparison of their photophysical properties is still lacking. Here, we compared the excited state behaviors of gold complexes, nanolcusters, and plasmonic nanoparticles, as well as small organic molecules by choosing four typical examples including the Au_{10} complex, Au_{25} nanocluster (1 nm metal core), 13 diameter Au nanoparticles, and Rhodamine B. To compare their photophysical behaviors, we performed steady-state absorption, photoluminescence, and femtosecond transient absorption spectroscopic measurements. It was found that gold nanoclusters behave somewhat like small molecules, showing both rapid internal conversion (<1 ps) and long-lived excited state lifetime (about 100 ns). Unlike the nanocluster form in which metal–metal transitions dominate, gold complexes showed significant charge transfer between metal atoms and surface ligands. Plasmonic gold nanoparticles, on the other hand, had electrons being heated and cooled (~100 ps time scale) after photo-excitation, and the relaxation was dominated by electron–electron scattering, electron–phonon coupling, and energy dissipation. In both nanoclusters and plasmonic nanoparticles, one can observe coherent oscillations of the metal core, but with different fundamental origins. Overall, this work provides some benchmarking features for organic dye molecules, organometallic complexes, metal nanoclusters, and plasmonic nanoparticles.

Keywords: gold nanomaterials; electron dynamics; phonon dynamics; optical properties

---

1. Introduction

Gold nanomaterials have attracted great interest in both fundamental science and practical applications such as sensing, catalysis, and optoelectronics owing to their unique properties [1–9]. For gold nanoparticles with diameters between 3–100 nm, the strong surface plasmon resonance (SPR) dominates in the absorption spectrum, which is caused by collective excitation of free electrons. In contrast, ultra-small gold nanoparticles consisting of tens to hundreds of gold atoms, often called gold nanoclusters, show multiple absorption bands spanning the UV-visible NIR range because of discrete energy levels [10,11]. The significant progress in ligand-protected (e.g., thiolate) gold nanoclusters has allowed atomically precise control of their size and structure [11–13]. The structure of a thiolate-protected gold nanocluster typically consists of a metal core with Au–Au bonds (~2.8 Å) and surface Au–S staple motifs [11]. The Au(I) complexes, on the other hand, do not have a metal core, albeit gold aurophilic interactions (Au···Au distance > 3 Å) often exist owing to the closed-shell d [10] configuration of Au(I) [14,15]. Typically, solutions of gold complexes are colorless because their absorption peaks lie in the ultraviolet (UV) range (shorter than 400 nm).

Plasmonic gold nanoparticles (AuNPs), gold nanoclusters, and Au(I) complexes have distinct optical features as a result of their differences in size, structure, and bonding. Therefore, understanding the photodynamics will help to deepen the understanding of their electronic structures and optical properties [2,7,9,12,16–18]. The electron dynamics of plasmonic gold nanoparticles has been intensively investigated [19,20]. Since the 1990s, El-Sayed and coworkers have probed the size and shape dependent electron dynamics of metallic AuNPs [21,22]. Hartland and Vallee groups have extensively investigated the phonon dynamics and electron–phonon coupling of different sized AuNPs [23,24]. In recent years, there has also been research on the excited state dynamics of ligand-protected gold nanoclusters [1,2,25–27]. The electron and phonon dynamics of the nanoclusters were found to be dependent on both size and structure [28–30]. Stamplecoskie and Kamat [31] found that the dynamics of Au(I) complexes are different from that of gold nanoclusters. Despite such progress, a systematic comparison of the photo-dynamics between them is still lacking.

Here, we chose Rhodamine B (RB), $Au_{10}(SR)_{10}$ complex, $Au_{25}(SR)_{18}$ nanocluster, and 13 nm diameter AuNPs protected by citrate (Figure 1) as typical examples to compare the photophysics of small molecules, gold complexes, nanoclusters, and nanoparticles. We employed time-resolved fluorescence and femtosecond transient absorption spectroscopy to probe their excited state behaviors. The electron and phonon dynamics are discussed and compared in detail. The obtained results are of great importance to understand their optical response and further promote their applications in sensing and optics.

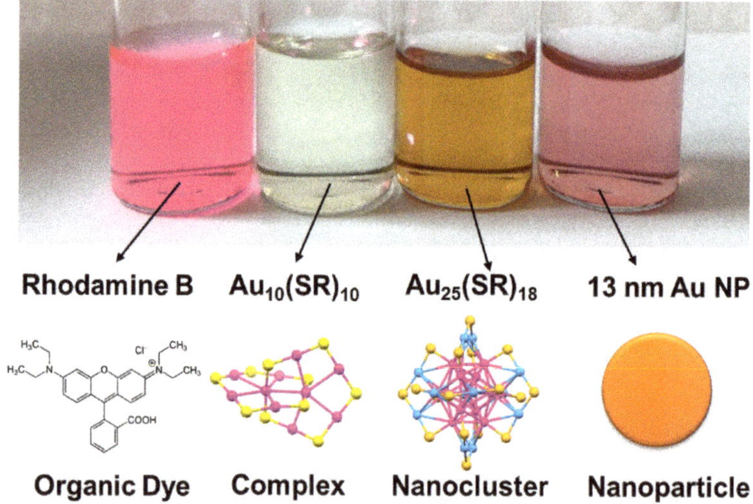

Figure 1. (Upper panel) Photograph of dilute solutions of Rhodamine B, sub-nanometer $Au_{10}(SR)_{10}$ complex, 1 nm $Au_{25}(SR)_{18}$ nanocluster, and ~13 nm diameter gold nanoparticles. (Lower panel) Molecular structures of Rhodamine B, $Au_{10}(SR)_{10}$ (where, R: Ph–^{t}Bu), and $Au_{25}(SR)_{18}$ (where, R: CH_2CH_2Ph). Color labels: yellow, S atoms; all other colors are for Au. The R groups and citrate stabilizers on ~13 nm Au nanoparticles are not shown.

2. Materials and Methods

Sample preparation: Rhodamine B was purchased from Sigma–Aldrich (St. Louis, MO, USA) and used as received. $Au_{10}(SR)_{10}$ (where, SR = 4-tert-butylbenzenethiolate), $Au_{25}(SR)_{18}$ (where, SR = 2-phenylethanethiol), and 13 nm diameter Au nanoparticles protected by trisodium citrate were prepared according to the literature [32,33].

Steady state absorption and photoluminescence: UV-vis absorption spectra were measured on a Shimadzu UV-3600plus spectrometer (Kyoto, Japan). Steady-state photoluminescence was measured on a Fluorolog-3 spectrofluorometer from Horiba Jobin Yvon (Piscataway, NJ, USA).

Time-resolved luminescence: fluorescence lifetimes were measured with a time-correlated single photon counting technique from Horiba Jobin Yvon (Piscataway, NJ, USA); a pulsed LED source (376 nm, 1.1 ns) was used as the excitation source.

Transient absorption spectroscopy: Femtosecond transient absorption spectroscopy was carried out using a commercial Ti:Sapphire laser system (SpectraPhysics, 800 nm, 100 fs, 3.5 mJ, 1 kHz) (Santa Clara, CA, USA). A pump pulse was generated using a commercial optical parametric amplifier (LightConversion, Vilnius, Lithuania). A small portion of the laser fundamental was focused into a sapphire plate to produce a supercontinuum in the visible region, which overlapped in time and space with the pump. Multi-wavelength transient spectra were recorded using dual spectrometers (i.e., signal and reference) (Thorlabs, Newton, NJ, USA) equipped with array detectors whose data rates exceed the repetition rate of the laser (1 kHz). Solution samples in 1 mm path length cuvettes were excited by the tunable output of the OPA (pump). Nanosecond transient absorption measurements were conducted using the same ultrafast pump pulses along with an electronically delayed supercontinuum light source with a sub-nanosecond pulse duration (EOS, Ultrafast Systems, Sarasota, FL, USA).

3. Results and Discussions

Steady-state absorption and photoluminescence spectra: From the steady-state spectra, one can clearly observe the differences between small molecules and different-sized gold nanomaterials. The RB dye showed significant absorption peaks at ~550 nm and a shoulder at ~520 nm (Figure 2A), which originated from 0–0 and 0–1 vibronic peaks of S_1 state, respectively. Other absorption peaks at shorter wavelengths were transitions from the ground state to higher excited states than S_1. The fluorescence spectrum of RB exhibited a mirror image to the S_0–S_1 absorption band, with a Stokes shift of 0.1 eV. The Au_{10} complex, on the other hand, showed absorption in the UV region only, with a peak at 346 nm and a shoulder at 380 nm (Figure 2B). The higher energy transitions ($\lambda < 300$ nm) originated from intraligand (IL) transitions, while the lower energy transitions (346 and 370 nm) should arise from ligand to metal or metal to ligand charge transfer (LM/MLCT) modified by Au(I)–Au(I) interactions [34,35]. The Au_{10} complex showed no observable photoluminescence (PL); however, some of the other gold complexes were reported to exhibit strong PL [36–38]. Unlike gold complexes, the Au_{25} nanocluster exhibited multiple absorption bands spanning the entire UV-Vis range (Figure 2C). Theoretical calculations revealed that the absorption band at 670 nm was from the *sp* to *sp* transition while absorption bands at shorter wavelengths involved both *sp* to *sp* and *d* to *sp* transitions [10]. The $Au_{25}(SR)_{18}$ exhibited weak photoluminescence (QY < 1%) related to the surface [39]. However, one can observe that the PL peak (centered at 750 nm) overlapped with the lowest absorption band, which indicates that the emission in the visible region may not be the intrinsic PL of Au_{25} nanoclusters. As the size of particles further increased, more and more gold atoms contributed to the electronic states, and finally SPR emerged [8] in the UV-vis spectrum, such as the ~13 nm Au nanoparticles (Figure 2D). Plasmonic gold nanoparticles had a continuous electronic band (i.e., the conduction band), and only had a very weak photoluminescence (QY = 10^{-4}) [40]. Below, we further discuss the excited state behaviors of these four species, from which one can find their molecular and plasmonic behaviors.

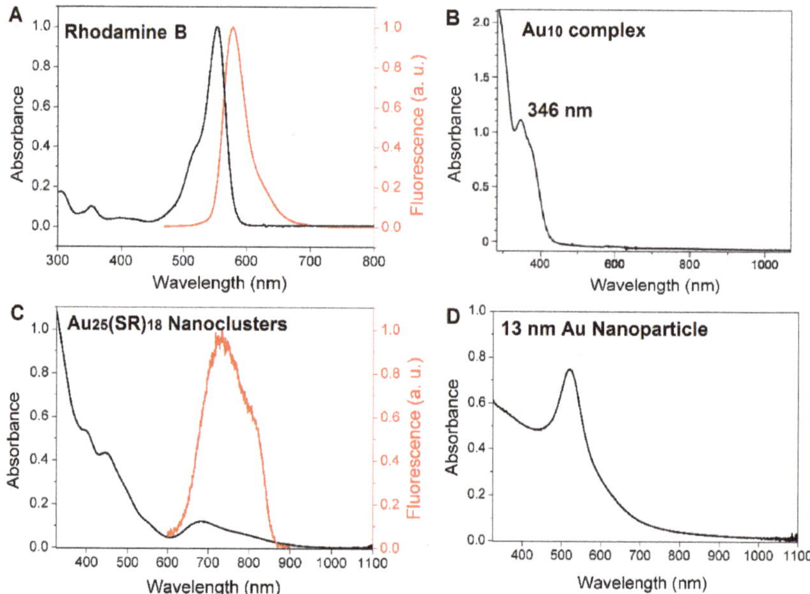

Figure 2. Steady-state optical spectra of (**A**) Rhodamine B; (**B**) complex $Au_{10}(SR)_{10}$; (**C**) nanocluster $Au_{25}(SR)_{18}$, and (**D**) 13 nm diameter Au nanoparticles.

Organic dyes (Rhodamine B): Before we discuss the photo-dynamics of different sized gold nanomaterials, we first discuss the excited state behavior of organic dyes such as Rhodamine B (RB) for an illustration of molecular behavior. The RB has been widely used as a probe in biological and synthetic polyelectrolyte systems [41]. The excited state behavior of organic dyes has been intensively investigated ever since the birth of time-resolved spectroscopy [42–44]. An aqueous solution of RB has strong luminescence (QY = 30%) and the PL lifetime is determined to be 1.7 ns in water (see Figure S1 in the Supplementary Materials). It is worth noting that the excited state lifetime of RB is highly dependent on solvent polarity (Figure S1), which has been reported previously [45]. In our current work, the transient absorption spectroscopic measurement on RB was performed with excitations at 360 nm and 560 nm, respectively, to excite the sample to the second and first singlet excited state. Upon photo-excitation, one can observe excited state absorption (ESA) around 450 nm, ground state bleaching (GSB), and stimulated emission (SE) between 500 nm and 700 nm (Figure 3A,B). With 360 nm excitation, one can observe an ultrafast decay (~100 fs) in GSB at 520 nm (Figure 3C, blue dip) as well as a rise in SE at 630 nm (Figure 3C, black). On the other hand, the ultrafast decay component was absent with excitation at 560 nm (Figure 3D); thus, the 100 fs component was assigned as internal conversion from S_2 to S_1 state. From the kinetic traces, ESA, GSB, and SE decay simultaneously from 1 ps to 3 ns and the transient absorption (TA) lifetime (1.7 ns) matched well with the PL lifetime (Table S1 in SI). The TA spectra and decay dynamics of RB we observed here matched well with the results of previous studies [44,45].

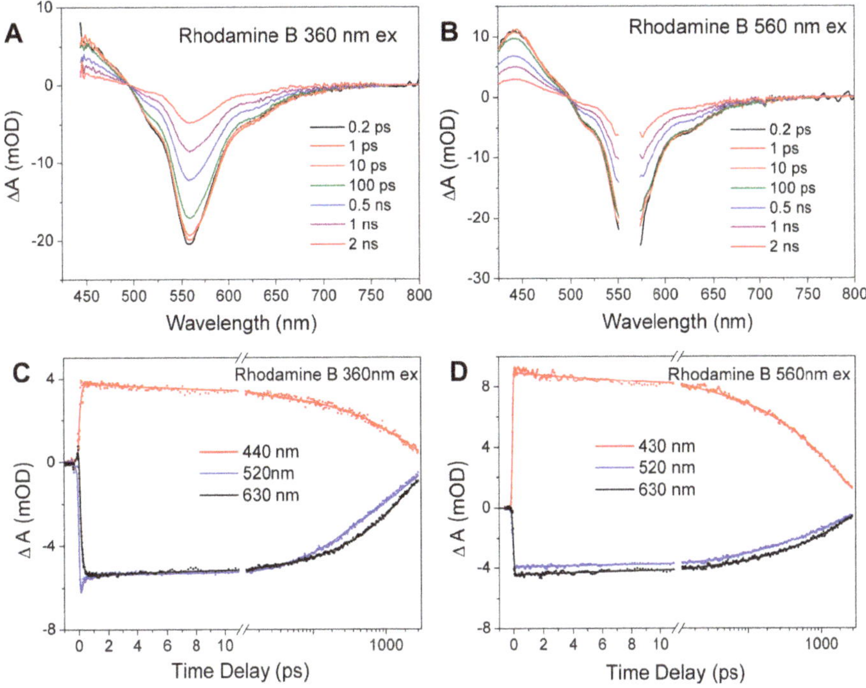

Figure 3. (**A**,**B**) Transient absorption spectra of Rhodamine B (dissolved in water) at different time delays with excitation of 360 nm and 560 nm, respectively. (**C**,**D**) Transient absorption decays at different probe wavelengths.

To sum up, the photophysics of RB and other typical organic dye molecules can be well explained by the Jablonski diagram and the relaxation follows the Kasha's rule [46], that is, the excited state molecules first experience a rapid non-radiative relaxation to the lowest singlet excited state and then emit photons to relax to the ground state.

Gold complexes ($Au_{10}(SR)_{10}$): Gold complexes are composed of one or a few gold atoms coordinated by ligands. The $Au_{10}(SR)_{10}$ complex is composed of two interlocked $Au_5(SR)_5$ rings and every gold atom is connected to two S atoms [32]. Upon photoexcitation at λ = 365 nm, broad positive signals were observed (Figure 4A) with no GSB signal, so the transient signal originates solely from ESA. Such an observation has also been reported in other Au(I) complexes [47]. In the initial 36 ps, the ESA1 around 780 nm decays to give rise to the ESA2 at 535 nm (Figure 4B). In the following 2.8 ns, ESA at all wavelengths decay dramatically by 90%. Decay associated spectra (DAS) obtained by global analysis and singular value decomposition (SVD) of the TA data exhibited three decaying components, 14 ps, 290 ps, and >1 ns (Figure 4C,D). Considering that gold–thiolate complexes show LM/MLCT characteristics in their steady-state UV-vis spectra, photo-excitation at 365 nm can directly generate LM/MLCT excited state [35,37,47]. The first 14 ps process can be assigned to the stabilization and equilibrium of LM/MLCT state. In a previous study, we probed the photodynamics of Au_{10} complexes dissolved in toluene and dichloromethane and found that the decay lifetime was dependent on solvent polarity [48]. Here, we repeated the TA spectra of Au_{10} dissolved in dichloromethane and the same processes can be observed with similar time constants to those in the previous study. Upon photoexcitation, intersystem crossing (ISC) from ^{1}LM/MLCT to ^{3}LM/MLCT occurred in less than 100 fs, which was not resolved in our TA measurement. The ^{3}LM/MLCT state was then stabilized in tens of picoseconds to form a ^{3}LM/MLCT* state, which decayed to the ground state (Figure 5).

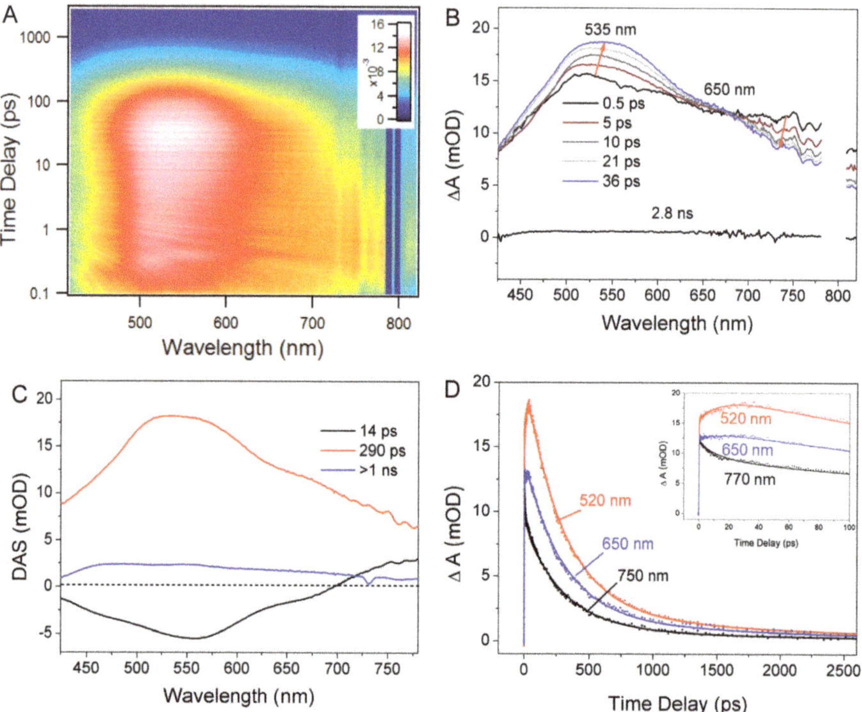

Figure 4. (**A**) Transient absorption (fs-TA) data of Au_{10} complex dissolved in dichloromethane. (**B**) Transient absorption spectra as a function of time delay from 0.5 ps to 36 ps and that at 2.8 ns. Scattering due to the laser pulse was excluded around 800 nm for (**A**,**B**). (**C**) Decay associated spectra (DAS) obtained from the global analysis of the TA data. (**D**) Kinetic traces (dots) and corresponding fit (solid line) at selected wavelengths, which shows the quality of the fitting.

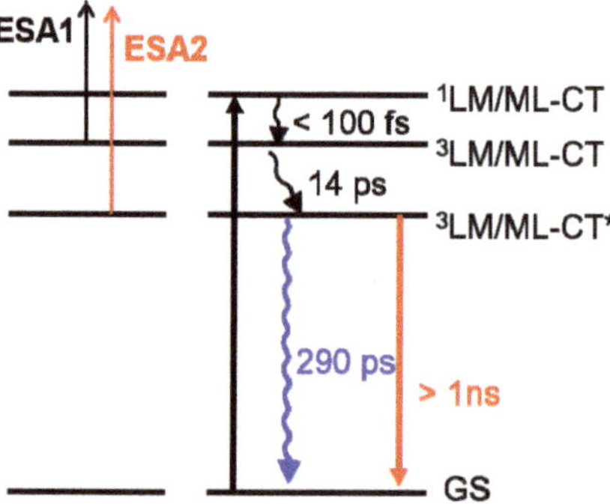

Figure 5. Diagram for the relaxation dynamics in the Au_{10} complex. The 290 ps process shows solvent dependency (dichloromethane in this study versus 350 ps in toluene in a previous study [48]).

Nanoclusters ($Au_{25}(SR)_{18}$ as the example): Unlike homoleptic gold complexes, thiolate-protected gold nanoclusters are composed of a well-defined metal core and surface staple motifs (e.g., –S–Au–S–) [11]. The relaxation dynamics of gold nanoclusters of different structures have been investigated by several groups and the relaxation model is more complicated as a result of multiple contributions of both Au and S atoms to the orbitals [27,28,49,50]. Here, $Au_{25}(SR)_{18}$ was chosen as an example to illustrate the photophysics. The $Au_{25}(SR)_{18}$ structure consisted of an icosahedral Au_{13} core and six $Au_2(SR)_3$ dimeric staple motifs for surface protection (Figure 1) [10]. With excitation at 490 nm, one can observe broad ESA overlapped with GSB peaks at 510 nm, 550 nm, and 675 nm (Figure 6A), which is a typical feature of gold nanoclusters [2,50,51]. During the first picosecond, one can observe a broad ESA band between 500 nm and 620 nm decaying rapidly and the time constant was ~600 fs (Figure 6B). The rapid relaxation was not observed under excitation of 800 nm (Figure 6C), which suggests that it should be internal conversion from higher to lower excited states. With excitation at 800 nm, the TA spectrum at ~0.3 ps was equal to the spectrum with excitation at 490 nm after 2 ps, which indicates that the 800 nm pulse excited Au_{25} directly to the lower excited state. The TA signal did not decay significantly between 2 ps and 3 ns, indicating a significantly longer excited state lifetime of Au_{25} than 3 ns. It was also interesting to see that the nanosecond TA and time resolved-PL give different lifetimes (Figure S2 in the Supplementary Materials). Compared with the dynamics of the Au_{10} complex, the excited state dynamics of the Au_{25} nanoclusters behaved more like that of small molecules.

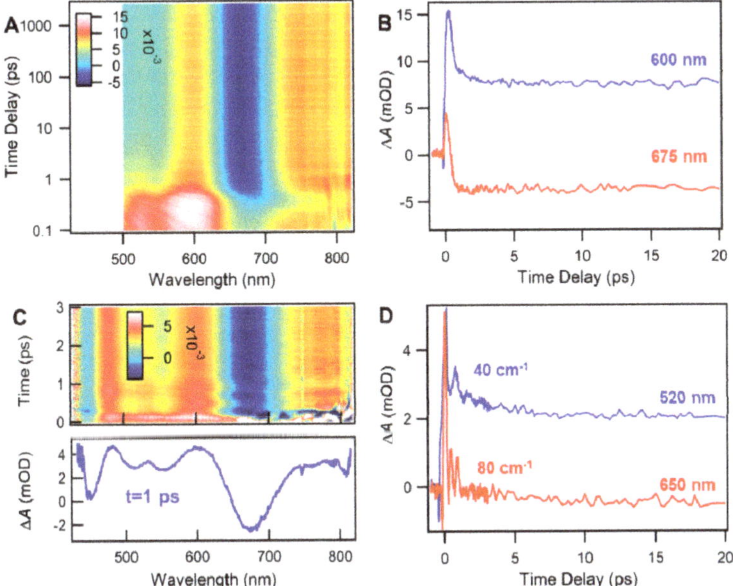

Figure 6. (**A**) Transient absorption data map of $Au_{25}(SR)_{18}$ with excitation at 490 nm; (**B**) kinetic traces probed at selected wavelengths with excitation at 490 nm. (**C**) Transient absorption data map and spectra probed at 1 ps with 800 nm excitation; (**D**) kinetic traces probe at 520 nm 650 nm with excitation at 800 nm.

With excitation at 800 nm, one can also observe coherent oscillations in the first few picoseconds (Figure 6C,D). Two frequencies (40 cm^{-1} and 80 cm^{-1}) were exhibited at different probe wavelengths (Figure 6D), which have been reported previously [51]. These oscillations observed in TA decays, also observed in other nanoclusters [52,53], were assigned to acoustic vibrations of the metal core and explained as displacive excitation [52,54], similar to that observed in semiconductors and small molecules [54,55].

Plasmonic nanoparticles (13 nm diameter AuNPs): As the particle size becomes larger, more gold atoms will contribute to the electronic states, and eventually the bandgap will disappear, giving rise to a transition from non-metallic to metallic [8]. The TA spectra of 13 nm AuNPs showed a significant bleaching signal at 520 nm and ESA on two sides (Figure 7A). We found that (i) the GSB became sharper from 0.2 ps to 5 ps and (ii) the higher pump fluence gave rise to a broader bleaching signal compared to that of the lower pump fluence (Figure 7A,B). The broadening of GSB in the initial time delay (full width at half maximum decreased from 50 nm to 30 nm in Figure 7A) is ascribed to the heating effect [20]. Upon photo-excitation, electrons in the AuNPs will be heated to a very high temperature (e.g., 1000 K, depending on the pump power). Heating the nanoparticle will result in broadening of the SPR peak as well as the GSB in the TA spectra. After the electrons are heated to a very high temperature, the excited state energy will first reach equilibrium via electron–electron scattering (manifested in the 100 fs rise of the kinetic traces in Figure 7C). Subsequently, the energy will be transferred from the electrons to the lattice through electron–phonon coupling (1–5 ps rapid decay in Figure 7C) [3,24]. Accompanied by the rapid decay, the GSB will also be narrowed during the electron–phonon coupling process (Figure 7A,B). Finally, the energy will dissipate into the environment by phonon–phonon relaxation, which is dependent on the surrounding medium (100 ps decay in Figure 7C). As there is no bandgap in plasmonic Au nanoparticles, there is no electron–hole separation or recombination process as in nanoclusters. Instead, the relaxation dynamics can be described by a well-established two-temperature model [21,56]. With the high pump fluence, one can observe a prominent oscillation with a frequency of 7.5 cm^{-1} (4 ps in periods) in the 13 nm AuNPs (Figure 7D). Unlike that of the gold nanoclusters, the oscillations in the metallic AuNPs originated from the periodic shift of the SPR band due to the lattice expansion induced by laser heating [20]. In addition, the oscillations in the metallic NPs can be well modelled by a classical continuum model (scaling as $1/D$, where D is the diameter), but this model fails in gold nanoclusters.

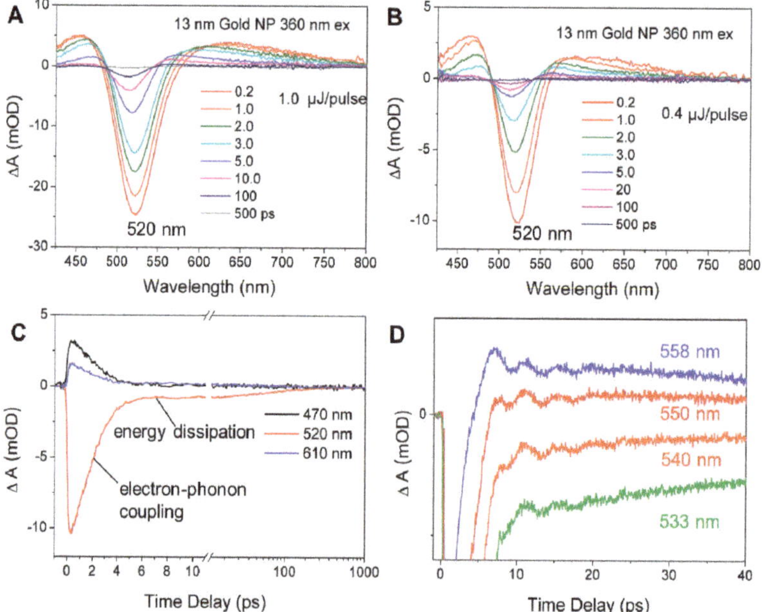

Figure 7. (**A,B**) Transient absorption spectra of 13 nm AuNPs at different time delays with an excitation of 360 nm under high and low pump fluences; (**C**) TA decay traces at selected probe wavelengths; (**D**) kinetic traces with prominent oscillation features.

A distinct feature of metallic gold nanoparticles is that the electron–phonon coupling is dependent on pump fluence [3,19]. When the pump fluence increased from 80 to 1800 uJ/cm^2, one can clearly observe that the electron–phonon coupling slowed down significantly (Figure 8A). After plotting the fitted time constants as a function of laser fluence, a linear relationship can be observed (Figure 8B), which agrees well with previous studies [3,56]. Such a power dependence is only observed in plasmonic nanoparticles, which can be well explained by the two-temperature model [21]. Compared to the plasmonic NPs, the TA measurements of RB, Au$_{10}$ complex, and Au$_{25}$ nanoclusters under different pump fluences exhibited no differences in the decay dynamics (Figures S3–S5 in the Supplementary Materials), hence, they were power independent. This serves as a signature of the non-metallic state. In Table 1, the photophysical features of the four types of materials are summarized. In Scheme 1, the relaxation processes as well as the time constants are illustrated.

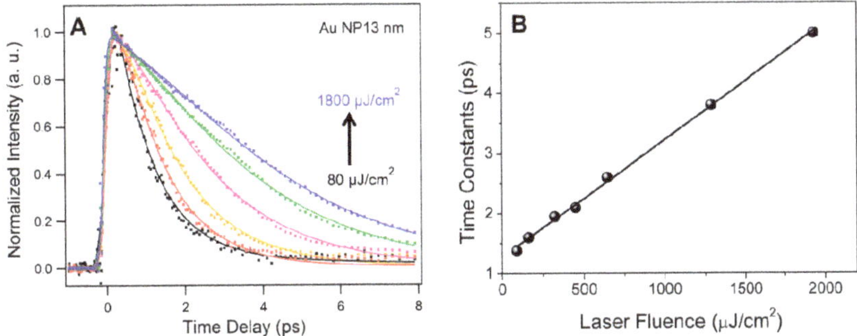

Figure 8. (**A**) Kinetic traces probe at 520 nm with different pump fluences; (**B**) electron–phonon time constants as a function of pump fluence.

Table 1. Characteristic features of molecules, complexes, nanoclusters, and plasmonic nanoparticles (PL: photoluminescence; ESA: excited state absorption; GSB: ground-state bleach; SE: stimulated emission; IC: internal conversion; ISC: intersystem crossing).

	Structure	E_g	Steady-State abs.	PL	Transient Absorption
Dye molecules e.g., Rhodamine B	No core	>1–2 eV	Multiple bands (e.g., π–π transition)	Yes	ESA + GSB + SE, IC, ISC, long lifetime (ns), power independence
Complexes, e.g., Au$_{10}$(SR)$_{10}$	No core	>2 eV	Multiple bands (charge transfer, CT)	Yes	ESA (predominant), ISC, long lifetime (µs–ns), power independence
Nanoclusters, e.g., Au$_{25}$(SR)$_{18}$	Core + surface (staple motifs)	~2.5 eV to zero	Multiple bands (metal core-based + metal ↔ ligand CT)	Yes	ESA + GSB, acoustic vibrations, IC, ISC, varying lifetime (ns–ps), power independence
Plasmonic NPs, e.g., 13 nm AuNPs	Core + surface	Zero	Single-band SPR (nanospheres)	Negligible	GSB, acoustic vibrations, short lifetime (ps), power dependence

$$\textbf{RhB: } S_n \xrightarrow{<100 \text{ fs}} S_1 \xrightarrow{1.7 \text{ ns}} G$$

$$\textbf{Au}_{10}\text{: } {}^1\text{LM/MLCT} \xrightarrow{<100 \text{ fs}} {}^3\text{LM/MLCT} \xrightarrow{14 \text{ ps}} {}^3\text{LM/MLCT}^* \xrightarrow{>1 \text{ ns}} G$$

$$\textbf{Au}_{25}\text{: higher excited state} \xrightarrow{<1 \text{ ps}} \text{lower excited state} \xrightarrow{\sim 100 \text{ ns}} G$$

$$\textbf{Au NP: } \text{e-e scattering} \xrightarrow{<100 \text{ fs}} \text{e-ph coupling} \xrightarrow{1\text{-}5 \text{ ps}} \text{energy dissipation} \xrightarrow{> 100 \text{ ps}}$$

Scheme 1. Relaxation pathways and time constants of four types of materials. G stands for ground state, e–e scattering stands for electron–electron scattering, e–ph coupling stands for electron–phonon coupling.

4. Conclusions

In summary, we have compared the photophysical properties of small organic molecules, gold complexes (no explicit core), nanoclusters (with a core and surface Au–S staples), and metallic-state nanoparticles by choosing four examples (Rhodamine B, $Au_{10}(SR)_{10}$, $Au_{25}(SR)_{18}$, and 13 nm diameter AuNPs). Femtosecond transient absorption spectroscopy was used to determine their relaxation time constants and photodynamics. Overall, gold nanoclusters behave similarly to small molecules, for example, showing a rapid internal conversion (<1 ps) and a long-lived excited state lifetime (~100 ns). In Au(I) complexes, LM/MLCT charge transfer states dominate the relaxation dynamics and no sub-picosecond relaxation was observed. On the other hand, the electron dynamics of plasmonic gold nanoparticles can be well explained by a two-temperature model, with no electron–hole separation or recombination being observed. Overall, the revealed features in photodynamics of the four different materials provide some benchmarking features and are expected to be of great importance for understanding their electronic structures and broadening their applications in various fields in future work.

Supplementary Materials: The Supplementary Materials are available online at http://www.mdpi.com/2079-4991/9/7/933/s1.

Author Contributions: R.J. designed the study; M.Z., C.Z., Q.L. and T.H. performed the experiments and analyzed the data. All authors discussed the results and commented on the manuscript.

Funding: This research was funded by the National Science Foundation (DMR-1808675).

Conflicts of Interest: The authors declare no conflict of interest.

References

1. Yau, S.H.; Varnavski, O.; Goodson, T. An ultrafast look at Au nanoclusters. *Acc. Chem. Res.* **2013**, *46*, 1506–1516. [CrossRef] [PubMed]
2. Zhou, M.; Higaki, T.; Hu, G.; Sfeir, M.Y.; Chen, Y.; Jiang, D.; Jin, R. Three-orders-of-magnitude variation of carrier lifetimes with crystal phase of gold nanoclusters. *Science* **2019**, *364*, 279–282. [PubMed]
3. Hartland, G.V. Optical studies of dynamics in noble metal nanostructures. *Chem. Rev.* **2011**, *111*, 3858–3887. [CrossRef] [PubMed]
4. Zhou, M.; Zeng, C.; Song, Y.; Padelford, J.W.; Wang, G.; Sfeir, M.Y.; Higaki, T.; Jin, R. On the non-metallicity of 2.2 nm $Au_{246}(SR)_{80}$ nanoclusters. *Angew. Chem. Int. Ed.* **2017**, *129*, 16475–16479. [CrossRef]
5. Bain, D.; Paramanik, B.; Patra, A. Silver(I)-induced conformation change of DNA: Gold nanocluster as a spectroscopic probe. *J. Phys. Chem. C* **2017**, *121*, 4608–4617. [CrossRef]
6. Paramanik, B.; Bain, D.; Patra, A. Making and breaking of DNA-metal base pairs: Hg^{2+} and Au nanocluster based off/on probe. *J. Phys. Chem. C* **2016**, *120*, 17127–17135. [CrossRef]

7. Guan, Z.; Gao, N.; Jiang, X.-F.; Yuan, P.; Han, F.; Xu, Q.-H. Huge enhancement in two-photon photoluminescence of au nanoparticle clusters revealed by single-particle spectroscopy. *J. Am. Chem. Soc.* **2013**, *135*, 7272–7277. [CrossRef]

8. Higaki, T.; Zhou, M.; Lambright, K.J.; Kirschbaum, K.; Sfeir, M.Y.; Jin, R. Sharp transition from nonmetallic Au_{246} to metallic Au_{279} with nascent surface plasmon resonance. *J. Am. Chem. Soc.* **2018**, *140*, 5691–5695. [CrossRef]

9. Russier-Antoine, I.; Bertorelle, F.; Vojkovic, M.; Rayane, D.; Salmon, E.; Jonin, C.; Dugourd, P.; Antoine, R.; Brevet, P.-F. Non-linear optical properties of gold quantum clusters. The smaller the better. *Nanoscale* **2014**, *6*, 13572–13578. [CrossRef]

10. Zhu, M.; Aikens, C.M.; Hollander, F.J.; Schatz, G.C.; Jin, R. Correlating the crystal structure of a thiol-protected Au_{25} cluster and optical properties. *J. Am. Chem. Soc.* **2008**, *130*, 5883–5885. [CrossRef]

11. Jin, R.; Zeng, C.; Zhou, M.; Chen, Y. Atomically precise colloidal metal nanoclusters and nanoparticles: Fundamentals and opportunities. *Chem. Rev.* **2016**, *116*, 10346–10413. [CrossRef] [PubMed]

12. Olesiak-Banska, J.; Waszkielewicz, M.; Samoc, M. Two-photon chiro-optical properties of gold Au_{25} nanoclusters. *Phys. Chem. Chem. Phys.* **2018**, *20*, 24523–24526. [CrossRef] [PubMed]

13. Zeng, C.; Chen, Y.; Kirschbaum, K.; Lambright, K.J.; Jin, R. Emergence of hierarchical structural complexities in nanoparticles and their assembly. *Science* **2016**, *354*, 1580–1584. [CrossRef] [PubMed]

14. Che, C.-M.; Lai, S.-W. Luminescence and photophysics of gold complexes. In *Gold Chemistry*; Wiley-VCH/Verlag GmbH & Co. KGaA: Weinheim, Germany, 2009; pp. 249–281.

15. Yam, V.W.-W.; Au, V.K.-M.; Leung, S.Y.-L. Light-emitting self-assembled materials based on d^8 and d^{10} transition metal complexes. *Chem. Rev.* **2015**, *115*, 7589–7728. [CrossRef] [PubMed]

16. Yu, K.; You, G.; Polavarapu, L.; Xu, Q.-H. Bimetallic Au/Ag core–shell nanorods studied by ultrafast transient absorption spectroscopy under selective excitation. *J. Phys. Chem. C* **2011**, *115*, 14000–14005. [CrossRef]

17. Olesiak-Banska, J.; Gordel, M.; Matczyszyn, K.; Shynkar, V.; Zyss, J.; Samoc, M. Gold nanorods as multifunctional probes in a liquid crystalline DNA matrix. *Nanoscale* **2013**, *5*, 10975–10981. [CrossRef] [PubMed]

18. Minutella, E.; Schulz, F.; Lange, H. Excitation-dependence of plasmon-induced hot electrons in gold nanoparticles. *J. Phys. Chem. Lett.* **2017**, 4925–4929. [CrossRef]

19. Link, S.; El-Sayed, M.A. Optical properties and ultrafast dynamics of metallic nanocrystals. *Annu. Rev. Phys. Chem.* **2003**, *54*, 331–366. [CrossRef]

20. Hartland, G.V. Coherent excitation of vibrational modes in metallic nanoparticles. *Annu. Rev. Phys. Chem.* **2006**, *57*, 403–430. [CrossRef]

21. Link, S.; Beeby, A.; FitzGerald, S.; El-Sayed, M.A.; Schaaff, T.G.; Whetten, R.L. Visible to infrared luminescence from a 28-atom gold cluster. *J. Phys. Chem. B* **2002**, *106*, 3410–3415. [CrossRef]

22. Link, S.; El-Sayed, M.A. Spectral properties and relaxation dynamics of surface plasmon electronic oscillations in gold and silver nanodots and nanorods. *J. Phys. Chem. B* **1999**, *103*, 8410–8426. [CrossRef]

23. Hartland, G.V. Coherent vibrational motion in metal particles: Determination of the vibrational amplitude and excitation mechanism. *J. Chem. Phys.* **2002**, *116*, 8048–8055. [CrossRef]

24. Del Fatti, N.; Bouffanais, R.; Vallée, F.; Flytzanis, C. Nonequilibrium electron interactions in metal films. *Phys. Rev. Lett.* **1998**, *81*, 922–925. [CrossRef]

25. Kwak, K.; Thanthirige, V.D.; Pyo, K.; Lee, D.; Ramakrishna, G. Energy gap law for exciton dynamics in gold cluster molecules. *J. Phys. Chem. Lett.* **2017**, *8*, 4898–4905. [CrossRef] [PubMed]

26. Zeng, C.; Chen, Y.; Kirschbaum, K.; Appavoo, K.; Sfeir, M.Y.; Jin, R. Structural patterns at all scales in a nonmetallic chiral $Au_{133}(SR)_{52}$ nanoparticle. *Sci. Adv.* **2015**, *1*, e1500045. [CrossRef] [PubMed]

27. Zhao, T.; Herbert, P.J.; Zheng, H.; Knappenberger, K. State-resolved metal nanoparticle dynamics viewed through the combined lenses of ultrafast and magneto-optical spectroscopies. *Acc. Chem. Res.* **2018**, *51*, 1433–1442. [CrossRef]

28. Zhou, M.; Zeng, C.; Sfeir, M.Y.; Cotlet, M.; Iida, K.; Nobusada, K.; Jin, R. Evolution of excited-state dynamics in periodic Au_{28}, Au_{36}, Au_{44}, and Au_{52} nanoclusters. *J. Phys. Chem. Lett.* **2017**, *8*, 4023–4030. [CrossRef] [PubMed]

29. Zhou, M.; Zeng, C.; Chen, Y.; Zhao, S.; Sfeir, M.Y.; Zhu, M.; Jin, R. Evolution from the plasmon to exciton state in ligand-protected atomically precise gold nanoparticles. *Nat. Commun.* **2016**, *7*, 13240. [CrossRef] [PubMed]

30. Mustalahti, S.; Myllyperkiö, P.; Malola, S.; Lahtinen, T.; Salorinne, K.; Koivisto, J.; Häkkinen, H.; Pettersson, M. Molecule-like photodynamics of $Au_{102}(pMBA)_{44}$ nanocluster. *ACS Nano* **2015**, *9*, 2328–2335. [CrossRef] [PubMed]

31. Stamplecoskie, K.G.; Kamat, P.V. Size-dependent excited state behavior of glutathione-capped gold clusters and their light-harvesting capacity. *J. Am. Chem. Soc.* **2014**, *136*, 11093–11099. [CrossRef]

32. Wiseman, M.R.; Marsh, P.A.; Bishop, P.T.; Brisdon, B.J.; Mahon, M.F. Homoleptic Gold Thiolate Catenanes. *J. Am. Chem. Soc.* **2000**, *122*, 12598–12599. [CrossRef]

33. Wu, Z.; Suhan, J.; Jin, R. One-pot synthesis of atomically monodisperse, thiol-functionalized Au_{25} nanoclusters. *J. Mater. Chem.* **2009**, *19*, 622–626. [CrossRef]

34. Yam, V.W.-W.; Cheng, E.C.-C.; Zhou, Z.-Y. A highly soluble luminescent decanuclear gold(I) complex with a propeller-shaped structure. *Angew. Chem. Int. Ed.* **2000**, *39*, 1683–1685. [CrossRef]

35. Forward, J.M.; Bohmann, D.; Fackler, J.P.; Staples, R.J. Luminescence studies of gold(I) thiolate complexes. *Inorg. Chem.* **1995**, *34*, 6330–6336. [CrossRef]

36. Konishi, K.; Iwasaki, M.; Shichibu, Y. Phosphine-ligated gold clusters with core+exo geometries: Unique properties and interactions at the ligand–cluster interface. *Acc. Chem. Res.* **2018**, *51*, 3125–3133. [CrossRef] [PubMed]

37. Wing-Wah Yam, V.; Kam-Wing Lo, K. Luminescent polynuclear d^{10} metal complexes. *Chem. Soc. Rev.* **1999**, *28*, 323–334.

38. Lei, Z.; Pei, X.-L.; Guan, Z.-J.; Wang, Q.-M. Full protection of intensely luminescent gold(I)–silver(I) cluster by phosphine ligands and inorganic anions. *Angew. Chem. Int. Ed.* **2017**, *56*, 7117–7120. [CrossRef] [PubMed]

39. Wu, Z.; Jin, R. On the ligand's role in the fluorescence of gold nanoclusters. *Nano Lett.* **2010**, *10*, 2568–2573. [CrossRef]

40. Link, S.; El-Sayed, M.A.; Schaaff, T.G.; Whetten, R.L. Transition from nanoparticle to molecular behavior: A femtosecond transient absorption study of a size-selected 28 atom gold cluster. *Chem. Phys. Lett.* **2002**, *356*, 240–246. [CrossRef]

41. Kim, H.N.; Lee, M.H.; Kim, H.J.; Kim, J.S.; Yoon, J. A new trend in rhodamine-based chemosensors: Application of spirolactam ring-opening to sensing ions. *Chem. Soc. Rev.* **2008**, *37*, 1465–1472. [CrossRef]

42. Sadkowski, P.J.; Fleming, G.R. Photophysics of acid and base forms of rhodamine-B. *Chem. Phys. Lett.* **1978**, *57*, 526–529. [CrossRef]

43. Moog, R.S.; Ediger, M.D.; Boxer, S.G.; Fayer, M.D. Viscosity dependence of the rotational reorientation of rhodamine B in mono- and polyalcohols. Picosecond transient grating experiments. *J. Phys. Chem.* **1982**, *86*, 4694–4700. [CrossRef]

44. Beaumont, P.C.; Johnson, D.G.; Parsons, B.J. Photophysical properties of laser dyes: Picosecond laser flash photolysis studies of rhodamine 6G, rhodamine B and rhodamine 101. *J. Chem. Soc. Faraday Trans.* **1993**, *89*, 4185–4191. [CrossRef]

45. Zhang, X.F.; Zhang, Y.K.; Liu, L.M. Fluorescence lifetimes and quantum yields of ten rhodamine derivatives: Structural effect on emission mechanism in different solvents. *J. Lumin.* **2014**, *145*, 448–453. [CrossRef]

46. Kasha, M. Characterization of electronic transitions in complex molecules. *Discuss. Faraday Soc.* **1950**, *9*, 14–19. [CrossRef]

47. Zhou, M.; Lei, Z.; Guo, Q.; Wang, Q.-M.; Xia, A. Solvent dependent excited state behaviors of luminescent gold(I)–silver(I) cluster with hypercoordinated carbon. *J. Phys. Chem. C* **2015**, *119*, 14980–14988. [CrossRef]

48. Zeng, C.-j.; Zhou, M.; Gayathri, C.; Gil, R.R.; Sfeir, M.Y.; Jin, R. $Au_{10}(TBBT)_{10}$: The beginning and the end of aun(tbbt)m nanoclusters. *Chin. J. Chem. Phys.* **2018**, *31*, 555–562. [CrossRef]

49. Devadas, M.S.; Kim, J.; Sinn, E.; Lee, D.; Goodson, T., 3rd; Ramakrishna, G. Unique ultrafast visible luminescence in monolayer-protected au_{25} clusters. *J. Phys. Chem. C* **2010**, *114*, 22417–22423. [CrossRef]

50. Zhou, M.; Tian, S.; Zeng, C.; Sfeir, M.Y.; Wu, Z.; Jin, R. Ultrafast relaxation dynamics of $Au_{38}(SC_2H_4Ph)_{24}$ nanoclusters and effects of structural isomerism. *J. Phys. Chem. C* **2017**, *121*, 10686–10693. [CrossRef]

51. Qian, H.; Sfeir, M.Y.; Jin, R. Ultrafast relaxation dynamics of $[Au_{25}(SR)_{18}]^q$ nanoclusters: Effects of charge state. *J. Phys. Chem. C* **2010**, *114*, 19935–19940. [CrossRef]

52. Varnavski, O.; Ramakrishna, G.; Kim, J.; Lee, D.; Goodson, T., III. Optically excited acoustic vibrations in quantum-sized monolayer-protected gold clusters. *ACS Nano* **2010**, *4*, 3406–3412. [CrossRef] [PubMed]

53. Zhou, M.; Jin, R.; Sfeir, M.Y.; Chen, Y.; Song, Y.; Jin, R. Electron localization in rod-shaped triicosahedral gold nanocluster. *Proc. Natl. Acad. Sci. USA* **2017**, *114*, E4697–E4705. [CrossRef] [PubMed]

54. Rafiq, S.; Scholes, G.D. Slow intramolecular vibrational relaxation leads to long-lived excited-state wavepackets. *J. Phys. Chem. A* **2016**, *120*, 6792–6799. [CrossRef] [PubMed]
55. Sagar, D.M.; Cooney, R.R.; Sewall, S.L.; Dias, E.A.; Barsan, M.M.; Butler, I.S.; Kambhampati, P. Size dependent, state-resolved studies of exciton-phonon couplings in strongly confined semiconductor quantum dots. *Phys. Rev. B* **2008**, *77*, 235321. [CrossRef]
56. Hodak, J.H.; Henglein, A.; Hartland, G.V. Electron-phonon coupling dynamics in very small (between 2 and 8 nm diameter) Au nanoparticles. *J. Chem. Phys.* **2000**, *112*, 5942–5947. [CrossRef]

 *nanomaterials*

Article

6-Aza-2-Thio-Thymine Stabilized Gold Nanoclusters as Photoluminescent Probe for Protein Detection

Hao-Hua Deng, Xiao-Qiong Shi, Paramasivam Balasubramanian, Kai-Yuan Huang, Ying-Ying Xu, Zhong-Nan Huang, Hua-Ping Peng and Wei Chen *

Fujian Key Laboratory of Drug Target Discovery and Structural and Functional Research, School of Pharmacy, Fujian Medical University, Fuzhou 350004, China; DHH8908@163.com (H.-H.D.); xiaoqiongshi@163.com (X.-Q.S.); chembala55@gmail.com (P.B.); 15980269577@163.com (K.-Y.H.); yingyingxu2010@hotmail.com (Y.-Y.X.); hzn651806692@163.com (Z.-N.H.); phpfjmu@126.com (H.-P.P.)
* Correspondence: Weichen@fjmu.edu.cn or chenandhu@163.com

Received: 31 December 2019; Accepted: 4 February 2020; Published: 7 February 2020

Abstract: This study puts forward an efficient method for protein detection in virtue of the tremendous fluorescence enhancement property of 6-aza-2-thio-thymine protected gold nanoclusters (ATT-AuNCs). In-depth studies of the protein-induced photoluminescence enhancement mechanism illustrate the mechanism of the interaction between ATT-AuNCs and protein. This new-established probe enables feasible and sensitive quantification of the concentrations of total protein in real samples, such as human serum, human plasma, milk, and cell extracts. The results of this proposed method are in good agreement with those determined by the classical bicinchoninic acid method (BCA method).

Keywords: gold nanocluster; 6-aza-2-thio-thymine; protein; fluorescence

---

1. Introduction

Proteins are the basic material of living organisms. There have been various methods for the accurate measurement of protein content, which are indispensable tools for biological researches, such as proteomics, molecular biology, cell biology, biochemistry, and neuroscience [1–4]. However, these methods have their own scope of application and limitations. Therefore, the development of a universal, precise, and rapid quantitative technique will provide a great deal of impetus for the molecular biology revolution and promote the progress of clinical diagnosis.

Owing to their unique physical and chemical properties caused by quantum confinement effect, gold nanoclusters (AuNCs) have emerged as an intensely pursued material for nanoscience research and biological applications in the past decade [5–9]. The molecule-like properties of AuNCs, such as discrete electronic states and size-dependent photoluminescence, makes them a distinguished material from larger nanoparticles and bulk metals [10–18]. As a matter of fact, the gradually extensive application of few-atom metal nanomaterials has promoted a depth exploration of nanotoxicology, and, correspondingly, highlighted the significance for the biosafety evaluation to be performed. Investigations of protein–nanoparticle interactions have been reported in several studies [19–23]. However, as far as we know, there has been a smaller number of literatures focusing on protein determination with the help of fluorescent AuNCs [24,25].

In our previous study, we found that the restriction of ligand motion can enhance the fluorescence quantum yield of 6-aza-2-thio-thymine protected gold nanoclusters (ATT-AuNCs) [8,26]. Inspired from this fact, we here test the influence of protein absorption on the fluorescence characteristics of ATT-AuNCs. The interaction between protein and ATT-AuNCs results in a dramatical enhancement on the fluorescence intensity of AuNCs. Based on this, we designed a universal method for detecting total protein (bovine serum albumin) in biological samples.

2. Materials and Methods

2.1. Chemical and Reagents

6-Aza-2-thiothymine (ATT) was purchased from Alfa Aesar Chemicals Co. Ltd. (Beijing, China). $HAuCl_4 \cdot 3H_2O$ was obtained from Aladdin Reagent Company (Shanghai, China). Bovine serum albumin (BSA) and NaOH was bought from Sinopharm Chemical Reagent Co. Ltd. (Shanghai, China). BCA kit was bought from Solarbio Science and Technology Co. Ltd. (Beijing, China). All chemicals and solvents were of analytical grade and commercially available. All solutions were prepared with deionized water (DIW).

2.2. Instruments

UV-2450 UV-Vis spectrophotometer (Shimadzu, Japan) and microplate reader (BioTek Instruments, Winooski, VT, USA) were used for UV-Vis measurements. The fluorescence spectra were recorded on a Cary Eclipse fluorescence spectrophotometer (Agilent, Santa Clara, CA, USA).

2.3. Preparation of ATT-AuNCs

Fluorescent ATT-AuNCs were prepared with a facile one-pot strategy as described previously [8]. In short, ATT (15 mL, 80 mM) containing 0.2 M NaOH was added into an aqueous solution of $HAuCl_4$ (15 mL, 10 mg/mL), and the mixture was continuously stirred for 1 h at room temperature. The ATT-AuNCs were purified by ultrafiltration (Millipore, Bedford, MA, USA, 50 kDa) and stored at 4 °C in the dark prior to use.

2.4. Sample Analysis

In a typical experiment, (a) to 200 µL of ATT-AuNCs stock solution (0.5 mg/mL, pH 5), 20 µL of different concentrations of BSA was added; the mixture was incubated at 33 °C for 25 min; the resulting reaction solution was measured by using fluorescence spectrometer under the excitation wavelength of 472 nm. (b) To 200 µL of BCA stock solution, 20 µL of different concentrations of BSA was added; the mixture was incubated at 37 °C for 25 min; the resulting reaction solution was measured by using a BioTek Microplate reader at 562 nm absorbance wavelength.

Human plasma and serum samples were donated by the Affiliated First Hospital of Fujian Medical University (Fuzhou, China). Our project has been approved by the Medical Ethics Committee of Fujian Medical University. We have obtained informed consent before the experiment, and the information of the volunteers was kept confidential. Milk samples were purchased from local supermarket. All samples were directly used for analysis.

HL60 cells were obtained from Shanghai Institute of Cell Biology and Biochemistry (Shanghai, China) and grown in RPMI 1640 medium (Hyclone) supplemented with 10% fetal calf serum.

For protein determination in plasma and serum samples, the samples were simply diluted by DIW three times, and determined according to the processes mentioned above. Standard addition experiments were further conducted by adding three different concentrations of BSA in the real plasma samples.

For protein determination in milk samples, the samples were simply diluted by DIW about three times. After completion of the reaction, 3 mM of EDTA was added to each sample to eliminate the possible interference of Ca^{2+}, and then determined according to the processes mentioned above.

For protein determination in cell extracts, HL60 cells were centrifuged and resuspended in 0.9% NaCl solution. Then, collected HL 60 cells were lysed by lysing buffer (50 mM Gly-NaOH, 0.15 M NaCl, 1% Triton X-100, 1 mM EDTA, 0.1% SDS). Cell extracts were collected by centrifugation, and then simply diluted with DIW three times. Protein content was determined according to the processes mentioned above. Results of the proposed method were also compared with the classical bicinchoninic acid (BCA) method.

3. Results and Discussion

3.1. Interaction between Protein and ATT-AuNCs

AuNCs were prepared via a simple one-pot strategy employing ATT as the reducing-cum-stabilizing/capping ligand [8]. The ATT-AuNCs in aqueous solution had a fluorescence emission band centered on 528 nm (photoluminescence quantum yield = 1.8%). As shown in Figure 1, interaction between BSA and ATT-AuNCs provokes a substantial enhancement (13.8-fold) of fluorescence intensity. This remarkable fluorescence change can be obviously observed by the naked eye under the irradiation of UV light (Figure 1 inset). Since BSA has no fluorescence in this spectral region, the enhancement of fluorescence can be reasonably attributed to the adsorption of ATT-AuNCs on the protein, which results in the restriction of rotation and vibration of the ligands.

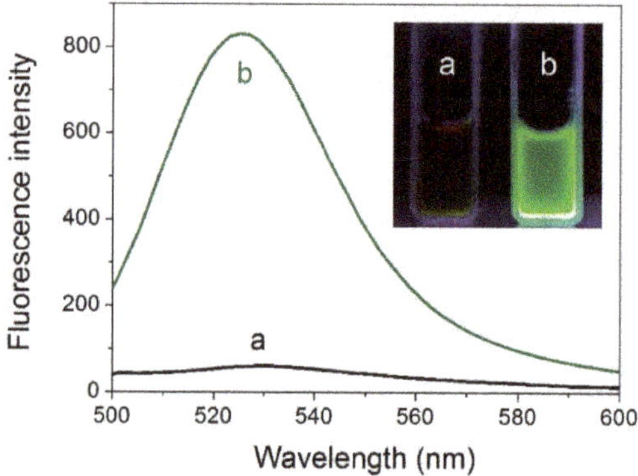

Figure 1. Fluorescence spectra of 6-aza-2-thio-thymine protected gold nanoclusters (ATT-AuNCs) (**a**) before and (**b**) after introducing bovine serum albumin (BSA). The concentrations of ATT-AuNCs and BSA were 0.5 mg/mL and 500 μg/mL, respectively. Inset: Photographs of ATT-AuNC and BSA/ATT-AuNC solutions under UV light.

Fluorescence spectroscopy is a most widely used experimental approach to obtain local conformational or dynamic changes of protein, in which tryptophan (Trp) and tyrosine (Tyr) are the main contributors to the endogenous fluorescence [27]. To investigate the interaction between ATT-AuNCs and BSA, fluorescence quenching of BSA upon the addition of ATT-AuNCs was performed. The photoemission intensity of BSA at 345 nm was suppressed progressively with increasing ATT-AuNCs concentration (Figure 2). Data were then analyzed with the Stern–Volmer equation (Equation (1)).

$$F_0/F = 1 + K_q\tau_0[Q] = 1 + K_{sv}[Q] \tag{1}$$

where F_0 and F represent the fluorescence intensities of BSA at 345 nm before and after the addition of AuNCs, respectively; [Q] is the concentration of AuNCs; K_{sv} is the Stern–Volmer fluorescence quenching constant; and K_q is the bimolecular dynamic fluorescence quenching rate constant. τ_0 is the lifetime of BSA without the addition of AuNCs, and is known to be approximately 5×10^{-9} s. From linear regression of the plot (r = 0.9987), $K_q = 5.05 \times 10^{15}$ L mol^{-1} s^{-1} was obtained. As the maximum value of K_q for a diffusion-controlled quenching process is about 2.0×10^{10} L mol^{-1} s^{-1}, we concluded a static quenching mechanism was responsible for the quenching process of BSA caused by ATT-AuNCs [28].

The forces of interaction between BSA and ATT-AuNCs were also investigated. Thermodynamic parameters were obtained from the static quenching equation (Equation (2)) and van 't Hoff equation (Equation (3))

$$lg[(F_0 - F)/F] = lgK + nlg[Q] \tag{2}$$

$$\Delta G = -RT\ln K = \Delta H - T\Delta S \tag{3}$$

As can be seen from Table 1, the negative ΔG values suggest that the interaction between ATT-AuNCs and BSA is spontaneous. The positive ΔH and ΔS values indicate that the interaction between ATT-AuNCs and BSA is mainly hydrophobic forces. It might be that when ATT-AuNCs are close to the binding sites of BSA, the hydration layer of BSA around the binding sites are partly destroyed, leading to heat absorption phenomenon [25]. We also found that the photoluminescence intensity of ATT-AuNCs was enhanced as the temperature increased in a certain range, which further validated that ATT-AuNCs approaching the binding sites of target protein is an endothermic reaction (Figure 3).

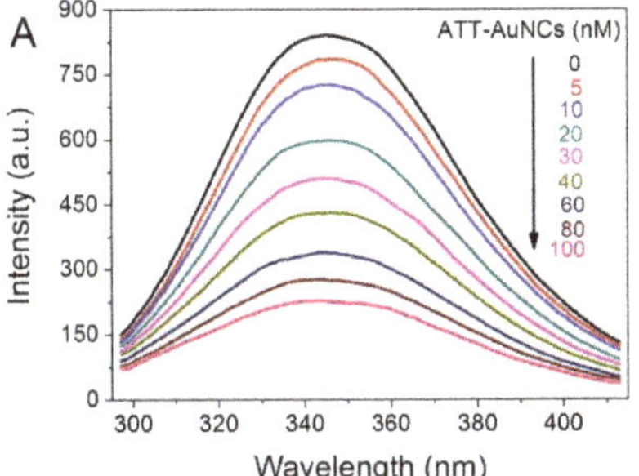

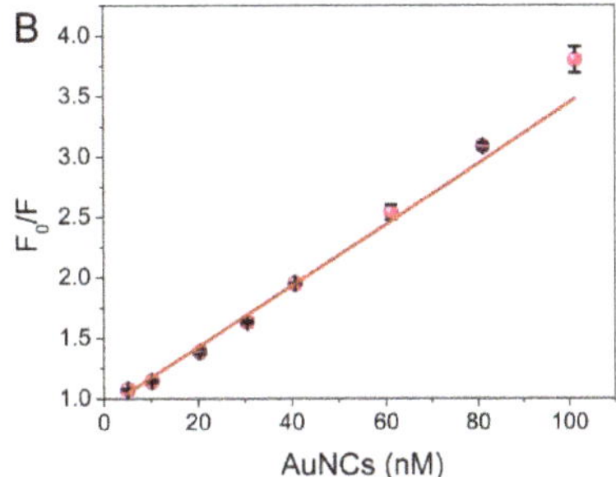

Figure 2. (**A**) Photoemission spectra of BSA (7.5×10^{-7} M, 50 µg/mL) with different concentrations of ATT-AuNCs (range from 0 to 100 nM) upon excitation at 280 nm at pH 5.0, 33 °C. (**B**) Stern–Volmer plot for fluorescence quenching of BSA by ATT-AuNCs. Error bars represent the standard deviations across three repetitive experiments.

Table 1. Calculated thermodynamic parameters for ATT-AuNCs-BSA composite.

T (°C)	K (10^8 L mol^{-1})	ΔH (kJ mol^{-1})	ΔG (kJ)	ΔS (J mol^{-1} K^{-1})	R	SD
21	5.70	17.583	−49.275	227.380	0.9996	0.97
33	7.38		−51.948		0.9978	2.41
38	8.51		−53.167		0.9997	0.70

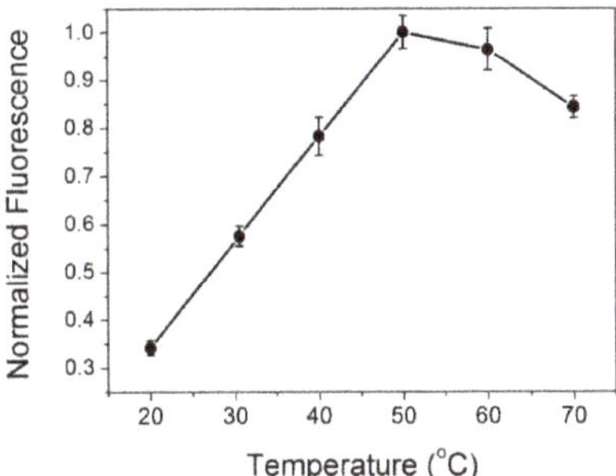

Figure 3. The thermodynamics curve of ATT-AuNCs (8 µM) after introducing BSA (35 µg/mL). Error bars represent the standard deviations across three repetitive experiments.

3.2. Analytical Performance

Fluorescence spectrophotometer was used to evaluate the BSA concentration in aqueous solution. In can be seen from Figure 4A, a progressive enhancement of photoluminescence intensity was obtained with the increased amount of BSA. The value of intensity exhibited a good linear correlation to BSA concentration in the range from 50 to 400 µg/mL (r = 0.9955). The detection limit of BSA was 0.88 µg/mL at an S:N ratio of 3. The relative standard deviation is 2.25% for the determination of 200 µg/mL BSA (n = 10).

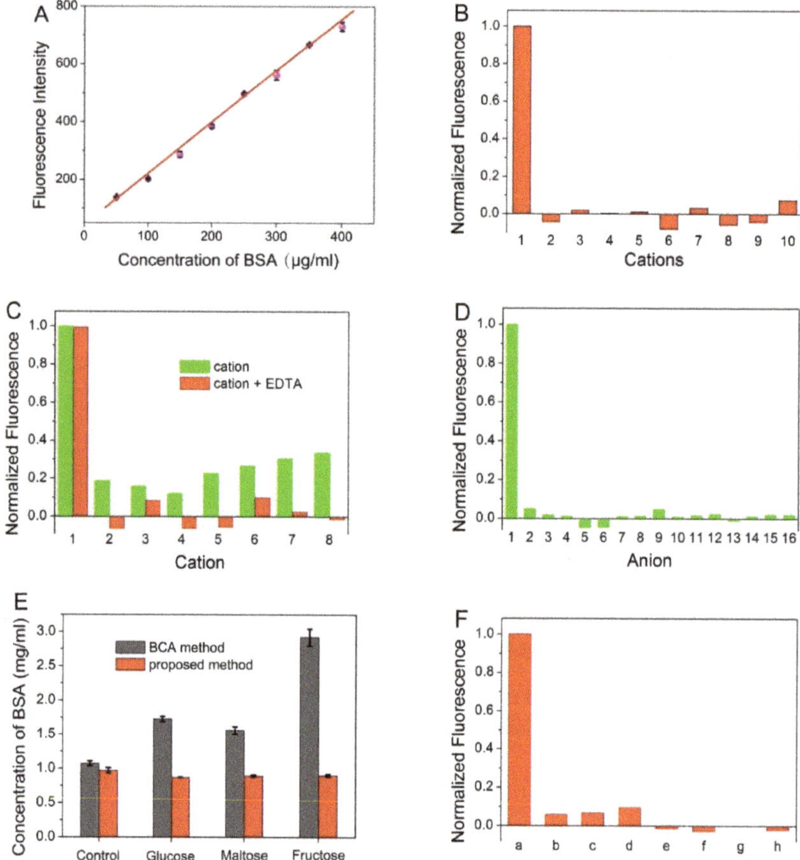

Figure 4. (**A**) The plot of the fluorescence intensity of ATT-AuNCs versus the concentration of BSA. Error bars represent the standard deviations across three repetitive experiments. (**B**) Fluorescence response of ATT-AuNCs to different cations. Samples marked 1 to 10 correspond to BSA (250 µg/mL), Ag^+, Cr^{3+}, K^+, Na^+, Hg^{2+}, NH_4^+, Fe^{3+}, Cu^{2+}, and Mn^{2+}, respectively. The concentration of each cation was 2 mM. (**C**) Fluorescence response of ATT-AuNCs to different cations in the presence of 2.5 mM of EDTA. Samples marked 1 to 8 correspond to BSA (250 µg/mL), Ni^{2+}, Ba^{2+}, Pb^{2+}, Zn^{2+}, Mg^{2+}, Ca^{2+}, Al^{3+}, respectively. The concentration of each cation was 2 mM. (**D**) Fluorescence response of ATT-AuNCs to different anions. Samples marked 1 to 16 correspond to BSA (250 µg/mL), $HCOO^-$, NO_2^-, NO_3^-, CN^-, IO_3^-, PO_4^{3-}, SO_4^{2-}, Br^-, SCN^-, I^-, BrO_3^-, ClO_4^-, F^-, Cl^-, and CO_3^{2-}, respectively. The concentration of each anion was 1 mM. (**E**) The detection results of BSA in the presence of different reducing sugar (5 M). Error bars represent the standard deviations across three repetitive experiments. (**F**) Fluorescence response of ATT-AuNCs to (a) 250 µg/mL BSA, (b) 0.5% SDS, (c) 20 mM creatine, (d) 0.5% Tween 20, (e) 150 mM urea, (f) 5% ethanol, (g) 20 mM EDTA, and (h) 0.5% TritonX-100. The normalized fluorescence intensity is calculated by $(I - I_0)/(I_{BSA} - I_0)$, where I_0, I, and I_{BSA} is the fluorescence intensity of ATT-AuNCs, ATT-AuNCs + interferent, ATT-AuNCs + BSA, respectively.

In order to examine the selectivity of the established protein assay, interferences from common ions were investigated. In addition, certain compounds often used during protein purification or cell lysis were also investigated. Figure 4B,D reveals that ATT-AuNCs were specific toward BSA over the tested cations or anions. Some of the tested cations that interfere with the protein assay can be eliminated to a certain extent by adding the chelator EDTA (Figure 4C). Figure 4E reveals that this gold nanoprobe based on PL enhancement exhibits the remarkable merits of tolerating reducing sugar, compared with the BCA method which gives false positive results. Figure 4F reveals that certain compounds often used during protein purification or cell lysis have little effect on the system.

3.3. Determination of Total Protein in Human Plasma and Serum

In order to elucidate the practical feasibility of the approach, a variety of biological samples were determined by the proposed method and compared with classical BCA method. As we know, it is of great importance to make a quick determination of total protein in plasma or serum content which is closely related to various diseases such as hypoalbuminemia [29] and hypertensive disease [30]. Figure 5 showed that human serum albumin (HSA) can also induce fluorescence enhancement of ATT-AuNCs and its efficiency is comparable to that of BSA. This result confirms that ATT-AuNC possesses the capacity to bind HSA. As shown in Tables 2 and 3, the results of total protein content measured by this method agree well with those obtained by the standard BCA method (Tables 2 and 3). In addition, we used BSA as a standard agent to conduct recovery experiments in human plasma samples. It was found that the recoveries were observed to range from 104% to 112%. Therefore, the proposed protein assay yields satisfactory results for total protein determination in human plasma and serum samples.

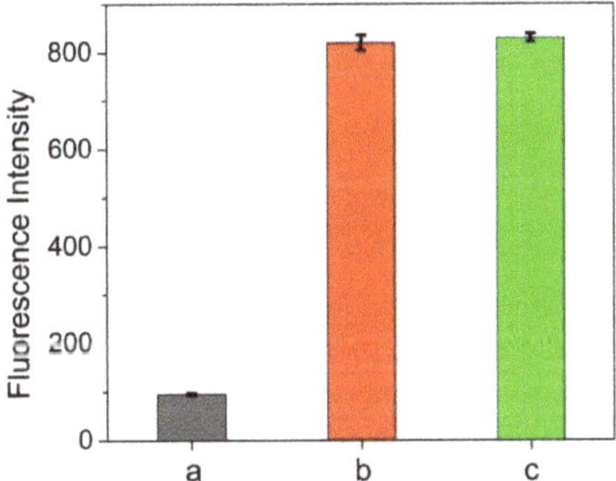

Figure 5. Fluorescence intensities of (a) 0.5 mg/mL ATT-AuNCs, (b) 0.5 mg/mL ATT-AuNCs + 500 µg/mL HSA, and (c) 0.5 mg/mL ATT-AuNCs + 500 µg/mL BSA. Error bars represent the standard deviations across three repetitive experiments.

Table 2. The analytical results of plasma total protein.

Sample	Proposed Method (mg/mL, n = 3)	BCA Method (mg/mL, n = 3)	F-Test [1]	t-Test [1]
1	62.60 ± 1.35	62.84 ± 1.27	1.13	0.22
2	61.73 ± 0.28	62.20 ± 1.17	17.46	0.68
3	58.08 ± 1.15	58.55 ± 1.26	1.21	0.48
4	58.95 ± 0.73	58.05 ± 0.78	1.14	1.46
5	65.18 ± 2.62	65.95 ± 1.42	3.40	0.45
6	63.26 ± 1.88	64.80 ± 1.66	1.28	1.07

[1] $F_{0.05,\,2,\,2} = 19.00$, $t_{0.05,\,4} = 2.776$.

Table 3. The results of serum total protein.

Sample	Proposed Method (mg/mL, n = 3)	BCA Method (mg/mL, n = 3)	F-Test [1]	t-Test [1]
1	59.15 ± 0.82	61.25 ± 2.78	11.49	1.26
2	65.28 ± 1.63	64.86 ± 1.96	1.45	0.29
3	64.81 ± 1.45	65.02 ± 0.52	7.78	0.24
4	70.56 ± 0.48	71.74 ± 1.93	16.17	1.03

[1] $F_{0.05, 2, 2} = 19.00$, $t_{0.05, 4} = 2.776$.

3.4. Determination of Proteins in Milk Samples

Food safety problems caused by the melamine incident have affected the entire food industry [31]. It can lead to different degrees of kidney failure or death, and induce bladder stones and urinary system diseases [32]. Classic food proteins determination (Kjeldahl methods) is cumbersome, time-consuming, and has a lack of specificity [33]. Thus, the demand for a rapid, accurate milk protein assay has increased dramatically. In this study, we chose three brands of pure milk in the supermarket randomly for protein assay. The results of total protein content obtained by the proposed method agree well with those measured by the BCA method (Table 4).

Table 4. The results of milk total protein.

Sample	Proposed Method (mg/mL, n = 3)	BCA Method (mg/mL, n = 3)	F-Test [1]	t-Test [1]
1	31.17 ± 0.18	32.29 ± 0.43	5.71	4.16
2	32.44 ± 0.19	32.01 ± 0.52	7.49	1.34
3	30.01 ± 0.16	30.05 ± 0.50	9.77	0.14

[1] $F_{0.05, 2, 2} = 19.00$, $t_{0.05, 4} = 2.776$.

3.5. Determination of Cellular Total Protein Concentration

Various kinds of cell protein factors were involved in complex biochemical processes like inflammatory responses, tumor growth, cell apoptosis and angiogenesis [34,35]. Thus, assay of cell total protein is of great significance. In the present study, we monitored cell total protein. HL 60 was used as a model cell. The results of total protein content obtained by the proposed method agree well with those measured by the BCA method (Table 5). It implies this method has potential application in the detection of total cellular protein.

Table 5. The results of cell total protein.

Sample	Proposed Method (mg/mL, n = 3)	BCA Method (mg/mL, n = 3)	F-Test [1]	t-Test [1]
1	5.52 ± 0.034	5.51 ± 0.026	1.71	0.40
2	6.77 ± 0.046	6.84 ± 0.072	2.45	1.42

[1] $F_{0.05, 2, 2} = 19.00$, $t_{0.05, 4} = 2.776$.

4. Conclusions

Herein, we present the BSA induced fluorescence enhancement of ATT-AuNCs through protein–nanocluster interactions, and it was successfully used to determine the total protein content of various biological samples. Fluorescence spectroscopy studies shed some light on the underlying mechanism of fluorescence enhancement of AuNCs upon protein adsorption. It turned out that the dominating quenching mechanism of protein–nanocluster interactions was the static quenching process. The experimental results indicate that this new method is better than the traditional BCA method in its compatibility with high concentrations of reducing sugar, which proposed an alternative scheme to determine the total protein content.

Author Contributions: Conceptualization, W.C. and H.-H.D.; Methodology, H.-P.P. and Y.-Y.X.; Investigation, X.-Q.S., K.-Y.H. and Z.-N.H.; Writing—Original Draft Preparation, H.-H.D. and X.-Q.S.; Writing—Review and Editing, P.B.; Funding Acquisition, W.C. All authors have read and agreed to the published version of the manuscript.

Funding: This research was funded by the National Natural Science Foundation of China, grant number 21675024; the Program for Innovative Leading Talents in Fujian Province, grant number 2016B016; and the Science and Technology Project of Fujian Province, grant number 2018L3008.

Conflicts of Interest: The authors declare no conflict of interest.

References

1. Smith, P.K.; Krohn, R.I.; Hermanson, G.T.; Mallia, A.K.; Gartner, F.H.; Provenzano, M.D.; Fujimoto, E.K.; Goeke, N.M.; Olson, B.J.; Klenk, D.C. Measurement of Protein Using Bicinchoninic Acid. *Anal. Biochem.* **1985**, *150*, 76–85. [CrossRef]

2. Lowry, O.H.; Nira, J.; Rosebrough, A.; Lewis, F.; Rose, J.R. Protein Measurement with the Folin Phenol Reagent. *J. Biol. Chem.* **1951**, *193*, 265–275. [PubMed]

3. Lorenzen, A.; Kennedy, S.W. A Fluorescence-Based Protein Assay for Use with a Microplate Reader. *Anal. Biochem.* **1993**, *214*, 346–348. [CrossRef] [PubMed]

4. Han, X.X.; Zhao, B.; Ozaki, Y. Surface-Enhanced Raman Scattering for Protein Detection. *Anal. Bioanal. Chem.* **2009**, *394*, 1719–1727. [CrossRef] [PubMed]

5. Xie, J.P.; Zheng, Y.G.; Ying, J.Y. Protein-Directed Synthesis of Highly Fluorescent Gold Nanoclusters. *J. Am. Chem. Soc.* **2009**, *131*, 888–889. [CrossRef] [PubMed]

6. Jin, R.C. Quantum Sized, Thiolate-Protected Gold Nanoclusters. *Nanoscale* **2010**, *2*, 343–362. [CrossRef] [PubMed]

7. Qian, H.F.; Zhu, M.Z.; Wu, Z.K.; Jin, R.C. Quantum Sized Gold Nanoclusters with Atomic Precision. *Acc. Chem. Res.* **2012**, *45*, 1470–1479. [CrossRef]

8. Deng, H.H.; Shi, X.Q.; Wang, F.F.; Peng, H.P.; Liu, A.L.; Xia, X.H.; Chen, W. Fabrication of Water-Soluble, Green-Emitting Gold Nanoclusters with a 65% Photoluminescence Quantum Yield Via Host-Guest Recognition. *Chem. Mater.* **2017**, *29*, 1362–1369. [CrossRef]

9. Peng, H.P.; Huang, Z.N.; Sheng, Y.L.; Zhang, X.P.; Deng, H.H.; Chen, W.; Liu, J.W. Pre-Oxidation of Gold Nanoclusters Results in a 66% Anodic Electrochemiluminescence Yield and Drives Mechanistic Insights. *Angew. Chem. Int. Ed.* **2019**, *58*, 11691–11694. [CrossRef] [PubMed]

10. Shang, L.; Dong, S.J.; Nienhaus, G.U. Ultra-Small Fluorescent Metal Nanoclusters: Synthesis and Biological Applications. *Nano Today* **2011**, *6*, 401–418. [CrossRef]

11. Lu, Y.Z.; Chen, W. Sub-Nanometre Sized Metal Clusters: From Synthetic Challenges to the Unique Property Discoveries. *Chem. Soc. Rev.* **2012**, *41*, 3594–3623. [CrossRef] [PubMed]

12. Cui, M.; Zhao, Y.; Song, Q. Synthesis, Optical Properties and Applications of Ultra-Small Luminescent Gold Nanoclusters. *TrAC-Trend Anal. Chem.* **2014**, *57*, 73–82. [CrossRef]

13. Shang, L.; Doerlich, R.M.; Brandholt, S.; Schneider, R.; Trouillet, V.; Bruns, M.; Gerthsen, D.; Nienhaus, G.U. Facile Preparation of Water-Soluble Fluorescent Gold Nanoclusters for Cellular Imaging Applications. *Nanoscale* **2011**, *3*, 2009–2014. [CrossRef] [PubMed]

14. Tian, D.H.; Qian, Z.S.; Xia, Y.S.; Zhu, C.Q. Gold Nanocluster-Based Fluorescent Probes for near-Infrared and Turn-on Sensing of Glutathione in Living Cells. *Langmuir* **2012**, *28*, 3945–3951. [CrossRef]

15. Deng, H.H.; Wu, G.W.; Zou, Z.Q.; Peng, H.P.; Liu, A.L.; Lin, X.H.; Xia, X.H.; Chen, W. pH-Sensitive Gold Nanoclusters: Preparation and Analytical Applications for Urea, Urease, and Urease Inhibitor Detection. *Chem. Comm.* **2015**, *51*, 7847–7850. [CrossRef]

16. Deng, H.H.; Zhang, L.N.; He, S.B.; Liu, A.L.; Li, G.W.; Lin, X.H.; Xia, X.H.; Chen, W. Methionine-Directed Fabrication of Gold Nanoclusters with Yellow Fluorescent Emission for Cu^{2+} Sensing. *Biosens. Bioelectron.* **2015**, *65*, 397–403. [CrossRef]

17. Martin-Barreiro, A.; Marcos, S.; Fuente, J.M.; Grazu, V.; Galban, J. Gold Nanocluster Fluorescence as an Indicator for Optical Enzymatic Nanobiosensors: Choline and Acetylcholine Determination. *Sens. Actuat. B-Chem.* **2018**, *277*, 261–270. [CrossRef]

18. Huang, K.Y.; He, H.X.; He, S.B.; Zhang, X.P.; Peng, H.P.; Lin, Z.; Deng, H.H.; Xia, X.H.; Chen, W. Gold Nanocluster-Based Fluorescence Turn-Off Probe for Sensing of Doxorubicin by Photoinduced Electron Transfer. *Sens. Actuat. B-Chem.* **2019**, *296*, 126656. [CrossRef]

19. Nel, A.E.; Madler, L.; Velegol, D.; Xia, T.; Hoek, E.M.V.; Somasundaran, P.; Klaessig, F.; Castranova, V.; Thompson, M. Understanding Biophysicochemical Interactions at the Nano-Bio Interface. *Nat. Mater.* **2009**, *8*, 543–557. [CrossRef]

20. Wang, J.; Jensen, U.B.; Jensen, G.V.; Shipovskov, S.; Balakrishnan, V.S.; Otzen, D.; Pedersen, J.S.; Besenbacher, F.; Sutherland, D.S. Soft Interactions at Nanoparticles Alter Protein Function and Conformation in a Size Dependent Manner. *Nano Lett.* **2011**, *11*, 4985–4991. [CrossRef]

21. Mahmoudi, M.; Lynch, I.; Ejtehadi, M.R.; Monopoli, M.P.; Bombelli, F.B.; Laurent, S. Protein-Nanoparticle Interactions: Opportunities and Challenges. *Chem. Rev.* **2011**, *111*, 5610–5637. [CrossRef] [PubMed]

22. Huang, Z.; Wang, H.; Yang, W. Gold Nanoparticle-Based Facile Detection of Human Serum Albumin and Its Application as an Inhibit Logic Gate. *ACS Appl. Mater. Inter.* **2015**, *7*, 8990–8998. [CrossRef] [PubMed]

23. Lacerda, S.H.; Paoli, D.; Park, J.J.; Meuse, C.; Pristinski, D.; Becker, M.L.; Karim, A.; Douglas, J.F. Interaction of Gold Nanoparticles with Common Human Blood Proteins. *ACS Nano* **2010**, *4*, 365–379. [CrossRef] [PubMed]

24. Shang, L.; Brandholt, S.; Stockmar, F.; Trouillet, V.; Bruns, M.; Nienhaus, G.U. Effect of Protein Adsorption on the Fluorescence of Ultrasmall Gold Nanoclusters. *Small* **2012**, *8*, 661–665. [CrossRef] [PubMed]

25. Xu, S.; Lu, X.; Yao, C.; Huang, F.; Jiang, H.; Hua, W.; Na, N.; Liu, H.; Ouyang, J. A Visual Sensor Array for Pattern Recognition Analysis of Proteins Using Novel Blue-Emitting Fluorescent Gold Nanoclusters. *Anal. Chem.* **2014**, *86*, 11634–11639. [CrossRef] [PubMed]

26. Deng, H.H.; Shi, X.Q.; Peng, H.P.; Zhuang, Q.Q.; Yang, Y.; Liu, A.L.; Xia, X.H.; Chen, W. Gold Nanoparticle-Based Photoluminescent Nanoswitch Controlled by Host-Guest Recognition and Enzymatic Hydrolysis for Arginase Activity Assay. *ACS Appl. Mater. Interfaces* **2018**, *10*, 5358–5364. [CrossRef]

27. Deng, H.H.; Wang, F.F.; Shi, X.Q.; Peng, H.P.; Liu, A.L.; Xia, X.H.; Chen, W. Water-Soluble Gold Nanoclusters Prepared by Protein-Ligand Interaction as Fluorescent Probe for Real-Time Assay of Pyrophospatase Activity. *Biosens. Bioelectron.* **2016**, *83*, 1–8. [CrossRef]

28. Kang, J.; Liu, Y.; Xie, M.; Li, S.; Jiang, M.; Wang, Y. Interactions of Human Serum Albumin with Chlorogenic Acid and Ferulic Acid. *BBA-Gen. Subj.* **2004**, *1674*, 205–214. [CrossRef]

29. Snyder, S.; Pendergraph, B. Detection and Evaluation of Chronic Kidney Disease. *Interventions* **2005**, *100*, 24–25.

30. Tibblin, G.; Bergentz, S.E.; Bjure, J.; Wilhelmsen, L. Hematocrit, Plasma Protein, Plasma Volume, and Viscosity in Early Hypertensive Disease. *Am. Heart J.* **1966**, *72*, 165–176. [CrossRef]

31. Ingelfinger, J.R. Melamine and the Global Implications of Food Contamination. *New Eng. J. Med.* **2008**, *359*, 2745–2748. [CrossRef] [PubMed]

32. Kobayashi, T.; Okada, A.; Fujii, Y.; Niimi, K.; Hamamoto, S.; Yasui, T.; Tozawa, K.; Kohri, K. The Mechanism of Renal Stone Formation and Renal Failure Induced by Administration of Melamine and Cyanuric Acid. *Urol. Res.* **2010**, *38*, 117–125. [CrossRef] [PubMed]

33. Finete, V.D.L.M.; Gouvêa, M.M.; de Carvalho Marques, F.F.; Netto, A.D.P. Is It Possible to Screen for Milk or Whey Protein Adulteration with Melamine, Urea and Ammonium Sulphate, Combining Kjeldahl and Classical Spectrophotometric Methods? *Food Chem.* **2013**, *141*, 3649–3655. [CrossRef]

34. Chiodoni, C.; Colombo, M.P.; Sangaletti, S. Matricellular Proteins: From Homeostasis to Inflammation, Cancer, and Metastasis. *Cancer Metastasis Rev.* **2010**, *29*, 295–307. [CrossRef] [PubMed]

35. De Mejia, E.G.; Dia, V.P. The Role of Nutraceutical Proteins and Peptides in Apoptosis, Angiogenesis, and Metastasis of Cancer Cells. *Cancer Metastasis Rev.* **2010**, *29*, 511–528. [CrossRef] [PubMed]

 *nanomaterials*

Article

A New Lamellar Gold Thiolate Coordination Polymer, [Au(m-SPhCO$_2$H)]$_n$, for the Formation of Luminescent Polymer Composites

Oleksandra Veselska [1], Nathalie Guillou [2], Gilles Ledoux [3], Chia-Ching Huang [4], Katerina Dohnalova Newell [4], Erik Elkaïm [5], Alexandra Fateeva [6] and Aude Demessence [1,*]

[1] Univ Lyon, Université Claude Bernard Lyon 1, Institut de Recherches sur la Catalyse et l'Environnement de Lyon (IRCELYON), UMR CNRS 5256, 69626 Villeurbanne, France; oleksandra.veselska@ircelyon.univ-lyon1.fr
[2] Institut Lavoisier de Versailles (ILV), UVSQ, Université Paris-Saclay, UMR CNRS 8180, 78035 Versailles, France; nathalie.guillou@uvsq.fr
[3] Univ Lyon, Université Claude Bernard Lyon 1, Institut Lumière Matière (ILM), UMR CNRS 5306, 69626 Villeurbanne, France; gilles.ledoux@univ-lyon1.fr
[4] Institute of Physics, University of Amsterdam, Science Park 904, 1098 XH Amsterdam, The Netherlands; c.huang@uva.nl (C.-C.H.); k.newell@uva.nl (K.D.N.)
[5] Beamline Cristal, Synchrotron Soleil, 91192 Gif-sur-Yvette, France; erik.elkaim@synchrotron-soleil.fr
[6] Univ Lyon, Université Claude Bernard Lyon 1, Laboratoire des Multimatériaux et Interfaces (LMI), UMR CNRS 5615, 69626 Villeurbanne, France; alexandra.fateeva@univ-lyon1.fr
* Correspondence: aude.demessence@ircelyon.univ-lyon1.fr

Received: 10 September 2019; Accepted: 30 September 2019; Published: 2 October 2019

Abstract: The photoluminescence of gold thiolate clusters brings about many potential applications, but its origin is still elusive because of its complexity. A strategy in understanding the structure–properties relationship is to study closely related neutral gold thiolate coordination polymers (CPs). Here, a new CP is reported, [Au(m-SPhCO$_2$H)]$_n$. Its structure is lamellar with an inorganic layer made of Au–S–Au–S helical chains, similar to the [Au(p-SPhCO$_2$H)]$_n$ analog. An in-depth study of its photophysical properties revealed that it is a bright yellow phosphorescent emitter with a band centered at 615 nm and a quantum yield (QY) of 19% at room temperature and in a solid state. More importantly, a comparison to the *para*-analog, which has a weak emission, displayed a strong effect of the position of the electron withdrawing group (EWG) on the luminescent properties. In addition, [Au(m-SPhCO$_2$H)]$_n$ CPs were mixed with organic polymers to generate transparent and flexible luminescent thin films. The ability to tune the emission position with the appropriate contents makes these nontoxic polymer composites promising materials for lighting devices.

Keywords: gold thiolate; coordination polymer; lamellar structure; luminescence; polymer composite

---

1. Introduction

Gold thiolate clusters, Au$_n$(SR)$_m$, are an intriguing family of materials that bridges the gap between molecular species and functionalized nanoparticles [1–4]. One of their attractive properties is their photoluminescence, which brings about many potential applications in areas such as chemical sensing, bioimaging, cell labeling, phototherapy, and drug delivery [5–8]. Nevertheless, the origin of their photoemission is still elusive, and the structure–properties relationship is difficult to predict. Indeed, in gold thiolate clusters, different parameters such as the gold core's composition and geometry, the functionality of the ligands, the length and organization of the Au(I)–SR shell, their aggregation, their rigidification, and the scale effect make those systems complicated to study and rationalize in terms of the effect of each parameter [9–11]. The limited amount of available crystallographic structures of luminescent gold thiolate clusters precludes an understanding of the structure–properties relationship. Our strategy to attempt to rationalize the photoluminescent properties of gold thiolate clusters has been to study their analogs, neutral gold thiolate coordination polymers (CPs) [Au(SR)]$_n$, which are also highly photoluminescent [12–18]. In crystalline CPs, all atoms are well organized on a micrometer scale, so their structure–properties relationships appear to be easier to correlate. In addition, gold thiolate CPs are the synthesis precursors of gold thiolate clusters and are consequently easier to obtain through their isolation before the addition of a reducing agent.

Even if gold thiolate CPs have been known for a long time, only four crystallographic structures have been reported so far [19]. Two are 1D CPs made of interpenetrated helical chains of Au–S–Au chains, and two are lamellar structures. The latter ones are 2D materials: in [Au(p-SPhCO$_2$H)]$_n$, the inorganic layers are made of helical gold–sulfur chains packed together through parallel aurophilic interactions [14], while in [Au(p-SPhCO$_2$Me)]$_n$, the Au–S chains have a zigzag shape and interact through perpendicular aurophilic bonds [13] (Figure S1a,b, respectively). This difference is explained by the presence of catemeric hydrogen bonds between the carboxylic acids of [Au(p-SPhCO$_2$H)]$_n$ (which implies highly bent S–Au–S angles) when such bonds are quasilinear in the ester analog. Consequently, this change in the lamellar structure induces different photoluminescent properties. [Au(p-SPhCO$_2$Me)]$_n$ exhibits a bright orange emission at 645 nm with a QY of around 70% at room temperature (RT). [Au(p-SPhCO$_2$H)]$_n$ has a double emission in blue (490 nm) and red (660 nm) at 260 K andits QY is below 1% at RT. The first emission band originates from the ligand, and the second one originates from a ligand-to-metal charge transfer (LMCT). Their intensity evolves differently with temperature and results in thermoluminochromism with good sensitivity for thermometry [14].

In order to evaluate the effect of the position of the substituent on the structure and the photoluminescent properties, *meta*-mercaptobenzoic acid (m-HSPhCO$_2$H) was used as a ligand to isolate [Au(m-SPhCO$_2$H)]$_n$. This new gold thiolate CP also has a 2D lamellar structure and exhibits bright photoemission. In addition, its easy integration into organic polymers shows its great potential for the fabrication of flexible and transparent luminescent devices.

2. Materials and Methods

2.1. Materials

Tetrachloroauric acid trihydrate (HAuCl$_4$·3H$_2$O, ≥49% Au basis) and tetrahydrofuran (THF) were purchased from Alfa Aesar (Haverhill, MA, USA), and 3-mercaptobenzoic acid (m-HSPhCO$_2$H) (>97%) was purchased from TCI (Portland, OR, USA). Ethanol and dimethylformamide (DMF) were purchased from VWR (Radnor, PA, USA). Polyvinylidene fluoride (PVDF) and poly(9-vinylcarbazole) (PVK) were purchased from Alfa Aesar and Sigma Aldrich (St. Louis, MO, USA), respectively. All reagents were used without further purification.

2.2. Synthesis of [Au(m-SPhCO$_2$H)]$_n$

A solution of HAuCl$_4$·3H$_2$O (100 mg, 0.25 mmol, 1 equiv.) in H$_2$O (10 mL) was added to m-HSPhCO$_2$H (178 mg, 1.15 mmol, 4.6 equiv.). The reaction was allowed to proceed for 18 h at 120 °C in a 20-mL sealed glass vial. White precipitate was obtained and washed with 40 mL of ethanol 3 times. The powder was recovered by centrifugation at 4000 rpm. The product was dried in air. The yield was 77% (68 mg); the chemical formula was C$_7$H$_5$AuO$_2$S; the molecular weight was 350.14; and the gold content from thermogravimetric analysis (TGA) (calculated) wt% was 56.4 (56.3).

2.3. Synthesis of the Polymer Composites

Here, a 0.5 to 15 wt% of CPs was mixed with the organic polymer PVDF or PVK and then dispersed in a solvent, DMF (8 mL) or THF (5 mL), respectively. The total mass of the polymer composite was 400 mg. The mixture was ultrasonicated until a homogeneous emulsion was obtained. Then it was deposited onto a flat substrate (i.e., a glass slide or a Petri dish), and the solvent was evaporated at 60 °C [20,21].

2.4. Structure Determination

The structural determination of $[Au(m\text{-}SPhCO_2H)]_n$ was carried out from high-resolution powder X-ray diffraction data. They were collected on a CRISTAL beamline at Soleil Synchrotron (Gif-sur-Yvette, France). A monochromatic beam was extracted from the U20 undulator beam by means of a Si(111) double monochromator. Its wavelength of 0.79276 Å was refined from a LaB_6 (NIST (National Institute of Standards and Technology) Standard Reference Material 660a) powder diagram recorded prior to the experiment. The X-ray beam was attenuated in order to limit radiation damage to the sample. A high angular resolution was obtained with (in the diffracted beam) a 21 perfect crystals Si(111) multi-analyzer. The sample was loaded in a 0.7-mm capillary (Borokapillaren, GLAS, Schönwalde, Germany) mounted on a spinner rotating at about 5 Hz to improve the particles' statistics. Diffraction data were collected in continuous scanning mode, and a diffractogram was obtained from the precise superposition and addition of the 21-channel data.

All calculations of structural investigation were performed with the TOPAS program [22]. The LSI (Least Squares Iterative) indexing method converged to two possible unit cells, a monoclinic C cell ($a = 6.8749(4)$, $b = 6.5715(3)$, $c = 16.3904(6)$ Å, $\beta = 91.199(7)°$, and $V = 740.3(5)$ Å^3 with $M_{20} = 57$) and a triclinic one ($a = 4.7567(5)$, $b = 4.7632(6)$, $c = 16.3975(5)$ Å, $\alpha = 90.780(8)°$, $\beta = 89.123(6)°$, $\gamma = 92.651(4)°$, and $V = 371.05(7)$ Å^3 with $M_{20} = 44$). The highest symmetry was first logically chosen to solve the structure. Nevertheless, no structural model could be initiated in the three space groups consistent with systematic extinctions (*C2*, *Cm*, and *C2/m*). Moreover, a careful examination of the triclinic solution showed that the unit cell parameters were very close to those of $[Au(p\text{-}SPhCO_2H)]_n$. The atomic coordinates of the gold and sulfur atoms of $[Au(p\text{-}SPhCO_2H)]_n$ were first directly used to initiate a Rietveld refinement, which did not converge. A structural investigation of $[Au(m\text{-}SPhCO_2H)]_n$ was then initialized in *P1* by using the charge flipping method, which allowed for locating two gold atoms. The direct space strategy was then used to complete the structural model, and two independent organic moieties were added to the fixed gold atomic coordinates and treated as rigid bodies in the simulated annealing process. Calculations converged to $R_B = 0.054$ and $R_{wp} = 0.113$. This structural model was used as the starting model in the Rietveld refinement. In its final stage, this involved the following structural parameters: 6 atomic coordinates, 6 translation and 6 rotation parameters for organic moieties as well as 4 distances and 2 torsion angles in the rigid bodies, 2 thermal factors, and 1 scale factor. The final Rietveld plot (Figure 1) corresponded with a satisfactory model indicator and profile factors (Table S1). A search for higher symmetry was undertaken by using Platon software [23], and no space group change was suggested. The crystal structure quoted the depository number CCDC-1952042.

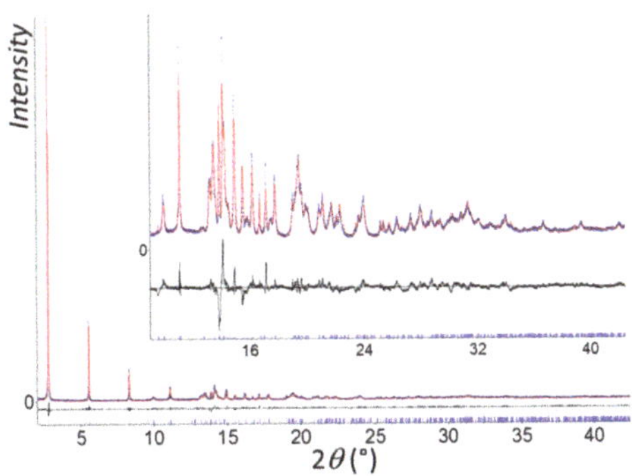

Figure 1. Final Rietveld plot of $[Au(m\text{-}SPhCO_2H)]_n$ showing observed (blue circles), calculated (red line), and difference (black line) curves. Zoomed-in high angles are shown as an inset.

2.5. Characterizations

Routine PXRD: Routine powder X-ray diffraction was carried out on a Bruker D8 Advance A25 diffractometer (Karlsruhe, Germany) using Cu Kα radiation equipped with a 1-dimensional position-sensitive detector (Bruker LynxEye). X-ray scattering was recorded between 4° and 90° (2θ) with 0.02° steps and 0.5 s per step (28 min for the scan). A divergence slit was fixed at 0.2°, and the detector aperture was fixed to 192 channels (2.95°).

SEM: SEM images were obtained with an FEI Quanta 250 FEG (Hillsboro, OR, USA) scanning electron microscope. Samples were mounted on stainless pads and sputtered with carbon to prevent charging during observation.

FTIR: Infrared spectra were obtained from a Bruker Vector 22 FTIR spectrometer (Richmond Scientific Ltd Unit 9; Chorley, UK) with KBr pellets at room temperature and were registered from 4000 cm^{-1} to 400 cm^{-1}.

TGA: Thermogravimetric analyses were performed with a TGA/DSC 1 STARe System from Mettler Toledo (Columbus, OH, USA). Around 5 mg of the sample was heated at a rate of 10 °C/min in a 70-µL alumina crucible under air (20 mL/min).

UV-VIS: A UV-VIS absorption spectrum was carried out with a LAMBDA 365 UV/VIS Spectrophotometer from Perkin Elmer (Waltham, MA, USA) in solid state at room temperature.

Photoluminescence (PL) excitation and emission spectra measurements were performed on a homemade apparatus. The sample was illuminated by an EQ99X laser-driven light source (Energetiq Technology Inc.; Woburn, MA, USA) filtered by a Jobin Yvon Gemini 180 monochromator (Paris, France). The exit slit from the monochromator was then reimaged on the sample by two MgF$_2$ lenses 100 m in focal length and 2 inch in diameter. The whole apparatus was calibrated by means of a Newport 918D (Irvine, CA, USA) Low power-calibrated photodiode sensor over the range 190–1000 nm. The resolution of the system was 4 nm. The emitted light from the sample was collected by an optical fiber connected to a Jobin-Yvon TRIAX320 monochromator equipped with a cooled charge coupled device (CCD) detector (Andor Newton 970; Oxford Instruments; Abingdon, UK). At the entrance of the monochromator, different long-pass filters could be chosen in order to eliminate the excitation light. The resolution of the detection system was better than 2 nm.

Temperature control over the sample was regulated with a THMS-600 heating stage with a T95-PE temperature controller from Linkam Scientific Instruments (Epsom, UK).

Luminescence lifetime measurements: Luminescence lifetime measurements were excited with a diode-pumped 50-Hz tunable optical parametric oscillator (OPO) laser from EKSPLA (Vilnuis, Lithuania). The luminescence of the sample was collected with an optical fiber and was afterwards filtered by a long-pass filter (FEL450 from Thorlabs; Newton, NJ, USA) and the monochromator described before and fed to an R943-02 photomultiplier tube from Hamamatsu (Hamamatsu, Japan). Photon arrival times were categorized by an MCS6A multichannel scaler from Fast ComTec (Munich, Germany).

The collected data cannot be fitted by a sum of simple exponential decays. For this reason, one stretched exponential decay is often used (Equation (1)) [24]:

$$I = a_1 e^{-\left(\frac{x}{t_1}\right)^{\beta_1}} + a_2 e^{-\frac{x}{t_2}} + a_3 e^{-\frac{x}{t_3}}, \tag{1}$$

with a_i and t_i being the amplitude and lifetime of a given component i, and $\beta_1 = [0, 1]$.

The distribution of a lifetime with a β factor is generally associated with the possibility of energy transfer toward a distribution of nonradiative centers through dipole–dipole/quadrupole–quadrupole, etc., interactions. Its value is defined by the types of interactions and the dimensionality of the system.

The average lifetime of the stretched exponential decay $\langle\tau_1\rangle$ is calculated through Equation (2). The procedure used for the fitting is described in Reference [25]:

$$\langle\tau_1\rangle = \tau_1 \cdot \frac{1}{\beta_1} \cdot \Gamma\left(\frac{1}{\beta_1}\right). \tag{2}$$

Determination of the QY with a standard integrating sphere (IS) (shown schematically in Figure S2a): To illuminate the sample, a stabilized Xenon lamp (Hamamatsu, L2273) coupled to a double-grating monochromator (Solar, MSA130; Minsk, Belarus) was used. The excitation beam was split using a spectrally broad bifurcated fiber (Ocean Optics, BIF600-UV-VIS; Duiven, The Netehrlands), where one part was used to monitor the fluctuations of the excitation intensity using a power meter (Ophir Photonics, PD300-UV; Jerusalem, Israel) and the other part was used to excite the sample. A collimator lens was used to reduce the spot size at the sample position to enable direct excitation of the sample ($F = 1$). The powder samples were placed in ethanol solution in a quartz cuvette and suspended using an aluminum holder in the center of the IS (internal walls made of Spectralon®) (PTFE), Newport, 70672) (with a diameter of 10 cm). The use of this type of holder has been verified using ray-tracing simulations [26]. Light was detected using a second spectrally broad optical fiber (Ocean Optics, QP1000-2-VIS-BX) coupled to a spectrometer (Solar, M266) equipped with a CCD (Hamamatsu, S7031-1108S). The sample (powder) did not dissolve completely in the solvent (ethanol), so the sample was shaken before the measurement and then measured. This was repeated several times to ensure reproducibility of the results. All measurements were corrected for the spectral response of the detection system, which we determined by illuminating the IS via the excitation port with a tungsten halogen calibration lamp (standard of spectral irradiance, Oriel, 63358; Irvine, CA, USA) for the visible range and a deuterium lamp (Oriel, 63945) for the UV range (<400 nm). The measured calibration spectrum was corrected for the spectrometer's stray-light contribution. Reabsorption effects were corrected for using the procedure described by Ahn et al. [27] by comparing the measured PL spectrum to that of a low-concentration sample for which reabsorption was negligible. Error estimates were obtained following Chung et al. [28]. QY was evaluated using Equation (3), where N_{em} and N_{abs} are numbers of emitted and absorbed photons, and N_S and N_{Ref} are numbers of photons transmitted through the sample (sample in solvent and cuvette) and reference (just cuvette with solvent) (star denotes emission spectral range, and no star means excitation spectral range). I_S and I_{Ref} are emission intensities detected within the sample and reference (Figure S3b), and C is a correction factor for the spectral response of the detection system:

$$QY = \frac{N_{em}}{N_{abs}} = \frac{N_S^* - N_{Ref}^*}{N_{Ref} - N_S} = \frac{\int^{em}\left[I_S(\lambda) - I_{Ref}(\lambda)\right]C(\lambda)d\lambda}{\int^{exc}\left[I_{Ref}(\lambda) - I_S(\lambda)\right]C(\lambda)d\lambda}. \tag{3}$$

3. Results and Discussion

3.1. Synthesis and Characterization

Pure and highly crystalline powder of [Au(m-SPhCO$_2$H)]$_n$ CP was synthesized under hydrothermal conditions. The procedure was similar to the one used for the *para*-substituted analog, [Au(p-SPhCO$_2$H)]$_n$ [14]. The PXRD pattern (Figure 1) showed predominant (00l) reflections underlining the lamellar structure of this material, with an interlamellar distance of 16.3 Å.

The morphology of the crystallites was thin pellets typical of lamellar compounds, as shown by SEM images (Figure S3). In the FTIR spectroscopy, the CO antisymmetric vibration band of the carboxylic acid was observed at 1683 cm^{-1}, the same position as for the free ligand, which showed that the carboxylic acids were not coordinated with the metal atoms. In addition, the broad ν(OH) bands between 2560 and 3065 cm^{-1} were consistent with the presence of hydrogen bonds between the carboxylic acid functions (Figure S4). The thermogravimetric analyses confirmed the purity of the compound with an expected 1:1 metal/organic content. This CP showed good thermal stability up to 300 °C under air (Figure S5), which was comparable to $[Au(p\text{-}SPhCO_2H)]_n$ (320 °C) [14].

The synthesis did not result in obtaining single crystals, and thus a structural determination from the high-resolution PXRD data was carried out.

3.2. Structure

The lamellar structure was triclinic (*P*1 space group) and was similar to the previously discussed $[Au(p\text{-}SPhCO_2H)]_n$ [4c]. It consisted of planes made of 1D helices of Au(I) atoms bridged by μ_2-S atoms (Figure 2). The Au–S distances were between 2.28(2) and 2.36(2) Å (Table S2). The Au–S–Au angles were 85.8(4)° and 90.5(4)°. The S–Au–S angles were 88.2(4)° and 151.1(5)°, far from the linear angle usually observed in dicoordinated gold atoms.

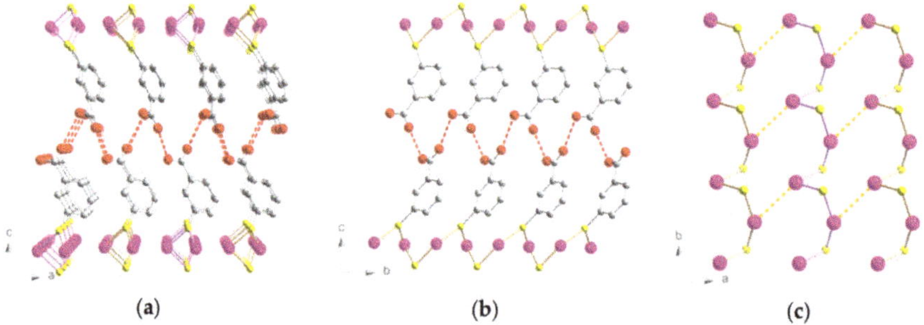

(**a**)   (**b**)   (**c**)

Figure 2. Structure representations of $[Au(m\text{-}SPhCO_2H)]_n$. Views of (**a**) (*ac*) and (**b**) (*bc*) planes. (**c**) View of the Au–S chains on the (*ab*) plane. Pink, Au; yellow, S; red, O; grey, C. Hydrogen atoms are omitted for clarity. Red and orange dotted lines represent the hydrogen bonds and interchain aurophilic interactions, respectively.

In the lamellar $[Au(p\text{-}SPhCO_2Me)]_n$ CP without hydrogen bonds, the S–Au–S angles were close to linear (177.5(1)°) [13]. Thus, the formation of such unusual S–Au–S angles in $[Au(m\text{-}SPhCO_2H)]_n$ was driven by the formation of both catemeric hydrogen bonds and aurophilic interactions, similarly to $[Au(p\text{-}SPhCO_2H)]_n$ [14]. The interchain Au–Au distances were slightly longer than in $[Au(p\text{-}SPhCO_2H)]_n$ (3.38(2) and 3.61(2) Å for the *meta* vs 3.36(1) and 3.42(1) Å for the *para*), while the intrachain Au–Au distances were shorter here than in the *para*-substituted counterpart (3.21(2) and 3.28(2) Å for the *meta* vs 3.59(1) and 3.73(1) Å for the *para*).

3.3. Photophysical Properties

The UV light excitation of the $[Au(m\text{-}SPhCO_2H)]_n$ resulted in a very bright yellowish-orange emission at room temperature.

The maximum absorption of $[Au(m\text{-}SPhCO_2H)]_n$ was at 320 nm (the optical band gap was 2.9 eV) (Figure S6). The absorption edge was red-shifted in comparison to the free ligand. The high-energy absorption was assigned to a π–π* transition of the phenyl group of the ligand [29].

At room temperature, the emission and excitation maxima were positioned at 615 and 352 nm, respectively (Figure 3a). Upon a temperature decrease, the emission intensity increased with a soft refinement of the width and a shift of the emission maximum to 595 nm at 93 K (Figure 3b). Upon a temperature increase, no important shift of emission relative to RT was observed. Meanwhile, the excitation maximum slightly shifted to 368 nm at 503 K (Figure S7).

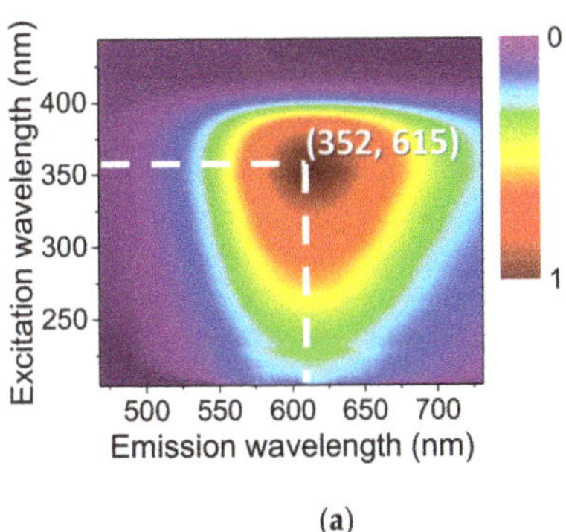
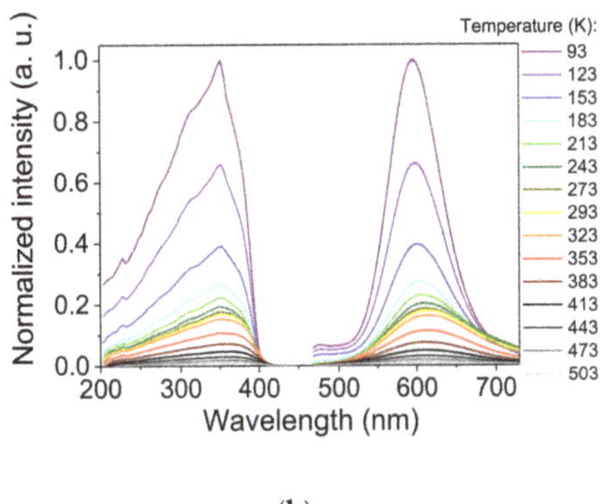

Figure 3. Luminescent properties of [Au(m-SPhCO$_2$H)]$_n$: (**a**) 2D map of the emission and excitation spectra conducted in a solid state at room temperature; (**b**) emission and excitation spectra (λ_{exc} = 352 nm, λ_{em} = 596 nm) in a solid state at varying temperatures.

Additionally, there was a low-intensity emission shoulder at 475 nm that appeared below 213 K. The free ligand exhibited a large band of emission centered at 507 nm under the same excitation wavelength. Thus, this high-energy emission could be assigned to a metal-perturbed intraligand (IL) transition. Similar behavior has been observed before in Au(I) phosphane complexes containing a 4-nitrophenylthiolate ligand [30] and dinuclear neutral thiolate Au(I) complexes with phenylene spacers [31].

The luminescence lifetime decay in solid state at room temperature was fitted with a triexponential curve with components of 0.31 µs (88%), 0.14 ms (10%), and 2.3 ms (2%) (Figures S8 and S9, Table S3). The lifetime decay study at various temperatures showed a shortening of the lifetime of the major components upon an increase of the temperature: at 93 K the lifetime was 0.87 µs, and at 503 K it dropped to 0.04 µs (Figure S10). Its contribution at low temperatures stayed at the level of 86%, and it grew up to 94% at high temperatures. Two smaller components experienced minor changes in lifetime in the studied temperature range. The contribution of both decreased upon temperature increase. The lifetime shortening could be attributed to thermal quenching. This results from thermal activation of the nonradiative decay pathways [32,33].

The long lifetime in the microsecond range and the large Stokes shift of 12,150 cm^{-1} (RT) were both characteristic of a phosphorescence process. The QY was 18.9% ± 0.2% in solid state at RT.

The close emission and excitation energies, the lifetime decay behavior, the low-intensity emission shoulder, and the high QY in solid state made the photophysical properties of [Au(m-SPhCO$_2$H)]$_n$ very similar to [Au(p-SPhCO$_2$Me)]$_n$ [13]. The overall luminescence process of [Au(p-SPhCO$_2$Me)]$_n$ was ascribed to a ligand–metal-to–ligand charge transfer transition (LMLCT: AuS → PhCO$_2$Me) based on Density Functional Theory calculations. Strong similarities in the properties of the two CPs suggested the same luminescence origin for [Au(m-SPhCO$_2$H)]$_n$ as the one proposed for [Au(p-SPhCO$_2$Me)]$_n$–LMLCT. This supports the hypothesis that the presence of an electron withdrawing group (EWG) precludes the possibility of LM(M)CT transitions, which is often ascribed to Au(I) compounds.

It is interesting to note that despite the fact that [Au(m-SPhCO$_2$H)]$_n$ was structurally related to [Au(p-SPhCO$_2$H)]$_n$, they had different luminescent properties. Moreover, as was shown above, the properties were closer to the ones of [Au(p-SPhCO$_2$Me)]$_n$, which differs structurally and possesses disparate Au–S arrangements (helices vs zigzags). The aurophilic interactions and substituents on the organic ligand are often the main parameters that can help to trace the origin of the photophysical properties [34]. The presence of aurophilic interactions is known to influence photoluminescent processes: their length, dimensionality (dimer or chain), and intra/interconnectivity play an intricate role in the wavelength and intensity of the emission [34,35].

Considering the Au–Au distances (Table S2), the intrachain Au–Au distances (two gold atoms bridged by one sulfur atom) were shorter in the *meta*-compound than in its *para*-analog: 3.21(2) and 3.28(2) Å versus 3.59(1) and 3.73(1) Å. Nevertheless, the interchain Au–Au distances generating chains in the case of [Au(*p*-SPhCO$_2$H)]$_n$ were slightly shorter (3.36(1) and 3.42(1) Å) than the ones of [Au(*m*-SPhCO$_2$H)]$_n$, for which there was one aurophilic interaction (3.38(2) Å) forming dimers, the second one being too long (3.61(2) Å) to be considered. In the case of [Au(*p*-SPhCO$_2$Me)]$_n$, the intrachain Au–Au distance was rather long (3.51(1) Å), and the interchain one was short (3.20(1) Å).

In the structural and photophysical data obtained for the three 2D gold thiolate CPs, an interesting trend could be seen. It appears that the two compounds with the shortest Au–Au distances, [Au(*m*-SPhCO$_2$H)]$_n$ (3.20(2) Å, intrachain) and [Au(*p*-SPhCO$_2$Me)]$_n$ (3.20(1) Å, interchain), exhibited higher energy emissions at 615 and 645 nm and higher QYs of 19% and 70%, respectively, than did [Au(*p*-SPhCO$_2$H)]$_n$, which had an Au–Au distance of 3.36(1) Å (interchain) and showed an emission at 665 nm with a QY of less than 1% at RT. Thus, an increase of the shortest Au–Au distances was followed by a decrease of the QY. In the literature, the effect of aurophilic interactions on the quantum yield is still under debate. Shorter aurophilic interactions are often associated with better photophysical properties [36]. However, sometimes the opposite is reported [37].

The difference in photophysical properties might also be explained by the change in substituent position on the phenyl ring and the stronger effect of the carboxylic acid EWG in the *meta*-position compared to the *para*-analog. EWG groups on the ligand stabilize the sulfur highest occupied molecular orbital (HOMO), making the ligand more difficult to oxidize, inducing a shift of the emission to higher energies [34,38]. Finally, it appears that the position change of an EWG for similar structures could induce different aurophilic interactions and a significant increase of the QY from ~1% for [Au(*p*-SPhCO$_2$H)]$_n$ to 19% for [Au(*m*-SPhCO$_2$H)]$_n$.

3.4. Polymer Composite Films

In order to integrate this bright luminescent CP in devices for lighting applications, films were elaborated [20,39]. Thus, the pellet-like crystals of [Au(*m*-SPhCO$_2$H)]$_n$ were used for the fabrication of homogeneous thin films that exhibited a bright luminescent response. Two organic polymers were chosen: polyvinylidene difluoride (PVDF), which is nonluminescent, and poly(9-vinylcarbazole) (PVK), which is blue-emissive.

PVDF films containing 0.5 wt%, 1 wt%, 2.5 wt%, and 5 wt% [Au(*m*-SPhCO$_2$H)]$_n$ were prepared. The PXRD showed that after film preparation, the CPs remained crystalline (Figure S11). SEM images indicated good dispersion of the CP crystallites in the PVDF following an increase in their density with an increase in the loading (Figure 4). In addition, the high QY of [Au(*m*-SPhCO$_2$H)]$_n$ allowed for the use of a small quantity of CPs, 0.5 wt%, when obtaining transparent, flexible, and luminescent films (Figure 5a–c). The intensity of the emission increased gradually with the CP loading (Figure 5d).

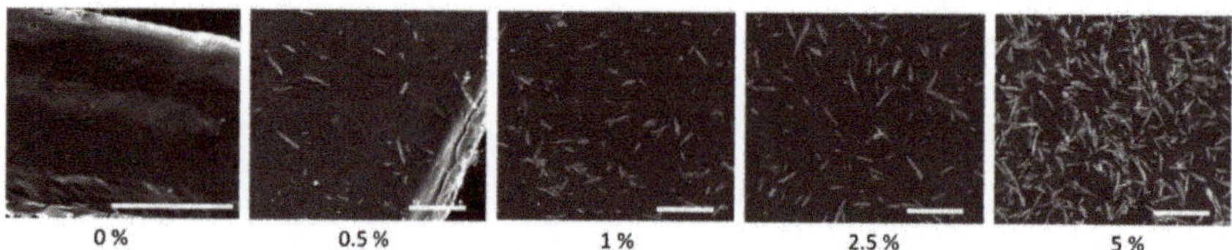

Figure 4. SEM images of the [Au(*m*-SPhCO$_2$H)]$_n$@ PVDF films with the different wt% loadings of the CP. The scale bar is 50 μm.

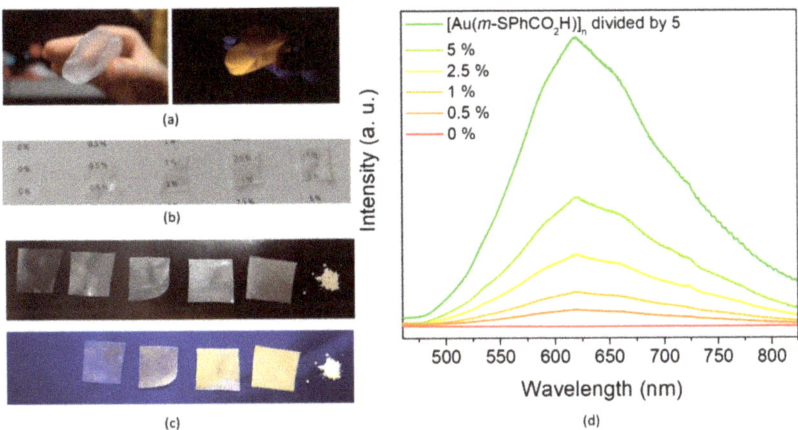

Figure 5. Photographs of (**a**) 2.5 wt% [Au(*m*-SPhCO$_2$H)]$_n$@PVDF film; and films with 0 wt%, 0.5 wt%, 1 wt%, 2.5 wt%, and 5 wt% [Au(*m*-SPhCO$_2$H)]$_n$@PVDF showing their (**b**) transparence and (**c**) luminescence. (**d**) Emission spectra (λ_{ex} = 380 nm) of these polymer composite films insolid state at room temperature.

An analogous approach was used for the synthesis of PVK films containing [Au(*m*-SPhCO$_2$H)]$_n$ (5 wt%, 10 wt%, and 15 wt% of loading). The PXRD patterns showed that the crystallinity of the CP was maintained in the polymer composites (Figure S12). PVK is a blue-emissive polymer. Thus, the mixture of PVK with yellowish-orange-emitting CP as [Au(*m*-SPhCO$_2$H)]$_n$ led to dual emission at 420 and 620 nm (Figure 6a). This allowed for fine-tuning the emission color from blue to orange by adjusting the quantity of the CP loading. This variable loading study specifically showed that flexible films with an emission close to white light could be obtained as shown in CIE 1931 (Comission Internationale de l'Eclairage) chromaticity diagram (Figure 6b). Given the very high interest in the development of white-light phosphors [40], formulating such composite CP/organic polymer films appears to be a valuable method for tuning the emission wavelength and generating nontoxic materials with white-light emission.

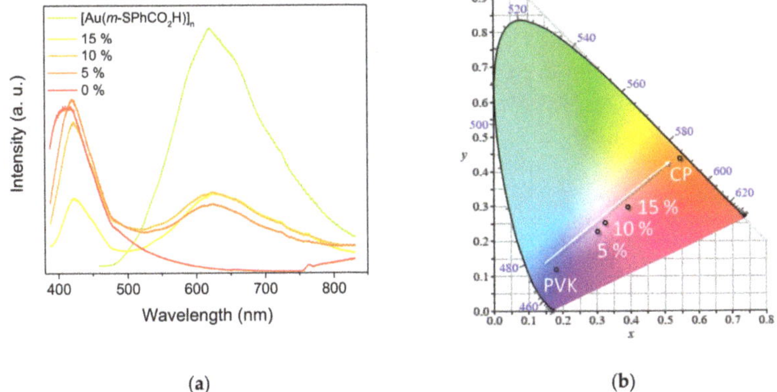

Figure 6. (**a**) Emission spectra of [Au(*m*-SPhCO$_2$H)]$_n$@PVK films at variable CP loadings; (**b**) CIE 1931 chromaticity diagram showing the luminescence color change of the [Au(*m*-SPhCO$_2$H)]$_n$@PVK films with variable CP loadings. Data were registered under an excitation of 370 nm in a solid state at room temperature.

4. Conclusions

Here, we report a new gold thiolate coordination polymer, [Au(*m*-SPhCO$_2$H)]$_n$. Its structure is lamellar, and its photophysical properties show a bright yellow emission up to 500 K with a QY of 19% in solid state and at room temperature. While its structure is similar to the analog [Au(*p*-SPhCO$_2$H)]$_n$, its luminescence is much more intense, displaying the effect of shorter Au–Au distances and the position of the EWG on the phenyl ring of the ligand. Related to the gold thiolate clusters, this shows that a subtle change in the structure can lead to tremendous change in the photophysical properties. Consequently, further studies on these basic CPs should be carried out to understand in-depth the photoemission of the more complicated clusters. In addition, this gold thiolate CP can be easily used along with organic polymers to formulate composite transparent, flexible, and luminescent films. In these latter materials, the emission color can be tuned by adjusting the CP loading. These abilities and their nontoxicity make gold thiolate CPs a good alternative to quantum dots for lighting devices.

Supplementary Materials: The following are available online at http://www.mdpi.com/2079-4991/9/10/1408/s1: Figure S1 Structure representations of (a) [Au(*p*-SPhCO$_2$H)]$_n$ [1] and (b) [Au(*p*-SPhCO$_2$Me)]$_n$ [2]. Pink, Au; yellow, S; red, O; grey, C. Hydrogen atoms are omitted for clarity. Red and purple dotted lines represent the hydrogen bonds and the aurophilic interactions, respectively. Figure S2 A scheme of a used standard integrating sphere setup was used (a) and a graphical representation of emission intensities detected with sample and reference (b). Figure S3 SEM image of [Au(m-SPhCO$_2$H)]n. Figure S4 FT-IR spectra of the free ligand (black) and [Au(m-SPhCO$_2$H)]n (red). Figure S5 The TGA conducted under air at 10 °C/min of [Au(m-SPhCO$_2$H)]n. Figure S6 UV-vis absorption spectra of [Au(m-SPhCO$_2$H)]n (red) and the free ligand (black) conducted in solid state at the room temperature. Figure S7 Normalized emission-excitation spectra (λexc = 352 nm, λem = 596 nm) of [Au(m-SPhCO$_2$H)]n conducted in solid state at variable temperatures. Figure S8 Temperature-dependent luminescence decay curves (λexc = 352 nm, λem = 610 nm) of [Au(m-SPhCO$_2$H)]n conducted in solid state at variable temperatures. Figure S9 Examples of lifetime decay fit of [Au(m-SPhCO$_2$H)]n at 93 K (a), 293 K (b) and 503 K (c). Figure S10 Temperature-dependent luminescence decays of [Au(m-SPhCO$_2$H)]n: values of the lifetime (a) and contributions (b) of different components at various temperatures. Figure S11 PXRD patterns of x %[Au(m-SPhCO$_2$H)]n@PVDF films. Figure S12 PXRD patterns of x %[Au(m-SPhCO$_2$H)]n@PVK films. Table S1 Crystallographic data and Rietveld refinement parameters for [Au(m-SPhCO$_2$H)]n. Table S2 Comparison of the main distances and angles in [Au(m-SPhCO$_2$H)]n and other lamellar [Au(SR)]n CPs, [Au(p-SPhCO$_2$H)]n [1] and [Au(p-SPhCO$_2$Me)]n [2]. Table S3 Luminescence lifetimes (τi), their pre-exponential factors (ai), parameters β1, average lifetimes (<τi>) and contributions of [Au(m-SPhCO$_2$H)]n at 93 to503 K temperature range.

Author Contributions: Conceptualization, O.V. and A.D.; methodology, O.V. and N.G.; validation, O.V. and N.G.; formal analysis, O.V. and G.L.; investigation, O.V., N.G., G.L., C.-C.H., A.F., and A.D.; resources, N.G., G.L., K.D.N., E.E., and A.D.; writing—original draft preparation, O.V.; writing—review and editing, A.D.; supervision, A.D.; funding acquisition, A.D.

Funding: This research was funded by SOLEIL synchrotron (Proposal 20150723) and by the Agence Nationale pour la Recherche (ANR) (MEMOL project ANR-16-JTIC-0004-01).

Acknowledgments: The authors acknowledge the CTμ (Centre Technologique des Microstructures of Lyon Univ) for providing the SEM. O.V. thanks the Auvergne-Rhône-Alpes region for her PhD grant.

Conflicts of Interest: The authors declare no conflict of interest.

References

1. Jin, R.; Zeng, C.; Zhou, M.; Chen, Y. Atomically Precise Colloidal Metal Nanoclusters and Nanoparticles: Fundamentals and Opportunities. *Chem. Rev.* **2016**, *116*, 10346. [CrossRef] [PubMed]
2. Chakraborty, I.; Pradeep, T. Atomically Precise Clusters of Noble Metals: Emerging Link between Atoms and Nanoparticles. *Chem. Rev.* **2017**, *117*, 8208–8271. [CrossRef] [PubMed]
3. Yao, Q.; Yuan, X.; Chen, T.; Leong, D.T.; Xie, J. Engineering Functional Metal Materials at the Atomic Level. *Adv. Mater.* **2018**, *30*, 1802751. [CrossRef] [PubMed]
4. Kang, X.; Zhu, M. Intra-cluster growth meets inter-cluster assembly: The molecular and supramolecular chemistry of atomically precise nanoclusters. *Coord. Chem. Rev.* **2019**, *394*, 1–38. [CrossRef]
5. Tao, Y.; Li, M.; Ren, J.; Qu, X. Metal nanoclusters: Novel probes for diagnostic and therapeutic applications. *Chem. Soc. Rev.* **2015**, *44*, 8636–8663. [CrossRef]
6. Song, X.-R.; Goswami, N.; Yang, H.-H.; Xie, J. Functionalization of metal nanoclusters for biomedical applications. *Analyst* **2016**, *141*, 3126–3140. [CrossRef] [PubMed]

7.   Zhang, X.-D.; Luo, Z.; Chen, J.; Shen, X.; Song, S.; Sun, Y.; Fan, S.; Fan, F.; Leong, D.T.; Xie, J. Ultrasmall Au10−12(SG)10−12 Nanomolecules for High Tumor Specificity and Cancer Radiotherapy. *Adv. Mater.* **2014**, *26*, 4565–4568. [CrossRef]

8.   Li, Q.; Pan, Y.; Chen, T.; Du, Y.; Ge, H.; Zhang, B.; Xie, J.; Yu, H.; Zhu, M. Design and mechanistic study of a novel gold nanocluster-based drug delivery system. *Nanoscale* **2018**, *10*, 10166–10172. [CrossRef]

9.   Cantelli, A.; Guidetti, G.; Manzi, J.; Caponetti, V.; Montalti, M. Towards Ultra-Bright Gold Nanoclusters. *Eur. J Inorg. Chem.* **2017**, *2017*, 5068–5084. [CrossRef]

10. Li, Q.; Zhou, M.; So, W.Y.; Huang, J.; Li, M.; Kauffman, D.R.; Cotlet, M.; Higaki, T.; Peteanu, L.A.; Shao, Z.; et al. A Mono-cuboctahedral Series of Gold Nanoclusters: Photoluminescence Origin, Large Enhancement, Wide Tunability, and Structure–Property Correlation. *J. Am. Chem. Soc.* **2019**, *141*, 5314–5325. [CrossRef]

11. Goswami, N.; Yao, Q.; Luo, Z.; Li, J.; Chen, T.; Xie, J. Luminescent Metal Nanoclusters with Aggregation-Induced Emission. *J. Phys. Chem. Lett.* **2016**, *7*, 962–975. [CrossRef] [PubMed]

12. Lavenn, C.; Okhrimenko, L.; Guillou, N.; Monge, M.; Ledoux, G.; Dujardin, C.; Chiriac, R.; Fateeva, A.; Demessence, A. A luminescent double helical gold(i)–thiophenolate coordination polymer obtained by hydrothermal synthesis or by thermal solid-state amorphous-to-crystalline isomerization. *J. Mater. Chem. C* **2015**, *3*, 4115–4125. [CrossRef]

13. Lavenn, C.; Guillou, N.; Monge, M.; Podbevšek, D.; Ledoux, G.; Fateeva, A.; Demessence, A. Shedding light on an ultra-bright photoluminescent lamellar gold thiolate coordination polymer [Au(p-SPhCO$_2$Me)]n. *Chem. Commun.* **2016**, *52*, 9063–9066. [CrossRef] [PubMed]

14. Veselska, O.; Okhrimenko, L.; Guillou, N.; Podbevšek, D.; Ledoux, G.; Dujardin, C.; Monge, M.; Chevrier, D.M.; Yang, R.; Zhang, P.; et al. An intrinsic dual-emitting gold thiolate coordination polymer, [Au(+I)(p-SPhCO$_2$H)]n, for ratiometric temperature sensing. *J. Mater. Chem. C* **2017**, *5*, 9843–9848. [CrossRef]

15. Veselska, O.; Podbevšek, D.; Ledoux, G.; Fateeva, A.; Demessence, A. Intrinsic triple-emitting 2D copper thiolate coordination polymer as a ratiometric thermometer working over 400 K range. *Chem. Commun.* **2017**, *53*, 12225–12228. [CrossRef] [PubMed]

16. Veselska, O.; Dessal, C.; Melizi, S.; Guillou, N.; Podbevšek, D.; Ledoux, G.; Elkaim, E.; Fateeva, A.; Demessence, A. New Lamellar Silver Thiolate Coordination Polymers with Tunable Photoluminescence Energies by Metal Substitution. *Inorg. Chem.* **2019**, *58*, 99–105. [CrossRef]

17. Gussenhoven, E.M.; Fettinger, J.C.; Pham, D.M.; Balch, A.L. A Reversible Polymorphic Phase Change Which Affects the Luminescence and Aurophilic Interactions in the Gold(I) Cluster Complex, [μ3-S(AuCNC7H13)3](SbF6) [J. *Am. Chem. Soc.* **2005**, *127*, 10838–10839. [CrossRef]

18. Cha, S.-H.; Kim, J.-U.; Kim, K.-H.; Lee, J.-C. Preparation and Photoluminescent Properties of Gold(I)−Alkanethiolate Complexes Having Highly Ordered Supramolecular Structures. *Chem. Mater.* **2007**, *19*, 6297–6303. [CrossRef]

19. Veselska, O.; Demessence, A. d^{10} Coinage Metal Organic Chalcogenolates: From Oligomers to Coordination Polymers. *Coord. Chem. Rev.* **2018**, *355*, 240–270. [CrossRef]

20. Troyano, J.; Castillo, O.; Martínez, J.I.; Fernández-Moreira, V.; Ballesteros, Y.; Maspoch, D.; Zamora, F.; Delgado, S. Reversible Thermochromic Polymeric Thin Films Made of Ultrathin 2D Crystals of Coordination Polymers Based on Copper(I)-Thiophenolates. *Adv. Funct. Mater.* **2018**, *28*, 1704040. [CrossRef]

21. Zhai, L.; Yang, Z.-X.; Zhang, W.-W.; Zuo, J.-L.; Ren, X.-M. Dual-emission and thermochromic luminescence alkaline earth metal coordination polymers and their blend films with polyvinylidene fluoride for detecting nitrobenzene vapor. *J. Mater. Chem. C* **2018**, *6*, 7030–7041. [CrossRef]

22. Bruker Axs. *Topas V4.2: General Profile and Structure Analysis Software for Powder Diffraction Data*; Bruker Axs Ltd: Karlsruhe, Germany, 2008.

23. Spek, A. Structure validation in chemical crystallography. *Acta Cryst.* **2009**, *D65*, 148–155. [CrossRef] [PubMed]

24. Klafter, J.; Shlesinger, M.F. On the relationship among three theories of relaxation in disordered systems. *Proc. Natl. Acad. Sci. USA* **1986**, *83*, 848–851. [CrossRef] [PubMed]

25. Mihalcescu, I. Analyse temporelle des mécanismes de luminescence du silicium poreux. Université Joseph Fourier-Grenoble 1, 1994.

26. Dam, B.V.; Bruhn, B.; Dohnal, G.; Dohnalová, K. Limits of emission quantum yield determination. *Aip Adv.* **2018**, *8*, 085313.

27. Ahn, T.-S.; Al-Kaysi, R.O.; Müller, A.M.; Wentz, K.M.; Bardeen, C.J. Self-absorption correction for solid-state photoluminescence quantum yields obtained from integrating sphere measurements. *Rev. Sci. Instrum.* **2007**, *78*, 086105. [CrossRef] [PubMed]

28. Chung, N.X.; Limpens, R.; Gregorkiewicz, T. *Investigating Photoluminescence Quantum Yield of Silicon Nanocrystals Formed in SiOx with Different Initial Si Excess*; SPIE: Bellingham, WA, USA, 2015; Volume 9562.

29. Lim, J.S.; Choi, H.; Lim, I.S.; Park, S.B.; Lee, Y.S.; Kim, S.K. Photodissociation Dynamics of Thiophenol-d1: The Nature of Excited Electronic States along the S–D Bond Dissociation Coordinate. *J. Phys. Chem. A* **2009**, *113*, 10410–10416. [CrossRef]

30. Li, C.-H.; Kui, S.C.F.; Sham, I.H.T.; Chui, S.S.-Y.; Che, C.-M. Luminescent Gold(I) and Copper(I) Phosphane Complexes Containing the 4-Nitrophenylthiolate Ligand: Observation of $\pi \rightarrow \pi^*$ Charge-Transfer Emission. *Eur. J. Inorg. Chem.* **2008**, *2008*, 2421–2428. [CrossRef]

31. Monzittu, F.M.; Fernández-Moreira, V.; Lippolis, V.; Arca, M.; Laguna, A.; Gimeno, M.C. Different emissive properties in dithiolate gold(i) complexes as a function of the presence of phenylene spacers. *Dalton Trans.* **2014**, *43*, 6212–6220. [CrossRef]

32. Huang, Y.; Wang, J.; Seo, H.J. The Irradiation Induced Valence Changes and the Luminescence Properties of Samarium Ions in Ba2SiO4. *J. Electrochem. Soc.* **2010**, *157*, J429–J434. [CrossRef]

33. Li, L.; Zhu, Y.; Zhou, X.; Brites, C.D.S.; Ananias, D.; Lin, Z.; Paz, F.A.A.; Rocha, J.; Huang, W.; Carlos, L.D. Visible-Light Excited Luminescent Thermometer Based on Single Lanthanide Organic Frameworks. *Adv. Funct. Mater.* **2016**, *26*, 8677–8684. [CrossRef]

34. Forward, J.M.; Bohmann, D.; Fackler, J.P.; Staples, R.J. Luminescence Studies of Gold(I) Thiolate Complexes. *Inorg. Chem.* **1995**, *34*, 6330–6336. [CrossRef]

35. Vogler, A.; Kunkely, H. Absorption and emission spectra of tetrameric gold(I) complexes. *Chem. Phys. Lett.* **1988**, *150*, 135–137. [CrossRef]

36. Kathewad, N.; Kumar, N.; Dasgupta, R.; Ghosh, M.; Pal, S.; Khan, S. The syntheses and photophysical properties of PNP-based Au(i) complexes with strong intramolecular Au··· Au interactions. *Dalton Trans.* **2019**, *48*, 7274–7280. [CrossRef] [PubMed]

37. Seki, T.; Sakurada, K.; Ito, H. Controlling Mechano- and Seeding-Triggered Single-Crystal-to-Single-Crystal Phase Transition: Molecular Domino with a Disconnection of Aurophilic Bonds. *Angew. Chem. Int. Ed.* **2013**, *52*, 12828–12832. [CrossRef] [PubMed]

38. Geoffroy, G.L.; Wrighton, M.S. *Organometallic Photochemistry*; Academic Press: New York, NY, USA, 1979.

39. Wang, Z.; Xiong, Y.; Kershaw, S.V.; Chen, B.; Yang, X.; Goswami, N.; Lai, W.-F.; Xie, J.; Rogach, A.L. In Situ Fabrication of Flexible, Thermally Stable, Large-Area, Strongly Luminescent Copper Nanocluster/Polymer Composite Films. *Chem. Mater.* **2017**, *29*, 10206–10211. [CrossRef]

40. Wang, M.-S.; Guo, G.-C. Inorganic–organic hybrid white light phosphors. *Chem. Commun.* **2016**, *52*, 13194–13204. [CrossRef] [PubMed]

 *nanomaterials*

Review

Origin of the Photoluminescence of Metal Nanoclusters: From Metal-Centered Emission to Ligand-Centered Emission

Tai-Qun Yang, Bo Peng, Bing-Qian Shan, Yu-Xin Zong, Jin-Gang Jiang, Peng Wu * and Kun Zhang *

Shanghai Key Laboratory of Green Chemistry and Chemical Processes, College of Chemistry and Molecular Engineering, East China Normal University, Shanghai 200062, China; tqyang@chem.ecnu.edu.cn (T.-Q.Y.); 51174300109@stu.ecnu.edu.cn (B.P.); 52184300051@stu.ecnu.edu.cn (B.-Q.S.); 51174300121@stu.ecnu.edu.cn (Y.-X.Z.); jgjiang@chem.ecnu.edu.cn (J.-G.J.)
* Correspondence: pwu@chem.ecnu.edu.cn (P.W.); kzhang@chem.ecnu.edu.cn (K.Z.)

Received: 23 December 2019; Accepted: 29 January 2020; Published: 4 February 2020

Abstract: Recently, metal nanoclusters (MNCs) emerged as a new class of luminescent materials and have attracted tremendous interest in the area of luminescence-related applications due to their excellent luminous properties (good photostability, large Stokes shift) and inherent good biocompatibility. However, the origin of photoluminescence (PL) of MNCs is still not fully understood, which has limited their practical application. In this mini-review, focusing on the origin of the photoemission emission of MNCs, we simply review the evolution of luminescent mechanism models of MNCs, from the pure metal-centered quantum confinement mechanics to ligand-centered p band intermediate state (PBIS) model via a transitional ligand-to-metal charge transfer (LMCT or LMMCT) mechanism as a compromise model.

Keywords: photoluminescence mechanism; metal nanoclusters; quantum confinement effect; ligand effect; p band intermediate state (PBIS); interface state; nanocatalysis

1. Introduction

Metal nanoclusters (Au, Ag, Pt, Cu) are a new class of attractive materials owing to their enhanced quantum confinement effect, which endows them with unusual optical and electronic properties [1–6]. Au and Ag NCs in particular have attracted tremendous interest because of their wide applications in single-molecule studies, sensing, biolabeling, catalysis, and biological fluorescence imaging [7–16]. The physical and chemical properties of metal nanoclusters (MNCs) are highly dependent on their size, shape, composition, and even assembly architecture [17–20]. These clusters are composed of a few to hundreds of atoms with a size approaching the Fermi wavelength of an electron (~0.5 nm for Au and Ag), and they possess discrete molecular-like electronic energy band structures, resulting in intense light absorption and emission. Luminescent MNCs exhibit outstanding optical properties including a large Stokes shift (generally large than 100 nm), large two-photon absorption cross-section, good photostability, good biocompatibility, and the emission wavelength could be easily adjusted by controlling their size and surface ligands, which are not present in conventional organic dyes [1,9,11,21]. Their excellent luminous performance is attractive for the applications in biomedicine, since it provides avenues for designing optical sensors, biolabeling, bioimaging, photosensitizers, and light-emitting devices. However, the luminous quantum yield of MNCs is still not competitive when compared to those organic dyes and semiconductor quantum dots (QDs), which always hinder their practical application. One primary reason is the lack of full understanding of the luminescence origin of MNCs at the molecular level, since its photoluminescence properties depend on many factors including cluster size, surface ligands architecture, cluster assembly structure, etc. The preparation and application of MNCs have been previously reviewed [2,8,10,14,15,22–25]. Herein, we are mainly focused on the photoluminescence emission origin of MNCs. It is imperative to clarify the origin of photoemission of MNCs and thus improve their luminous efficacy. Two major explanations about the photoluminescence mechanism have become gradually accepted by researchers. One is the pure metal quantum confinement effect [1,26–28]. That is, as the size of MNCs approaches the Fermi wavelength of metals (usually <1 nm), the continuous band of energy level becomes discrete, leading to the emerging of molecule-like properties. The emission of MNCs was originated from intraband (sp–sp) and interband (sp–d) transitions [29,30]. The other explanation is the charge transfer on the shell of metal clusters due to the interaction between the functional ligands and the metal core, i.e., ligand-to-metal charge transfer (LMCT) [31] or ligand-to-metal–metal charge transfer (LMMCT: emphasize the gold–gold interactions resulted from the relativistic effect) [32] mechanism borrowing from the concept of PL emission of the transition organometallic complexes. Both of these suggest that the intraband (sp–sp) and interband (sp–d) transitions from metal NCs probably act an intermediate state (or dark state) to determine the electron transfer [33]. These two explanations are proposed to explain the size-dependent and ligand-dependent emission phenomenon respectively, but they are self-contradicted to illustrate the size-dependent and size-independent emission. In addition, more and more studies show that nonluminescent groups could serve as chromophores in a certain solvent condition or confined nanospaces. In addition, more and more strong evidences from recent studies tend to prove the limitation of metal-centered LMCT and LMMCT models to explain the photoluminescence (PL) properties of metal NCs.

So far, two main approaches—the bottom–up and top–down strategies—have been developed for the synthesis of emission wavelength adjustable luminescent MNCs with high quantum yield (QY) in large quantities, as illustrated in Figure 1 [8–10,14,15,34]. The size effect (including metal corearchitecture) [35–40], ligand effect (charge-donating capability [33], chirality [41,42], hydrophily [43]), valance state of surface metals [33,44], and superlattice structures of assembly of metal clusters [45,46] have been intensively studied.

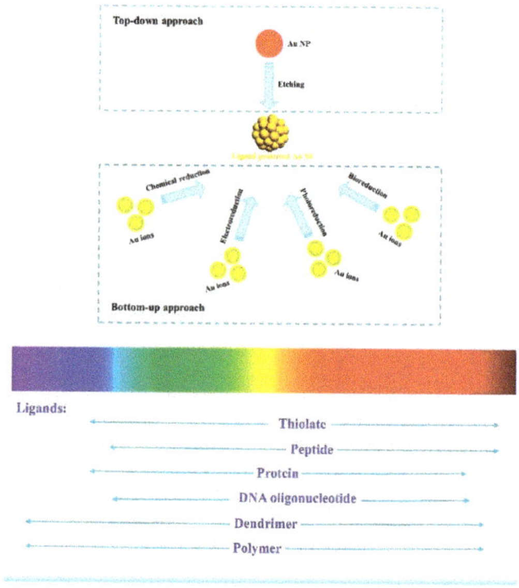

Figure 1. Schematic illustration of synthetic strategies of luminescent AuNCs and effect of ligands on their photoluminescence. Reprinted with permission from Ref. [14]. Copyright (2017) Elsevier.

It is important to note that especially for the bottom–up synthetic strategy, a common feature of luminescence metal NCs is that the templates used for the synthesis as protective ligands, such as nonconjugated polymers, proteins, amino acid molecules, and even small chemical molecules, always contain electron-rich heteroatoms, including oxygen (O), nitrogen (N), sulfur (S), phosphorus (P), and others, emphasizing the dominant role of the surface ligand to tune the PL properties. However, the exact role of the surface ligand and the dynamic and/or kinetic of electron relax at the nanoscale interface between core–shell structured MNCs was not clarified. In this report, we simply review the present fundamental understanding of the photoluminescence origin of MNCs, especially our recently developed the ligand-centered p band intermediate state (PBIS) model to elucidate the abnormal PL emission phenomena. Finally, we proposed some forward-looking perspectives to the future developments of MNCs, in particular, for the application of MNCs in the nanocatalysis science.

2. Photoemission Mechanism of Metal Nanoclusters

2.1. Size-Dependent Photoemission

As the saying goes, "gold will glitter forever!" The first experimental observation of photoluminescence from bulk gold metal can date back to as early as 1969 [29]. Mooradian reported that bulk gold and copper film could generate photoemission when excited under a high-energy laser. The emission was arising from the interband transitions between electrons in conduction-band states below the Fermi level and holes in the d bands generated by optical excitation, as shown in Figure 2. This is the first proposed energy band structure to explain the photoemission mechanism of transition bulk metal. The energy gap was elucidated to 2.0 eV for copper and 2.2 eV for gold with the help of emission spectra. While photoemission was observed from bulk metal, the emission was generated under an extreme condition, and the quantum yield was very low (10^{-10}), which was not suitable for practical application. Metal nanoparticles with a size range from 2 to 50 nm exhibit intense colors due to the surface plasmon resonance (SPR); in some cases, photoluminescence was also observed in these small metal nanoparticles, which contrasts the conclusion that MNCs with SPR effect generally do not produce fluorescence [2]. In 1998, Wilcoxon et al. observed relatively intense photoluminescence when the size of the metal nanocluster was sufficiently small [47]. Nonluminescent gold particles can also become luminescent by partial etching with potassium cyanide (KCN). The photoemission with large Stokes shift is ascribed to the electron and hole interband recombination process. When the particle size was further reduced to 2 nm or less, the continuous band structure will be broken down into discrete energy levels according to quantum confinement mechanics, exhibiting molecular-like properties and not exhibiting typical plasmonic properties. Indeed, these very small transition metal clusters with precisely defined architectures are first studied in heterogeneous catalysis [48], which did not attract much more attention from photochemists and photophysicists. With the rapid development of synthetic nanochemistry, more and more people focused on the peculiar optoelectronic properties of these tiny metal clusters, even for still low quantum yield (QY) confined in a gas matrix at low temperature [49,50]. By combining solid-state principles as well as a molecular model, Link et al. improved the energy band structure model to explain the electron transition process and subsequent PL emission, which was attributed to the intraband (sp–sp) and interband (sp–d) transitions as illustrated in Figure 3 [30].

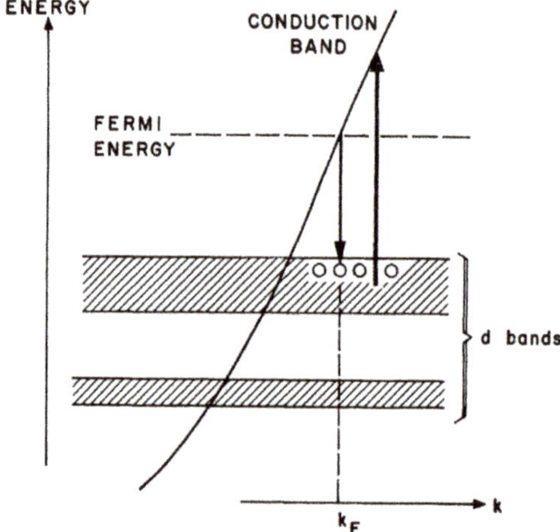

Figure 2. Schematic band structure of a noble metal showing the excitation and recombination transitions. Reprinted with permission from Ref. [29]. Copyright (1969) American Physical Society.

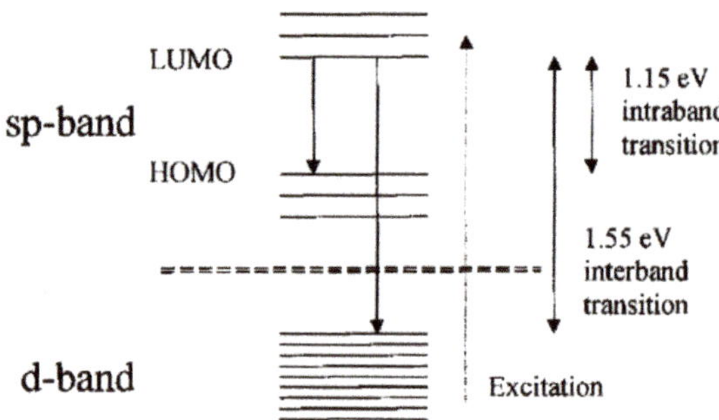

Figure 3. Solid-state model for the origin of the two luminescence bands: The high-energy band is proposed to be due to the radiative interband recombination between the sp and d-bands, while the low-energy band is thought to originate from radiative intraband transitions within the sp-band across the HOMO-LUMO gap (highest occupied molecular orbital and lowest unoccupied molecular orbital). Note that intraband recombination has to involve the prior nonradiative recombination of the hole in the d-band created after excitation with an (unexcited) electron in the sp-band. Reprinted with permission from Ref. [30]. Copyright (2002) American Chemical Society.

With the progress of wet chemical synthesis strategy, luminescent water-soluble metal nanoclusters with different sizes were successfully synthesized with improved PL emission efficiency and tunable colors [26,51–54]. The photoemission wavelength could be varied from the ultraviolet (UV) to near-infrared (NIR) region by controlling the sizes of metal NCs [1]. In the early 2000s, Dickson et al. first achieved the preparation of water-soluble Au NCs with high fluorescence in solution using a biocompatible dendrimer (OH-terminated poly(amidoamine) (PAMAM)) as a capping agent [26,51,52]. The emission wavelength of as-prepared Au NCs could be adjusted from UV to the NIR region by changing the metal/polymer ratio during the synthesis. Benefiting from the progress of soft-ionization mass spectrometry (MS) technology, the cluster size could be precisely identified. By ingeniously creating the linear correlation between the emission intensity at a certain wavelength and the intensity of mass spectra at a certain cluster size, the emission wavelength of different clusters was determined. The correlation between the size of Au nanodots and emission energies was established and fit a simple scaling relation of $E_{\text{Fermi}}/N^{1/3}$ (where E_{Fermi} is the Fermi energy of bulk gold and N is the number of Au atoms) [26], as shown in Figure 4. The size-dependent scaling of excitation and emission energies with $E_{Fermi}/N^{1/3}$ confirms the good matching with the jellium model where the gap between the discrete 5d valence band and the 6sp conduction band decreases with the increasing cluster size [1]. If this model is rational, the coinage metals at the same family with the similar size or the close atom numbers should show very similar PL emission. However, the size distribution and optical properties of stable "magic" number Ag NCs were found to be obviously different to the gold NCs. Recently, Udaya Bhaskara Rao et al. successfully synthesized Ag NCs with very similar particle size (Ag7 and Ag8) using mercaptosuccinic acid (H2MSA) as capping ligands; they showed distinguished PL emission at approximately 440 nm and 650 nm for Ag7 and Ag8 NCs, respectively. Only one silver atom difference in cluster structures could lead to such a big change of emission color from blue to red [55]. This could not be simply exchanged only by a size-dependent quantum confinement effect (QCE).

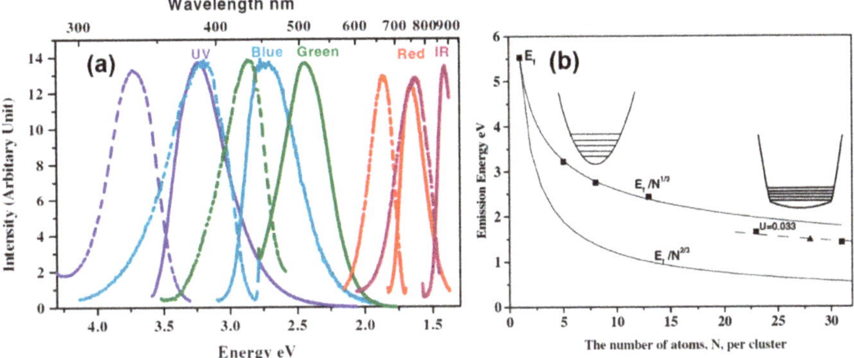

Figure 4. (a) Excitation (dashed) and emission (solid) spectra of different sizes of PAMAM-Au NCs. Excitation and emission maxima shift to longer wavelengths with increasing initial Au concentration, suggesting that increasing the nanocluster size leads to lower energy emission. (b) Correlation of the number of Au atoms per cluster (N) with the photoemission energy of Au NCs. Reprinted with permission from Ref. [26]. Copyright (2004) American Physical Society.

Even though MNCs shows strong size-dependent PL emission properties, the quantum lifetime and QY calculated by theoretical calculation are much less than the experimental observed values, and if we are only considering the contribution of metal to PL emission, the wavelength of PL emissions calculated by the theoretical model should lie in a near-infrared range, instead of the observed visible wavelength emission. These disagreements between experimental and theoretical data indicate the invalidity of metal-centered quantum confinement mechanics.

2.2. Size-Independent Photoemission

With sophisticated synthetic techniques, atom-precise MNCs were achieved for a number of metal nanoclusters (referred to as M_nL_m, where n and m are the numbers of metal atoms and ligands in the cluster). Very recently, Luo and his co-workers synthesized a series of glutathione (GSH) protected Au NCs with precisely atomic compositions (Au29SG27, Au30SG28, Au36SG32, Au39SG35, Au43SG37), and, to one's surprise, they all exhibited completely the same emission at approximately 610 nm [56], which is completely contradictory to the explanation of the classical QCE mechanism. Furthermore, the emission wavelength could be readily regulated from 600 nm to 810 nm by just fine-tuning the surface ligands' coverage [57], suggesting the pivotal role of surface-protective ligands to tune the PL of Au NCs. In addition, the Au NCs with the same number of core atoms but different protecting ligands show different PL properties [33,58,59]. These results clearly indicate that the metal core is not the only determining factor for the photoemission of MNCs. Other components of MNCs including the nature of coordinate ligands, valence states of surface metal atoms, and assembly architectures of nanoclusters also make a significant difference on the emission properties of MNCs.

2.2.1. Ligand Effect

It is well accepted that the organic/inorganic scaffolds and/or surface protective ligands were used to stabilize the metal nanoclusters, since naked MNCs would strongly interact with each other and aggregate irreversibly as to reduce their surface energy. Solid matrices including noble gas matrices [60], glasses [61], and zeolites [62–64] are widely selected for incubating silver nanoclusters with only a few atoms. However, the large size of solid matrices impedes their further application in biomedicine science. The organic templates that encapsulated MNCs were first achieved using water-soluble dendrimer (PAMAM) as the scaffold by Dickson et al. in 2002 [51]. Since then, various synthesis strategies including the template-assisted method and ligand-induced etching have been developed to fabricate organic ligands-capped luminescent MNCs [10,14]. The formation of MNCs with different capping ligands in solution have been accomplished in various ways, as shown in Figure 5 [34]. The synthesis parameters, such as the reduction method of metal ions, the initial ratio of reactants, and the reaction temperature have a profound effect on the generation of luminescent MNCs.

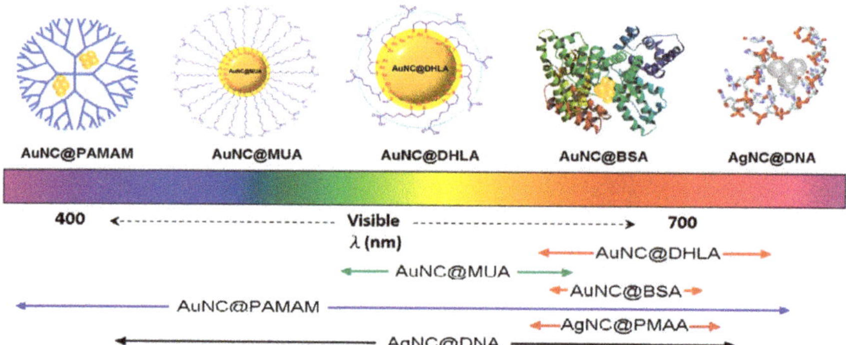

Figure 5. Representative luminescent noble–metal nanoclusters scaled as a function of their emission wavelength superimposed over the spectrum. Protected molecules show different capabilities to tune the emission wavelength of metallic nanoclusters from current reports.

Prominently, the emission wavelength of the ligand-protected MNCs exhibits a strong ligand-dependent effect. For example, the emission wavelength of DNA oligomer-capped Ag NCs could be varied from the blue to NIR region by playing with the nucleotide sequence [65]. In addition, the photoluminescence intensity is also dependent on the nucleotide sequence. The photoluminescence of DNA-AgNCs can be enhanced 500-fold when placed in proximity to guanine-rich DNA sequences. Based on this discovery, Yeh et al. designed a fluorescent probe to detect specific nucleic acid targets [66]. This fluorescence probe can easily reach high signal-to-background ratios (S/B ratios) (>100) upon target binding, since it does not rely on Forster energy transfer for quenching. Wu et al. systematically studied the ligand's role in the fluorescence of gold nanoclusters [33]. They found that the photoluminescence intensity was scaled with the electron donation capacity of thiol ligands. The photoluminescence was attributed to the charge transfer from the ligands to the metal nanoparticle core (i.e., LMNCT) through the Au-S bonds. These ligands with electron-rich atoms (e.g., N, O) or groups (e.g., –COOH, NH_2) can largely promote fluorescence via surface interactions, as shown in Figure 6. Luo et al. demonstrated that even only the aggregated Au(I)–thiolate complexes induced by solvents or cations could generate strong photoluminescence. The formation of aurophilic bonds provided the impetus for aggregation; then, denser and more rigid aggregates were formed. The emission from the aggregates was attributed to ligand-to-metal charge transfer (LMCT) or ligand-to-metal–metal charge transfer

(LMMCT) from the sulfur atom in the thiolate ligands to the Au atoms, and subsequent radiative relaxation, which was most likely via a metal-centered triplet state. Pyo et al. reported an efficient strategy to enhance the photoluminescence of gold clusters ($Au_{22}(SG)_{18}$) by rigidifying its gold shell with bulky tetraoctylammonium (TOA) cations [67]. The luminous quantum yield could improve up to approximately 60% at room temperature. The luminescence was ascribed to the LMMCT triplet state from the gold shell, which exhibits a strong rigidifying effect. They conclude that the inter-complex aurophilic Au(I)···Au(I) interactions are unlikely in the well-separated dimeric gold shell of the $Au_{25}(SG)_{18}$ cluster encapsulated in dendrimer (PAMAM), since these as-synthesized AuNCs were well-separated by the cavity of the dendrimer, and the luminescence can be ascribed to the ligand-to-metal charge transfer (LMCT) effect, rather than the LMMCT effect [26]. Londono-Larrea et al. successfully synthesized water-soluble naked gold nanoclusters ($AuNC_{naked}$) by using only NaOH (the reductant) and $HAuCl_4$ [68]. They found that the nonluminescent $AuNC_{naked}$ could be transferred to strong luminescent when passivated with different thiols and adenosine monophosphate. The photoluminescence of the passivated NCs was clearly attributed to the ligand–AuNC surface interaction.

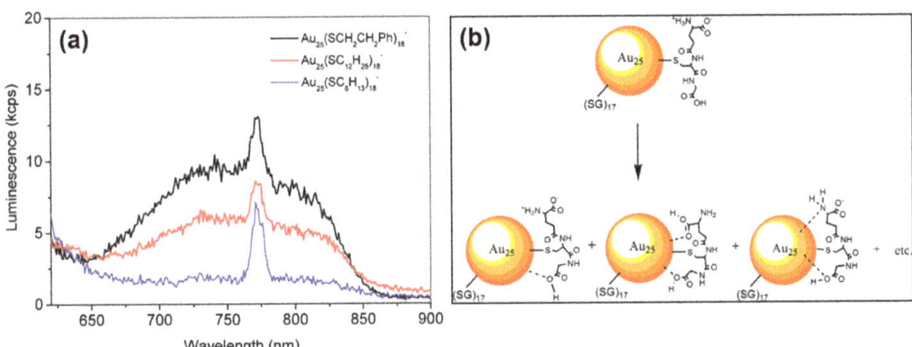

Figure 6. (**a**) Weak fluorescence of [$Au_{25}(SR)_{18}$]- with different R groups ($-C_2H_4Ph$, $-C_{12}H_{25}$, and $-C_6H_{13}$). (**b**) Possible Interactions of amine and carboxyl groups of glutathione ($-SG$) ligands to the gold surface. Reprinted with permission from Ref. [33]. Copyright (2010) American Chemical Society.

These results indicated that the surface ligands of MNCs were also pivotal to their photoluminescence properties. Ligand-to-metal charge transfer (LMCT) and ligand-to-metal–metal charge transfer (LMMCT or LMNCT) mechanisms were proposed to understand the principle of the luminescence process. However, some issues are still difficult to understand. For example, these MNCs with identical size could be luminescent or nonluminescent, and they are very sensitive to the delicate change of reaction parameters.

2.2.2. Metal Valence State Correlated Photoemission

In addition to the cluster size and surface-capping ligands, a surface metal valence state was also a considerable factor for synthesizing highly luminescent MNCs. Zhou et al. observed that the completely different luminescence properties of glutathione protected gold nanoparticles (GS-Au NPs) with identical size but different in metal valence states [44]. The luminescent GS-Au NPs with a core size of approximately 2 nm become nonluminescent after treatment by $NaBH_4$, noting that the core size was sustained. If this is true (same size), obviously the metal centered quantum confinement effect (QCE) can not answer this peculiar observation. The authors concluded that the high content of gold(I) (40–50%) in the luminescent nanoparticles verified by X-ray photoelectron spectroscopy (XPS) was responsible for the unique optical properties of the luminescent gold nanoparticles. The luminescence lifetime highly depends on the excitation wavelength; when excited at approximately 420 nm, it emits orange colors with a long lifetime of several microseconds at approximately 565 nm, while, if excited at 530 nm, the same color is emitted with a short lifetime with nanoseconds (Figure 7). This interesting observation of dual PL emissions with different lifetimes indicates the presence of two completely different chromophores. The authors pointed out that the degeneration in energy of the triplet and singlet excited states in the luminescent gold nanoparticles led to the same wavelength emissions due to the change of metal charge valence state. Wu et al. also observed that the fluorescence of $Au_{25}(SC_2H_4Ph)_{18}$ could be largely enhanced by increasing their oxidation charge state (from −1 up to +2) using oxidants such as O_2, H_2O_2, $Ce(SO_4)_2$, etc. [69,70]. In these reports, all the evidence confirmed the influence of the valence state of a metal ion or core on the PL emissions, and the PL emission of MNCs was originated from the interband or intraband transitions of the outmost shell electrons. However, it is important to note that the increasing of the electropositivity of the metal core, i.e., the high charge state of metals, can also promote the interactions between the surface ligands and metal core, which cannot completely exclude the role of ligand assembly to tune the PL emission.

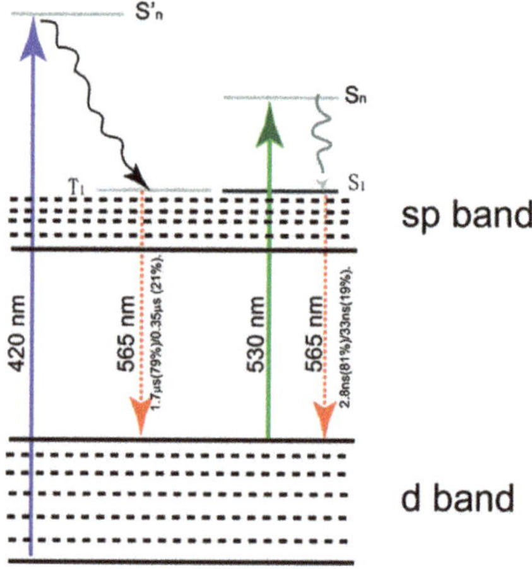

Figure 7. A possible optical transition scheme for orange-emitting GS-AuNPs where the luminescence originates from transitions between d and sp bands. When the NPs are excited at 420 nm, the electrons will be relaxed from triplet states in sp bands to some ground states in d bands, leading to microsecond emission. Once the excitation wavelength is shifted to 530 nm, the electrons will be decayed from singlet excited states in the sp band to singlet states in ground states and give out a nanosecond emission. The triplet and singlet excited states in the luminescent gold nanoparticles are degenerate in energy. Reprinted with permission from Ref. [44]. Copyright (2010) American Chemical Society.

2.2.3. Self-Assembly Governed Photoemission

The intrinsic defect of luminous MNCs was their limited quantum yield, since these ultra-small clusters with high surface-to-volume ratio easily suffer from the solvent molecules and oxygens, which could quench their photoluminescence [71]. Many MNCs only exhibit photoluminescence under low temperature and are nonluminescent at room temperature, which largely limits their practical application [72]. Aggregation-induced emission (AIE) is an efficient strategy for improving the photoluminescence performance of lumigens. It has attracted tremendous attention since the first observation in organic chromophores by Tang's group in 2001 [73]. Generally, the aggregation of organic dyes could lead to photoluminescence quenching due to the formation of detrimental species such as excimers [74,75]. In contrast, the luminescence of (luminigens) AIEgens is conspicuously enhanced via aggregation due to the strong restriction of the intramolecular vibrations (RIV) and rotations (RIR) [76]. Recently, this fascinating AIE effect has been regarded as an efficient strategy for optimizing the photoemission performance of nanoclusters by directing their spatial construction [45,56,77,78]. The alternation of nanoarchitectures greatly influences the charge transfer and energy conduction process of MNCs, and thus their photoelectric performance. However, the physical origin of AIE effect is not clearly addressed. In fact, if we carefully checked the structures and PL properties of AIEgens, they showed very similar PL properties with MNCs, such as a very strong solvent effect and large Stock shift. It is possible that their PL origins come from the same physical principle.

Jianping Xie's group and Erkang Wang's group first reported that the nonluminescent oligomeric Au(I)−thiolate complexes and weak luminescent CuNCs in aqueous solution could generate strong luminescence upon aggregation induced by weak polar solvents or divalent cations [56,79], and their luminescence enhancement was assigned to the AIE effect. Very recently, Jia et al. demonstrated that the photophysical features of the thiolated AgNCs were dependent on their morphology, which was controlled by the solvents-induced aggregation [41]. The ordered structure of AgNCs assemblies was well defined by XRD and TEM techniques. Subsequently, Benito et al. observed the mechanochromic luminescence properties in copper iodide clusters [46]. Two kinds of crystalline polymorphs with green and yellow emission were obtained. Upon mechanical grinding, the green emissive polymorph exhibits great modification of its emission from green to yellow, as shown in Figure 8. XRD analyses indicated that the crystalline packing was damaged, and an almost complete amorphous state was formed, which implies an assembly architecture-dependent emission property of CuNCs. Obviously, the change of PL properties triggered by simply grounding cannot be simply answered by metal-centered QCE, since individual CuNCs remains unchanged. The authors suggested that the Cu-Cu interactions were responsible for the luminescence properties. Upon mechanical stimulation, destruction of the crystalline structure led to shortening of the Cu−Cu bond in the cluster core and resulting red shift of emission wavelength from green to yellow. These results definitely established the important role of cuprophilic interactions in the mechanochromic mechanism of CuNCs. A similar mechanism was used by Zhang to explain the PL emission of metal nanoclusters (Cu, Au, Ag) self-assemblies with different assembly morphology [45,78,80–83]. These as-synthesized nonluminescent individual CuNCs could generate strong photoemission when assembled to a crystalline packing structure, similar to the AIE effect, noting that the emission wavelength could be easily adjusted by controlling the assembly structure. As shown in Figure 9, CuNCs exhibit yellow-green emission in nanosheets but blue emission in nanoribbons. A mechanochromic property was also observed in which the blue emissive nanoribbon could transfer to yellow emissive with less crystallinity.

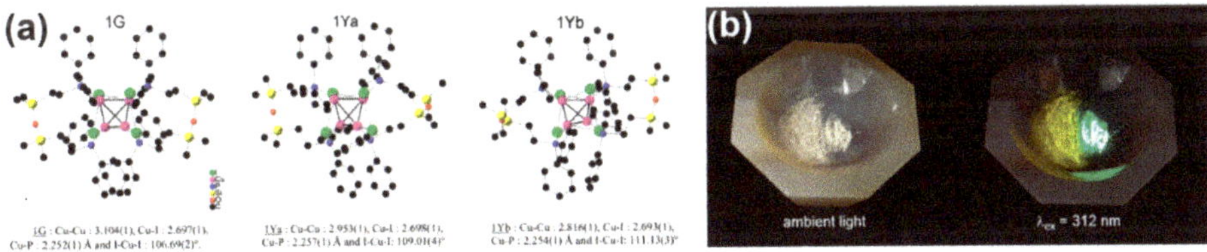

Figure 8. (**a**) Molecular structure of clusters 1G and 1Y and the mean of selected bond lengths. Hydrogen atoms have been omitted for clarity. (**b**) Photos of the ground (left) and intact (right) crystalline powder of 1G under ambient light and under UV irradiation at 312 nm (UV lamp) at room temperature. Reprinted with permission from Ref. [46]. Copyright (2014) American Chemical Society.

The relationship between the compactness of assemblies and the emission was summarized as follows. (1) "High compactness reinforces the cuprophilic Cu(I)···Cu(I) interaction of inter- and intra-NCs, and meanwhile, it suppresses intramolecular vibration and rotation of the capping ligand of 1-Dodecanethiol (DT), thus enhancing the emission intensity of Cu NCs. (2) The emission energy depends on the distance of Cu(I)···Cu(I); the improved compactness increases the average Cu(I)···Cu(I) distance by inducing additional inter-NCs cuprophilic interaction, and therewith leads to the blue shift of NCs emission" [45]. Noting that it is counterintuitive for improving the compactness could lead to the increase of Cu(I)···Cu(I) distance. The authors explain this abnormal phenomenon by inducing additional inter-NCs cuprophilic interaction between the neighboring NCs. However, it needs to be reiterated that cuprophilic interactions could only generate when the adjacent Cu⋯Cu distances are in the range of the van der Walls interaction distance (generally less than 3.6 Å); however, the distance between these two neighboring NCs is at a nanometer scale, which is far beyond the effective distance of metallophilic interactions [32,84,85]. Thus, the rationality of the LMCT and/or LMMCT model was challenged. However, the fact remains that the change of spacing distance between adjacent surface ligands in varied morphologies is definitively confirmed, which further evidenced the paramount role of surface ligand packing to tune the PL properties.

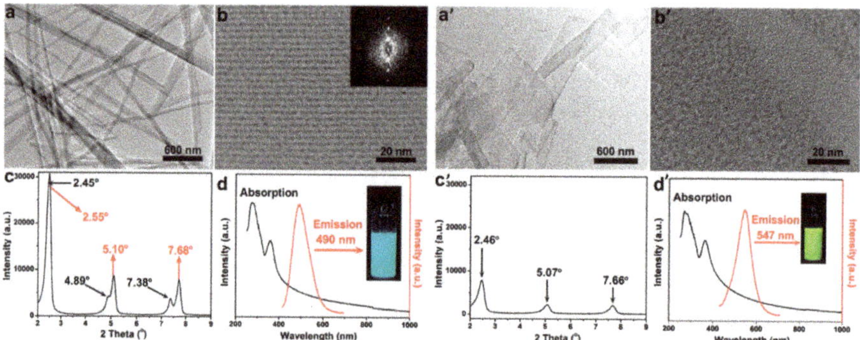

Figure 9. (**a**) TEM image of the ribbons from Cu NCs' self-assembly. (**b**) HRTEM image of the ribbons. Inset: the Fourier transform image. (**c**) Small-angle region of XRD pattern. (**d**) Steady-state absorption (black) and emission (red) spectra of the ribbons in chloroform. Inset: the fluorescent image with 365 nm excitation. (**a′**) TEM image of the sheets from Cu NCs' self-assembly. (**b′**) HRTEM image of the sheets. (**c′**) Small-angle region of XRD pattern. (**d′**) Steady-state absorption (black) and emission (red) spectra of the sheets in chloroform. Inset: the fluorescent image with 365 nm excitation. Reprinted with permission from Ref. [45]. Copyright (2015) American Chemical Society.

3. Our P Band Intermediate State (PBIS) Model Dominates the Photoluminescence Emission of MNCs

Even though the metal-centered free-electron model based on the QCE and, subsequent LMCT and/or LMMCT mechanism, in some sense, explained some important PL emission phenomena, the elucidation and origin of optoelectronic properties is diverse and contradictory, at every point inviting inquiry and debate over a decade. Since 2014, this continuous and long-term study in my group and collaborators has been carried out to understand the nature of the photoluminescence emission of metal nanoclusters and related quantum nanostructures. We first provided the key evidences that the distribution of surface-protecting ligands on the metal core played a paramount role to tune the optoelectronic properties of noble metal NCs. However, the basic chemical principle hidden behind abnormal optical phenomena has been troubling us, such as its room temperature phosphorescence enhancement, surface ligand selectivity, unprecedented large Stokes Shift, tunable optical absorption, newborn electronic band structure, etc. Very recently, using metal NCs as a model system, by judiciously manipulating the delicate surface ligand interactions at the nanoscale interface of a single metal nanocluster, superlattice and mesoporous materials, together with a careful control of the solvophobicity and solvophilicity of the ligands, the resulting interplay of various noncovalent interactions can lead to the precise modulation of optoelectronic properties of metal NCs. A completely new p band intermediate state (PBIS) model was proposed to understand the origin of PL emission of all related quantum nanodots, which completely challenges the metal-centered quantum mechanics for the elucidation of PL emission of metal NCs. We definitively identify that the PBIS stems from the overlapping of p orbitals of the paired or more adjacent heteroatoms (O and S) from the surface-protected ligands on the metal NCs, which can be considered as a dark state at the metal–NCs interface to activate the triplet site of the surface chromophores.

In our earlier investigations, in order to fully understand the structure of luminescent MNCs, we used water-soluble polymers poly(methacrylic acid) (PMAA) as templates to encapsulate AgNCs with small size (2−5 nm) [86]. By precisely designing the experiments' parameters, we found that the photoluminescence of AgNCs showed high selectivity on the surface-anchoring ligands and strong dependence on the valence state of surface metal. That was, the strong fluorescence of carboxyl-protected AgNCs disappeared when the surface functional ligands were changed to sulfonic acid groups, even though the metal core was sustained, indicating the key role of ligand type on the regulation of PL properties. Based on these experimental results, a core-shell structural model was proposed to understand the nature of photoluminescence of Ag NCs. As shown in Figure 10, the fluorescence from the AgNCs was attributed to ligand-to-metal−metal charge transfer (LMMCT) from Ag(I)-carboxylate complexes (the oxygen atom in the carboxylate ligands to the Ag(I) ions) to the Ag atoms and subsequent radiative relaxation. In this report, we also followed the classical QCE to elucidate the PL emission of MNCs. However, we highlighted the pivotal role of surface ligands to regulate the PL properties of MNCs.

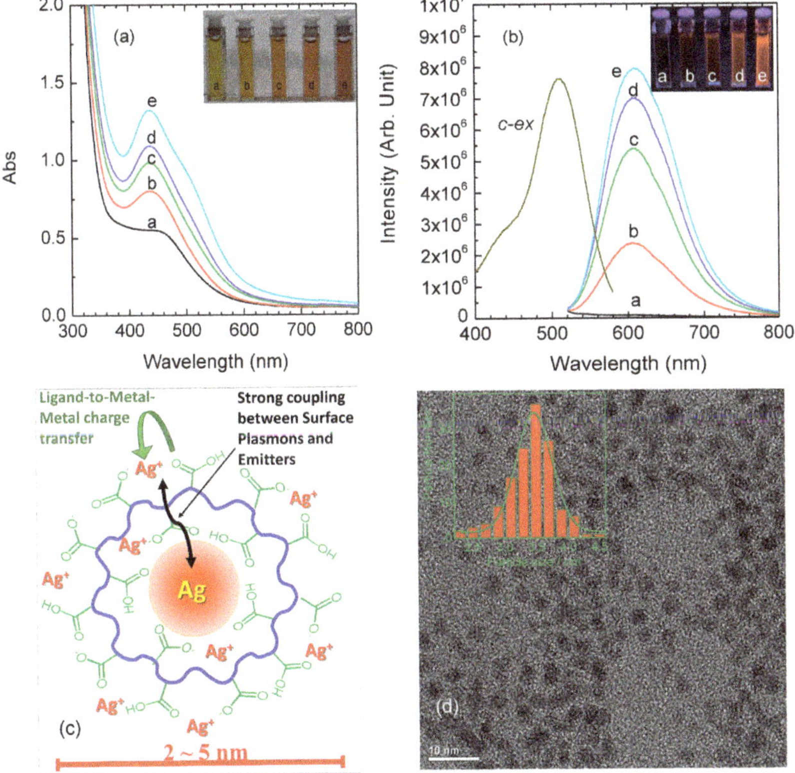

Figure 10. (**a**) UV−vis absorption of the freshly prepared Ag-carboxylate NCs using PMAA as scaffold at various radiation times. (**b**) PL spectra of the Ag-carboxylate NCs. Inset images show the corresponding photographs of Ag-carboxylate NCs under room and UV light exposure at $\lambda = 365$ nm. (**c**) Schematic illustration of the structures of our luminescent Ag-carboxylate with Ag⁺-carboxylate complexes shell. (**d**) TEM image of freshly prepared Ag-carboxylate NCs (histogram describes the statistical distribution of the particle size, which is approximately 3.5 nm). Reprinted with permission from Ref. [86]. Copyright (2014) American Chemical Society.

To further clarify the emission origin of MNCs, we separately discussed the functions of surface ligands and metal core to the photoluminescence. Metal-centered emission (MCE) and ligand-centered emissions (LCE) were simultaneously observed in AgNCs [87]. Luminescent water-soluble AgNCs protected by sulfydryl ligands with different functional groups (carboxyl, amino, alkyl) were successfully fabricated using a modified cyclic reduction–decomposition approach. Two distinct photoemissions with emission peaks at approximately 580 and 665 nm were observed, as shown in Figure 11. The emission at approximately 580 nm with a small Stokes shift (30 nm), narrow peak width (full width at half maxima (FWHM) approximately 30 nm), short lifetime (1.6 ns), and low quantum yield (<1%) was ligand-independent, since it could be observed in all of these as-synthesized AgNCs with different capping ligands. In contrast, the emission at approximately 665 nm with a large Stokes shift (>200 nm), broad peak width (FWHM approximately 100 nm), relative longer lifetime (180 ns), and higher quantum yield (approximately 10%) was ligand-dependent, and it could only be observed in AgNCs protected by carboxyl. They were attributed to MCE and LCE, respectively. The MCE was highly dependent on the metal core, and the LCE was highly related to the surface ligands and solvent environment. Accordantly, the emission at approximately 580 nm was pH independent, and the emission at approximately 665 nm exhibits a strong pH-dependent effect (Figure 12a–c). Based on these observations, a new ligand synergistic emission effect was proposed to understand the PL emission of MNCs. That is, the amino-correlated $n\pi^*$ state provides a pivot to bridge the carboxyl correlated $\pi\pi^*$ and $n\pi^*$ states to enhance the charge transfer efficiency between different surface electronic states. Consequently, the photoluminescence quantum yields were significantly improved (approximately 1% to 10%), as shown in Figure 12. Very recently, our unpublished results showed that the assignment on the emission at approximately 580 nm is probably not right. Thus, the ligand synergistic emission mechanism needs further optimization.

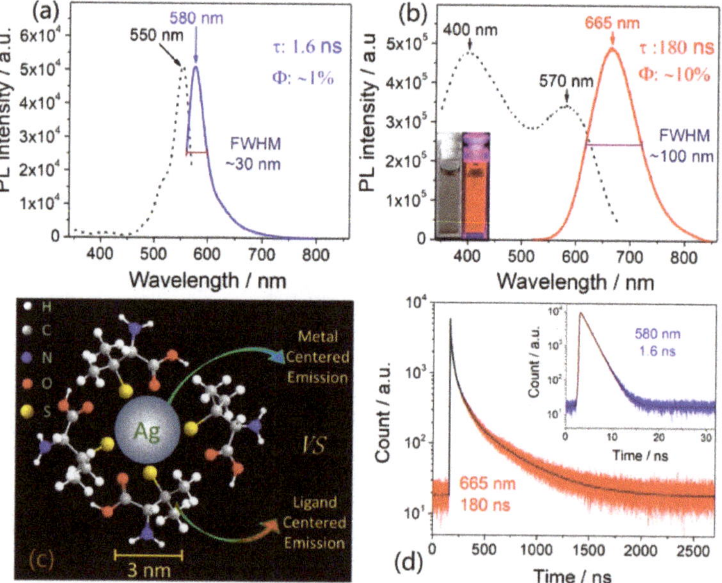

Figure 11. Excitation and emission spectra of the narrow approximately 580 nm emission (**a**) and broad approximately 665 nm emission (**b**). Inset images show the corresponding photographs of D-penicillamine-capped silver nanoclusters (DPA-AgNCs) under room and UV light exposure at λ = 365 nm. (**c**) Schematic illustration of the metal-centered and ligand-centered emission mechanisms of DPA-capped AgNCs. (**d**) Time-resolved luminescence decay profiles of DPA-AgNCs in water solution measured at 665 and 580 nm (inset), excited at 400 and 550 nm, respectively. Reprinted with permission from Ref. [87]. Copyright (2019) American Chemical Society.

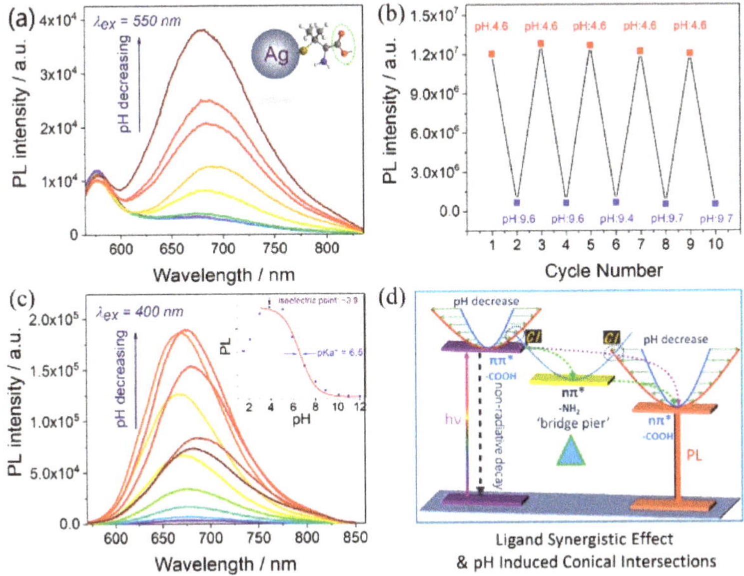

Figure 12. (**a**) Photoluminescence spectra of as-synthesized DPA-AgNCs in different pH conditions. (**b**) PL intensity of DPA-AgNCs upon cyclic switching of the pH between 4.6 and 9.7. (**c**) PL spectra and PL intensity (inset) of DPA-AgNCs with changing pH values. PL spectra were excited at 400 nm; pH value over the range 1.5–12. (**d**) Schematic illustration of the ligand synergistic effect and pH-induced conical intersections. Reprinted with permission from Ref. [87]. Copyright (2019) American Chemical Society.

If the metal-centered QCE dominates the PL emission of MNCs, the optical properties should not be significantly affected by the solvent effect. However, it is not the case. Very recently, we unexpectedly observed a strong solvent-induced enhancement effect of carboxyl-protected water-soluble AgNCs, indicating the invalidity of the well-accepted metal centered QCE model [88]. The emission QY of AgNCs could be largely improved from approximately 1% to 40% by a simple solvent-stimulated strategy (Figure 13). The fluorescence-phosphorescence dual solvoluminescence (SL) of water-soluble metal nanoclusters (NCs) at room temperature was observed. The photoluminescence was originated from the self-assembly of surface ligands, which was induced by solvent stimulation, that is, the clustering of surface ligands, as illustrated in Figure 14. The clustering of surface carbonyl groups was promoted in two different ways: (1) the strong interaction between carboxylate ligands and the metal core, and (2) strong n → π* interactions between these two adjacent carbonyl groups [89–91]. The clustering of carbonyl groups could lead to the extension of conjugation and efficient delocalization of electrons in overlapped C=O double bonds between neighboring carbonyl groups. In addition, the molecular conformation becomes more rigid, which could largely restrict the intramolecular vibrations (RIV) and rotations (RIR). Finally, the photoluminescence was largely enhanced. We first proposed that the exact chromophore of metal NCs for aggregation-induced emission (AIE) mechanics was originated from the clustering carbonyl groups. This interpretation gives a completely new insight into the luminescent principle of MNCs, but we cannot completely preclude the role of the metal core for the PL emission.

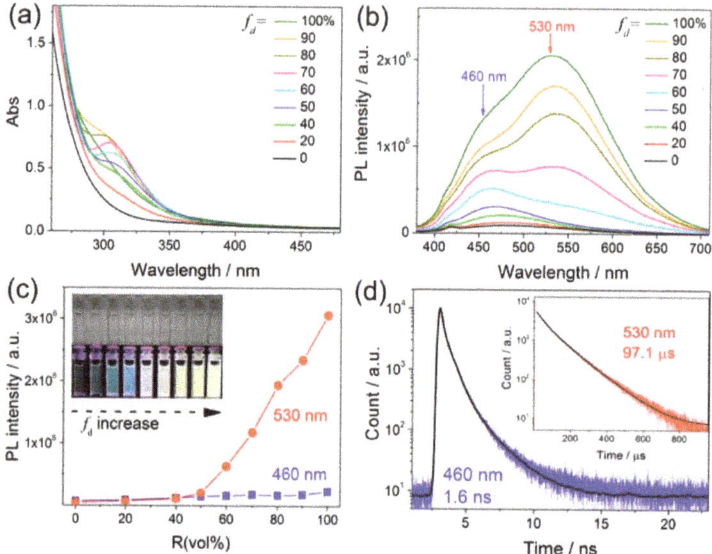

Figure 13. (**a**) UV–vis absorption and (**b**) photoluminescence spectra of polymethyl vinyl ether-alt-maleic acid-capped silver nanoclusters (PMVEM-Ag NCs) in the varied volume fraction f_d of DMSO in the mixed solvent ($f_d = V_{DMSO}/V_{DMSO+water}$). Photoluminescence was excited at 365 nm. (**c**) Correlations of the emission intensities of the two peaks centered at approximately 460 and approximately 530 nm versus f_d. (Inset) Photographs of PMVEM-Ag NCs at different f_d under visible (top row) and UV (bottom row) light. (**d**) Time-resolved luminescence decay profiles of PMVEM-Ag NCs in DMSO measured at 460 and 530 nm (inset), respectively. Reprinted with permission from Ref. [88]. Copyright (2017) American Chemical Society.

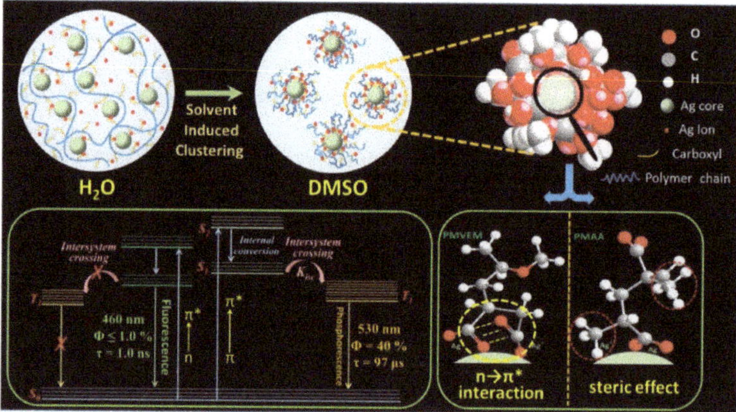

Figure 14. Schematic illustration of the solvent-induced clustering process of Ag NCs (top) depending on the unique molecular structure of the polymer used as a template for the synthesis of Ag NCs (bottom-right) and the energy-level structure of Ag NCs in water solution and DMSO solution (bottom-left). Reprinted with permission from Ref. [88]. Copyright (2017) American Chemical Society.

Very most recently, based on the complimentary characterizations of steady-state absorption, excitation, and time-resolved PL spectroscopy, we definitively identify that the induced p band intermediate state (PBIS) stems from the overlapping of p orbitals of the paired or more adjacent heteroatoms (O and S) from the surface-protected ligands on the metal NCs, which can be considered as a dark state at the metal–NCs interface to activate the triplet site of the surface chromophores by enhancing intersystem crossing [43]. Figure 15 showed that the water-soluble and oil-soluble MNCs with identical size synthesized by a ligand-exchange strategy (denoted as Au NCs@GSH and Au NCs@DT, respectively) exhibited the strong solvent-enhanced PL emission behavior, and that it is very interesting that the number and intensity of PL emission was strongly dependent on the type of used surface-protective ligands, even though the metal core remains intact. The more convincing and powerful evidence for the PBIS model comes from the luminescent mesoporous silica nanoparticles functionalized by nonluminescent organosilanes with amino and carbonyl groups free of any metals, which bears the same spectroscopic properties as metal NCs. These surface-modified MSNs with target organic functions showed a very strong and tunable photoluminescence emission due to the clustering of nonluminescent chromophores in the confined nanospace [92–99]. It is importantly noted that the selectively used organosilane in the functional groups contains heteroatoms with unpaired lone electrons, such as oxygen (O), nitrogen (N), and sulfur (S), as used protecting ligand molecules for the synthesis of metal NCs [100–102]. These nonluminescent functional groups assembled on a non-metallic surface generated the same photoluminescence emission with MNCs, as shown in Figure 16. We definitively confirmed that the pairing of surface ligands could serve as an exact chromophore for PL emission, which has nothing to do with the metal core.

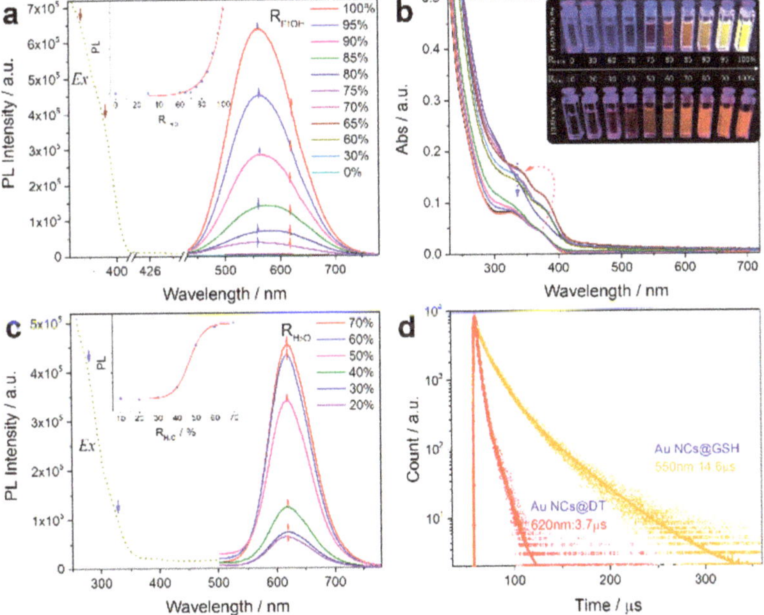

Figure 15. Solvent-induced ligand-dependent optical absorption and emissions. Photoemission and excitation spectra of water-soluble Au NCs@GSH (**a**) and oil-soluble Au NCs@DT (**c**) in mixed solvents with different volume fractions of R (Inset, relationship between the luminescence intensity and R (R_{EtOH} = Vol$_{EtOH}$/Vol$_{EtOH + H2O}$, R_{H2O} = Vol$_{H2O}$/Vol$_{EtOH + H2O}$), the spectra were recorded at 0.5 h after the sample preparation). (**b**) Ultraviolet-visible (UV–vis) absorption spectra of Au NCs@GSH in mixed solvents with different R_{EtOH}. Inset shows the digital photos of water-soluble Au NCs@GSH and oil-soluble Au NCs@DT in mixed solvents of ethanol and water with varied volume fractions of R_{EtOH} and R_{H2O} under UV light. (**d**) Time-resolved luminescence decay profiles of solvent-induced luminescent Au NCs@GSH and Au NCs@DT. Reprinted with permission from Ref. [43]. Copyright (2019) Springer Nature. GSH: glutathione.

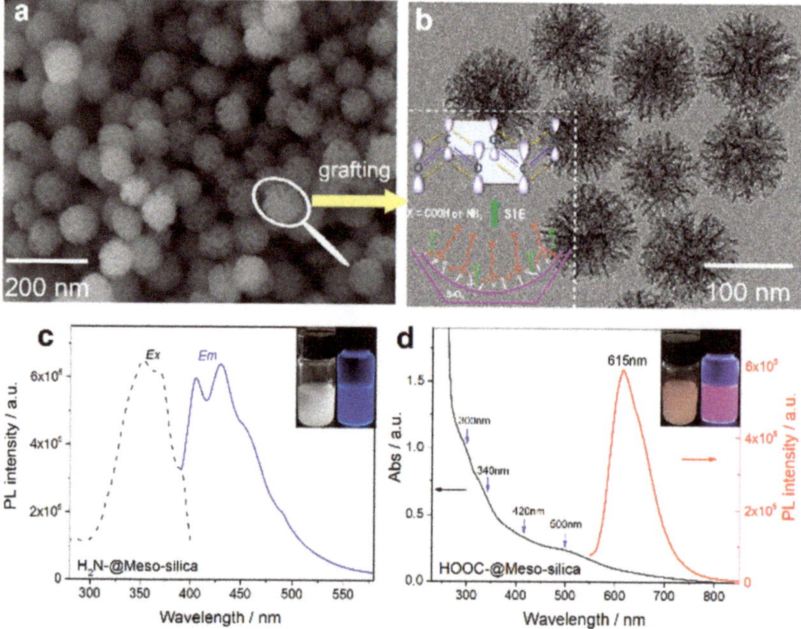

Figure 16. Ligand assembly in the mesoporous silica nanoparticles (MSNs) free of metals and their tunable luminescent properties. Scanning electron microscopy (SEM) (**a**) and TEM (**b**) images of as-synthesized fluorescent mesoporous silica nanoparticles. The inset shows the assembly of amino and carbonyl groups in the confined nanopores. (**c**) Excitation and emission spectra of aminopropyl-functionalized MSNs. (**d**) Absorption and emission spectra of propylsuccinic-functionalized MSNs. Reprinted with permission from Ref. [43]. Copyright (2019) Springer Nature.

A descriptive ligand-assembly-mediated interfacial p band intermediate state (PBIS) model was proposed to understand the origin of the PL emission of MNCs, as shown in Figure 17. At the nanoscale interface of the metal nanocluster core or in the confined nanomesopores, the adjacent ligands locally interact with each other to form Rydberg matter-like clusters by the overlapping of p-orbitals from O, N, S, and P on proximal carbonyl/thiol groups with high-energy lone-pair electrons [103,104]. The delocalization of the high-energy electrons in coupled p orbitals in ligand-directed molecular architectures produces a new overall lower energy state, the so-called PBIS, which acts as an intermediate (or dark) state to tune PL emissions by intersystem crossing where the energy gap between the singlet excited state and the intermediate p band state governs the direction and the extent of the electron transfer. Time-dependent density functional theory (TD-DFT) calculation indicated that the energy level of the p-band center is very sensitive to the local proximity ligand chromophores at heterogeneous interfaces, which further confirmed the validity of the PBIS model.

Based on this model, which is shown here as an atypical example, the abnormal polymorph-dependent emission wavelength phenomenon in anisotropic superlattices could be readily understood [45]. As illustrated in Figure 18, in the highly compact structure (nanoribbon), due to the strong intercalation between DT molecules in the layers, the d spacing between two DT molecules is larger than that in nanosheets; then, the overlapping or the interactions between paired or neighboring DT molecules are weakened, which answers that the blue-shift emission of the nanoribbons is due to the declined electron delocalization between adjacent DT molecules.

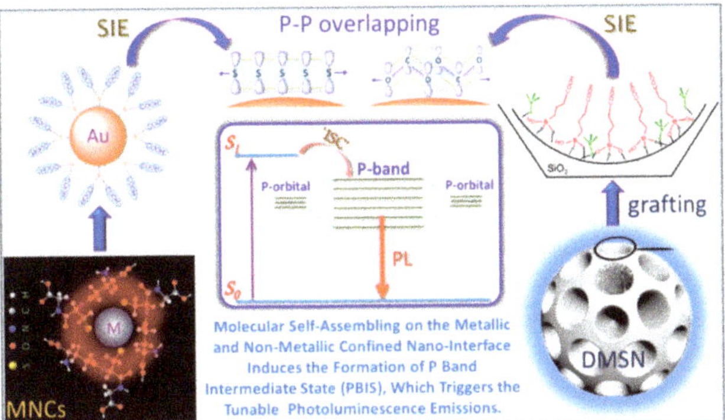

Figure 17. Ligand-assembly-mediated p band intermediate state (PBIS) dominates photoluminescence emission. Schematic illustration of the ligand exchange process and solvent-induced emission (SIE) properties of Au NCs (inset: the energy-level structure of Au NCs in water and ethanol mixed solution). The p band formed by the overlapping of p orbitals of electron-rich sulfur and oxygen heteroatoms of well-organized surface ligands is used as an intermediate state or dark state to tune the optoelectronic properties. Please see more details on the pioneering conceptual PBIS model. Reprinted with permission from Ref. [43]. Copyright (2019) Springer Nature.

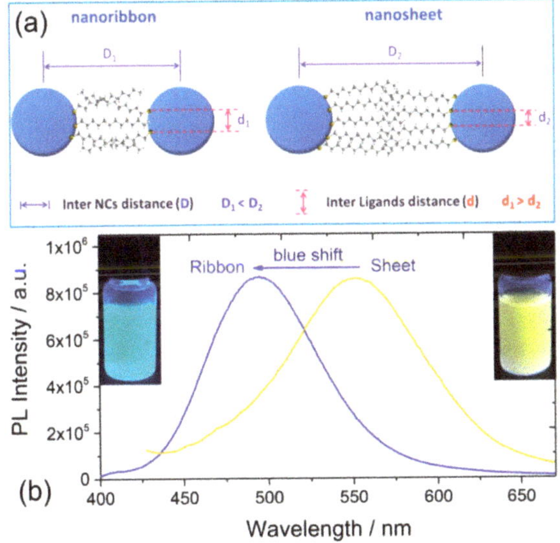

Figure 18. (**a**) Schematic structures of a Cu superlattice with different nanoribbon and nanosheet morphology. (**b**) Corresponding emission spectra of Cu NCs with nanoribbon and nanosheet morphology. The real distance between two thiol groups in the nanoribbon Cu superlattice is larger than that in the nanosheet Cu superlattice, indicating the loose packing of DT molecules on the metal core. The emission peak exhibits a blue shift from 550 nm to 490 nm when the morphology of the Cu superlattice changed from a nanosheet to nanoribbon, which implies a change of the packing model of DT molecules. Reprinted with permission from Ref. [43]. Copyright (2019) Springer Nature.

The PBIS model was also successful to explain the long-term debated and self-contradictory size-dependent and size-independent PL phenomena of MNCs: the smaller the metal nanoparticle size, the more coordinated unsaturated metal atoms are exposed with increased surface-to-volume rations, resulting in the strong binding and high surface coverage of surface coating ligands, consequently emitting enhanced PL intensity with low energy due to the maximum overlapping of p orbitals from surface ligands, where the MNCs provided an ideal nanoscale interface for the strong binding interactions of adsorbed molecules, for example, the surface-protective ligands. When the nanoparticle size is fixed, the intensity and wavelength of PL are dependent on ligand coverage or densities: a high Au-S coordination number (CN) and high surface coverage results in stronger PL emission at long wavelengths because of the close packing of surface ligands, whereas a low Au-S CN and a low surface coverage make weak PL emissions with high energy. Obviously, the PL emission of metal nanoclusters is strongly dependent on the close packing or self-assembly of targeted surface ligands; thus, the used template for MNCs must contain electron-rich heteroatoms, including oxygen (O), nitrogen (N), sulfur (S), and phosphorus (P).

4. Summary and Outlooks

Numerous and complicated superficial factors, such as the size and type of MNCs, the surface ligands, the valence state of surface metal, the self-assembly, and even the solvent effect affect the luminescent properties of metal nanoclusters, which interferes with the understanding of the luminescent nature of metal nanoclusters. Using metal NCs as a model system, by judiciously manipulating the delicate surface ligand interactions at the nanoscale interface together with a careful calculation of time-dependent density functional theory (TD-DFT), we proposed a completely new p band intermediate state (PBIS) model to elucidate the origin of PL emission of MNCs, which is completely different from the metal-centered quantum confinement mechanics (QCE) for the elucidation of PL emission of metal NCs. This mechanism could be universal to explain the PL emission nature of other quantum dot systems, including carbon and graphene nanodots, transition metal dichalcogenide (TMDC) nanomaterials, luminescent metal organic framework (MOFs), luminescent perovskites, and even organometallic complexes [105–110], because the overlapping of p orbitals at the confined nanointerface and nanospace could not be precluded [111]. Even though the PBIS model could qualitatively describe the PL properties, the exact dynamics and kinetics of electron transition of the excited state were not clear. In the near future, ultrafast time-resolved technology such as transient absorption and fluorescence up-conversion at the femtosecond scale will provide some reliable measurement tools to describe the relaxation and decay pathways of excited electrons. Most importantly, with respect to the paramount importance and universality of ligand and interface interaction at the nanoscale interface, the proof of concept of PBIS not only elucidates the fundamental physical principles driving nanoscale PL emission phenomena, but it also provides completely new insights to understand the interface-confined nanocatalysis on the molecule level. The presence of a new interface state with a low energy level could act as an extra channel to promote election transfer, which reduces the energy barrier of redox reaction, similar to the role of the transition state called in the heterogeneous catalysis [112,113]. At the same time, if the atomic orbitals of reactants interact with the PBIS, the binding strength of the reactant with the metal active site could be optimized (the classical Sabatier principle), which will improve the selectivity and lifetime of catalysts [114]. Of course, PBIS could be self-spontaneously formed between two adsorbed neighboring reactants at the heterogeneous nanoscale interface, which will also accelerate the chemical reaction depending the surface coverage. Hence, this significant conceptual advance will be of immediate interest to a broad readership of researchers in the nanocatalysis science.

Author Contributions: Conceptualization, T.-Q.Y. and K.Z.; writing—original draft preparation, T.-Q.Y., B.P., B.-Q.S., Y.-X.Z., J.-G.J., P.W. and K.Z.; writing—review and editing, T.-Q.Y., P.W. and K.Z.; supervision, P.W. and K.Z.; project administration, K.Z.; funding acquisition, P.W. and K.Z.; All authors have read and agreed to the published version of the manuscript.

Funding: This research was funded by the NSFC (21872053, 21573074 and 21872052), the Science and Technology Commission of Shanghai Municipality (19520711400), the CAS key laboratory of Low-Coal Conversion Science & Engineering (KLLCCSE-201702), and the JORISS program, the Postdoctoral Science Foundation of China (2018M640360).

Acknowledgments: K.Z. thanks ENS de Lyon for a temporary position as an invited professor in France. This paper is dedicated to Ming-Yuan He in the occasion of his 80th birthday.

Conflicts of Interest: The authors declare no conflict of interest.

References

1. Zheng, J.; Nicovich, P.R.; Dickson, R.M. Highly fluorescent noble-metal quantum dots. *Ann. Rev. Phys. Chem.* **2007**, *58*, 409–431. [CrossRef] [PubMed]

2. Diez, I.; Ras, R.H. Fluorescent silver nanoclusters. *Nanoscale* **2011**, *3*, 1963–1970. [CrossRef] [PubMed]

3. Zheng, J.; Zhou, C.; Yu, M.; Liu, J. Different sized luminescent gold nanoparticles. *Nanoscale* **2012**, *4*, 4073–4083. [CrossRef] [PubMed]

4. Yu, P.; Wen, X.; Toh, Y.-R.; Ma, X.; Tang, J. Fluorescent Metallic Nanoclusters: Electron Dynamics, Structure, and Applications. *Part. Part. Syst. Charact.* **2015**, *32*, 142–163. [CrossRef]

5. Higaki, T.; Li, Q.; Zhou, M.; Zhao, S.; Li, Y.; Li, S.; Jin, R. Toward the Tailoring Chemistry of Metal Nanoclusters for Enhancing Functionalities. *Acc. Chem. Res.* **2018**, *51*, 2764–2773. [CrossRef]

6. Nieto-Ortega, B.; Burgi, T. Vibrational Properties of Thiolate-Protected Gold Nanoclusters. *Acc. Chem. Res.* **2018**, *51*, 2811–2819. [CrossRef]

7. He, X.; Yam, V.W.W. Luminescent gold(I) complexes for chemosensing. *Coord. Chem. Rev.* **2011**, *255*, 2111–2123. [CrossRef]

8. Shang, L.; Dong, S.; Nienhaus, G.U. Ultra-small fluorescent metal nanoclusters: Synthesis and biological applications. *Nano Today* **2011**, *6*, 401–418. [CrossRef]

9. Choi, S.; Dickson, R.M.; Yu, J. Developing luminescent silver nanodots for biological applications. *Chem. Soc. Rev.* **2012**, *41*, 1867–1891. [CrossRef]

10. Guével, X.L. Recent Advances on the Synthesis of Metal Quantum Nanoclusters and Their Application for Bioimaging. *IEEE J. Top. Quant.* **2014**, *20*, 6801312.

11. Palmal, S.; Jana, N.R. Gold nanoclusters with enhanced tunable fluorescence as bioimaging probes. *WIREs Nanomed. Nanobiotechnol.* **2014**, *6*, 102–110. [CrossRef] [PubMed]

12. Zhang, L.; Wang, E. Metal nanoclusters: New fluorescent probes for sensors and bioimaging. *Nano Today* **2014**, *9*, 132–157. [CrossRef]

13. Chen, L.Y.; Wang, C.W.; Yuan, Z.; Chang, H.T. Fluorescent gold nanoclusters: Recent advances in sensing and imaging. *Anal. Chem.* **2015**, *87*, 216–229. [CrossRef]

14. Zheng, Y.; Lai, L.; Liu, W.; Jiang, H.; Wang, X. Recent advances in biomedical applications of fluorescent gold nanoclusters. *Adv. Colloid Interface Sci.* **2017**, *242*, 1–16. [CrossRef] [PubMed]

15. Li, C.; Chen, H.; Chen, B.; Zhao, G. Highly fluorescent gold nanoclusters stabilized by food proteins: From preparation to application in detection of food contaminants and bioactive nutrients. *Crit. Rev. Food Sci. Nutr.* **2018**, *58*, 689–699. [CrossRef] [PubMed]

16. Bonacic-Koutecky, V.; Kulesza, A.; Gell, L.; Mitric, R.; Antoine, R.; Bertorelle, F.; Hamouda, R.; Rayane, D.; Broyer, M.; Tabarin, T.; et al. Silver cluster-biomolecule hybrids: From basics towards sensors. *Phys. Chem. Chem. Phys.* **2012**, *14*, 9282–9290. [CrossRef]

17. Goswami, N.; Yao, Q.; Luo, Z.; Li, J.; Chen, T.; Xie, J. Luminescent Metal Nanoclusters with Aggregation-Induced Emission. *J. Phys. Chem. Lett.* **2016**, *7*, 962–975. [CrossRef]

18. New, S.Y.; Lee, S.T.; Su, X.D. DNA-templated silver nanoclusters: Structural correlation and fluorescence modulation. *Nanoscale* **2016**, *8*, 17729–17746. [CrossRef]

19. Li, D.; Chen, Z.; Mei, X. Fluorescence enhancement for noble metal nanoclusters. *Adv. Colloid Interface Sci.* **2017**, *250*, 25–39. [CrossRef]

20. Musnier, B.; Wegner, K.D.; Comby-Zerbino, C.; Trouillet, V.; Jourdan, M.; Hausler, I.; Antoine, R.; Coll, J.L.; Resch-Genger, U.; Le Guevel, X. High photoluminescence of shortwave infrared-emitting anisotropic surface charged gold nanoclusters. *Nanoscale* **2019**, *11*, 12092–12096. [CrossRef]

21. Peric, M.; Sanader Marsic, Z.; Russier-Antoine, I.; Fakhouri, H.; Bertorelle, F.; Brevet, P.F.; le Guevel, X.; Antoine, R.; Bonacic-Koutecky, V. Ligand shell size effects on one- and two-photon excitation fluorescence of zwitterion functionalized gold nanoclusters. *Phys. Chem. Chem. Phys.* **2019**, *21*, 23916–23921. [CrossRef] [PubMed]

22. Ishida, Y.; Corpuz, R.D.; Yonezawa, T. Matrix Sputtering Method: A Novel Physical Approach for Photoluminescent Noble Metal Nanoclusters. *Acc. Chem. Res.* **2017**, *50*, 2986–2995. [CrossRef] [PubMed]

23. Tao, Y.; Li, M.; Ren, J.; Qu, X. Metal nanoclusters: Novel probes for diagnostic and therapeutic applications. *Chem. Soc. Rev.* **2015**, *44*, 8636–8663. [CrossRef] [PubMed]

24. Yu, H.; Rao, B.; Jiang, W.; Yang, S.; Zhu, M. The photoluminescent metal nanoclusters with atomic precision. *Coord. Chem. Rev.* **2019**, *378*, 595–617. [CrossRef]

25. Kang, X.; Zhu, M. Tailoring the photoluminescence of atomically precise nanoclusters. *Chem. Soc. Rev.* **2019**, *48*, 2422–2457. [CrossRef] [PubMed]

26. Zheng, J.; Zhang, C.; Dickson, R.M. Highly Fluorescent, Water-Soluble, Size-Tunable Gold Quantum Dots. *Phys. Rev. Lett.* **2004**, *93*, 077402. [CrossRef]

27. LEE, T.-H.; Gonzalez, J.I.; Zheng, J.; Dickson, R.M. Single-Molecule Optoelectronics. *Acc. Chem. Res.* **2005**, *38*, 534–541. [CrossRef]

28. Jin, R. Quantum sized, thiolate-protected gold nanoclusters. *Nanoscale* **2010**, *2*, 343–362. [CrossRef]

29. Mooradian, A. Photoluminescence of Metals. *Phys. Rev. Lett.* **1969**, *22*, 185–187. [CrossRef]

30. Link, S.; Beeby, A.; FitzGerald, S.; El-Sayed, M.A.; Schaaff, T.G.; Whetten, R.L. Visible to Infrared Luminescence from a 28-Atom Gold Cluster. *J. Phys. Chem. B* **2002**, *106*, 3410–3415. [CrossRef]

31. Forward, J.M.; Bohmann, D.; John, P.; Fackler, J.; Staples, R.J. Luminescence Studies of Gold(I) Thiolate Complexes. *Inorg. Chem.* **1995**, *34*, 6330–6336. [CrossRef]

32. Cha, S.-H.; Kim, J.-U.; Kim, K.-H.; Lee, J.-C. Preparation and Photoluminescent Properties of Gold(I)-Alkanethiolate Complexes Having Highly Ordered Supramolecular Structures. *Chem. Mater.* **2007**, *19*, 6297–6303. [CrossRef]

33. Wu, Z.; Jin, R. On the ligand's role in the fluorescence of gold nanoclusters. *Nano Lett.* **2010**, *10*, 2568–2573. [CrossRef] [PubMed]

34. Lin, C.A.J.; Li, J.K.; Lee, C.H.; Shen, J.L.; Hsieh, J.T.; Chan, W.H.; Wang, H.-H.; Yeh, H.I.; Chang, W.H. Synthesis of Fluorescent Metallic Nanoclusters towards Biomedical Application: Recent Progress and Present Challenges. *J. Med. Biol. Eng.* **2009**, *29*, 276–283.

35. Jadzinsky, P.D.; Calero, G.; Ackerson, C.J.; Bushnell, D.A.; Kornberg, R.D. Structure of a Thiol Monolayer–Protected Gold Nanoparticle at 1.1 Å Resolution. *Science* **2007**, *318*, 430–433. [CrossRef]

36. Zhu, M.; Aikens, C.M.; Hollander, F.J.; Schatz, G.C.; Jin, R. Correlating the Crystal Structure of A Thiol-Protected Au_{25} Cluster and Optical Properties. *J. Am. Chem. Soc.* **2008**, *130*, 5883–5885. [CrossRef]

37. Yang, H.; Wang, Y.; Lei, J.; Shi, L.; Wu, X.; Makinen, V.; Lin, S.; Tang, Z.; He, J.; Hakkinen, H.; et al. Ligand-stabilized $Au_{13}Cu_x$ (x = 2, 4, 8) bimetallic nanoclusters: Ligand engineering to control the exposure of metal sites. *J. Am. Chem. Soc.* **2013**, *135*, 9568–9571. [CrossRef]

38. Aikens, C.M. Electronic and Geometric Structure, Optical Properties, and Excited State Behavior in Atomically Precise Thiolate-Stabilized Noble Metal Nanoclusters. *Acc. Chem. Res.* **2018**, *51*, 3065–3073. [CrossRef]

39. Li, Q.; Zhou, M.; So, W.Y.; Huang, J.; Li, M.; Kauffman, D.R.; Cotlet, M.; Higaki, T.; Peteanu, L.A.; Shao, Z.; et al. A Mono-cuboctahedral Series of Gold Nanoclusters: Photoluminescence Origin, Large Enhancement, Wide Tunability, and Structure-Property Correlation. *J. Am. Chem. Soc.* **2019**, *141*, 5314–5325. [CrossRef]

40. Khatun, E.; Bodiuzzaman, M.; Sugi, K.S.; Chakraborty, P.; Paramasivam, G.; Dar, W.A.; Ahuja, T.; Antharjanam, S.; Pradeep, T. Confining an Ag_{10} Core in an Ag_{12} Shell: A Four-Electron Superatom with Enhanced Photoluminescence upon Crystallization. *ACS Nano* **2019**, *13*, 5753–5759. [CrossRef]

41. Jia, X.; Li, J.; Wang, E. Supramolecular self-assembly of morphology-dependent luminescent Ag nanoclusters. *Chem. Commun.* **2014**, *50*, 9565–9568. [CrossRef]

42. Bertorelle, F.; Russier-Antoine, I.; Calin, N.; Comby-Zerbino, C.; Bensalah-Ledoux, A.; Guy, S.; Dugourd, P.; Brevet, P.F.; Sanader, Z.; Krstic, M.; et al. $Au_{10}(SG)_{10}$: A Chiral Gold Catenane Nanocluster with Zero Confined Electrons. Optical Properties and First-Principles Theoretical Analysis. *J. Phys. Chem. Lett.* **2017**, *8*, 1979–1985. [CrossRef] [PubMed]

43. Yang, T.; Shan, B.; Huang, F.; Yang, S.; Peng, B.; Yuan, E.; Wu, P.; Zhang, K. P band intermediate state (PBIS) tailors photoluminescence emission at confined nanoscale interface. *Commun. Chem.* **2019**, *2*, 132. [CrossRef]

44. Zhou, C.; Sun, C.; Yu, M.; Qin, Y.; Wang, J.; Kim, M.; Zheng, J. Luminescent Gold Nanoparticles with Mixed Valence States Generated from Dissociation of Polymeric Au(I) Thiolates. *J. Phys. Chem. C* **2010**, *114*, 7727–7732. [CrossRef]

45. Wu, Z.; Liu, J.; Gao, Y.; Liu, H.; Li, T.; Zou, H.; Wang, Z.; Zhang, K.; Wang, Y.; Zhang, H.; et al. Assembly-Induced Enhancement of Cu Nanoclusters Luminescence with Mechanochromic Property. *J. Am. Chem. Soc.* **2015**, *137*, 12906–12913. [CrossRef] [PubMed]

46. Benito, Q.; Le Goff, X.F.; Maron, S.; Fargues, A.; Garcia, A.; Martineau, C.; Taulelle, F.; Kahlal, S.; Gacoin, T.; Boilot, J.P.; et al. Polymorphic copper iodide clusters: Insights into the mechanochromic luminescence properties. *J. Am. Chem. Soc.* **2014**, *136*, 11311–11320. [CrossRef] [PubMed]

47. Wilcoxon, J.P.; Martin, J.E.; Parsapour, F.; Wiedenman, B.; Kelley, D.F. Photoluminescence from nanosize gold clusters. *J. Chem. Phys.* **1998**, *108*, 9137–9143. [CrossRef]

48. Yan, J.; Teo, B.K.; Zheng, N. Surface Chemistry of Atomically Precise Coinage-Metal Nanoclusters: From Structural Control to Surface Reactivity and Catalysis. *Acc. Chem. Res.* **2018**, *51*, 3084–3093. [CrossRef]

49. Harbich, W.; Fedrigo, S.; Buttet, J.; Lindsay, D.M. Deposition of mass selected gold clusters in solid krypton. *J. Chem. Phys.* **1992**, *96*, 8104–8108. [CrossRef]

50. Fedrigo, S.; Harbich, W.; Buttet, J. Optical response of Ag_2, Ag_3, Au_2, and Au_3 in argon matrices. *J. Chem. Phys.* **1993**, *99*, 5712–5717. [CrossRef]

51. Zheng, J.; Dickson, R.M. Individual Water-Soluble Dendrimer-Encapsulated Silver Nanodot Fluorescence. *J. Am. Chem. Soc.* **2002**, *124*, 13982–13983. [CrossRef] [PubMed]

52. Zheng, J.; Petty, J.T.; Dickson, R.M. High Quantum Yield Blue Emission from Water-Soluble Au8 Nanodots. *J. Am. Chem. Soc.* **2003**, *125*, 7780–7781. [CrossRef] [PubMed]

53. Angel, L.A.; Majors, L.T.; Dharmaratne, A.C.; Dass, A. Ion Mobility Mass Spectrometry of $Au_{25}(SCH_2CH_2Ph)_{18}$ Nanoclusters. *ACS Nano* **2010**, *4*, 4691–4700. [CrossRef] [PubMed]

54. Jin, R.; Qian, H.; Wu, Z.; Zhu, Y.; Zhu, M.; Mohanty, A.; Garg, N. Size Focusing: A Methodology for Synthesizing Atomically Precise Gold Nanoclusters. *J. Phys. Chem. Lett.* **2010**, *1*, 2903–2910. [CrossRef]

55. Udaya Bhaskara Rao, T.; Pradeep, T. Luminescent Ag_7 and Ag_8 clusters by interfacial synthesis. *Angew. Chem. Int. Ed.* **2010**, *49*, 3925–3929. [CrossRef]

56. Luo, Z.; Yuan, X.; Yu, Y.; Zhang, Q.; Leong, D.T.; Lee, J.Y.; Xie, J. From aggregation-induced emission of Au(I)-thiolate complexes to ultrabright Au(0)@Au(I)-thiolate core-shell nanoclusters. *J. Am. Chem. Soc.* **2012**, *134*, 16662–16670. [CrossRef]

57. Liu, J.; Duchesne, P.N.; Yu, M.; Jiang, X.; Ning, X.; Vinluan, R.D., 3rd; Zhang, P.; Zheng, J. Luminescent Gold Nanoparticles with Size-Independent Emission. *Angew. Chem. Int. Ed.* **2016**, *55*, 8894–8898. [CrossRef]

58. Xie, J.; Zheng, Y.; Ying, J.Y. Protein-Directed Synthesis of Highly Fluorescent Gold Nanoclusters. *J. Am. Chem. Soc.* **2009**, *131*, 888–889. [CrossRef]

59. Wu, Z.; Wang, M.; Yang, J.; Zheng, X.; Cai, W.; Meng, G.; Qian, H.; Wang, H.; Jin, R. Well-defined nanoclusters as fluorescent nanosensors: A case study on $Au_{25}(SG)_{18}$. *Small* **2012**, *8*, 2028–2035. [CrossRef]

60. Brewer, L.; King, B.A.; Wang, J.L.; Meyer, B.; Moore, G.F. Absorption Spectrum of Silver Atoms in Solid Argon, Krypton, and Xenon. *J. Chem. Phys.* **1968**, *49*, 5209–5213. [CrossRef]

61. Simo, A.; Polte, J.; Pfander, N.; Vainio, U.; Emmerling, F.; Rademann, K. Formation mechanism of silver nanoparticles stabilized in glassy matrices. *J. Am. Chem. Soc.* **2012**, *134*, 18824–18833. [CrossRef]

62. Cremer, G.D.; Coutiño-Gonzalez, E.; Roeffaers, M.B.J.; Moens, B.; Ollevier, J.; Auweraer, M.V.d.; Schoonheydt, R.; Jacobs, P.A.; Schryver, F.C.D.; Hofkens, J.; et al. Characterization of Fluorescence in Heat-Treated Silver-Exchanged Zeolites. *J. Am. Chem. Soc.* **2009**, *131*, 3049–3056. [CrossRef]

63. Grandjean, D.; Coutiño-Gonzalez, E.; Cuong, N.T.; Fron, E.; Baekelant, W.; Aghakhani, S.; Schlexer, P.; D'Acapito, F.; Banerjee, D.; Roeffaers, M.B.J.; et al. Origin of the bright photoluminescence of few-atom silver clusters confined in LTA zeolites. *Science* **2018**, *361*, 686–690. [CrossRef]

64. Fenwick, O.; Coutiño-Gonzalez, E.; Grandjean, D.; Baekelant, W.; Richard, F.; Bonacchi, S.; De Vos, D.; Lievens, P.; Roeffaers, M.; Hofkens, J.; et al. Tuning the energetics and tailoring the optical properties of silver clusters confined in zeolites. *Nature Mater.* **2016**, *15*, 1017–1022. [CrossRef]

65. Richards, C.I.; Choi, S.; Hsiang, J.-C.; Antoku, Y.; Vosch, T.; Bongiorno, A.; Tzeng, Y.-L.; Dickson, R.M. Oligonucleotide-Stabilized Ag Nanocluster Fluorophores. *J. Am. Chem. Soc.* **2008**, *130*, 5038–5039. [CrossRef]

66. Yeh, H.C.; Sharma, J.; Han, J.J.; Martinez, J.S.; Werner, J.H. A DNA-silver nanocluster probe that fluoresces upon hybridization. *Nano Lett.* **2010**, *10*, 3106–3110. [CrossRef]

67. Pyo, K.; Thanthirige, V.D.; Kwak, K.; Pandurangan, P.; Ramakrishna, G.; Lee, D. Ultrabright Luminescence from Gold Nanoclusters: Rigidifying the Au(I)-Thiolate Shell. *J. Am. Chem. Soc.* **2015**, *137*, 8244–8250. [CrossRef]

68. Londoño-Larrea, P.; Vanegas, J.P.; Cuaran-Acosta, D.; Zaballos-García, E.; Pérez-Prieto, J. Water-Soluble Naked Gold Nanoclusters Are Not Luminescent. *Chem. Eur. J.* **2017**, *23*, 8137–8141. [CrossRef]

69. Zhu, M.; Eckenhoff, W.T.; Pintauer, T.; Jin, R. Conversion of Anionic [$Au_{25}(SCH_2CH_2Ph)_{18}$]$^-$ Cluster to Charge Neutral Cluster via Air Oxidation. *J. Phys. Chem. C* **2008**, *112*, 14221–14224. [CrossRef]

70. Negishi, Y.; Chaki, N.K.; Shichibu, Y.; Whetten, R.L.; Tsukuda, T. Origin of Magic Stability of Thiolated Gold Clusters: A Case Study on $Au_{25}(SC_6H_{13})_{18}$. *J. Am. Chem. Soc.* **2007**, *129*, 11322–11323. [CrossRef]

71. Shang, L.; Dong, S. Facile preparation of water-soluble fluorescent silver nanoclusters using a polyelectrolyte template. *Chem. Commun.* **2008**, 1088–1090. [CrossRef]

72. Xie, Z.; Sun, P.; Wang, Z.; Li, H.; Yu, L.; Sun, D.; Chen, M.; Bi, Y.; Xin, X.; Hao, J. Metal-Organic Gels from Silver Nanoclusters with Aggregation-Induced Emission and Fluorescence-to-Phosphorescence Switching. *Angew. Chem. Int. Ed.* **2019**, *58*, 1–6.

73. Luo, J.; Xie, Z.; Lam, J.W.Y.; Cheng, L.; Tang, B.Z.; Chen, H.; Qiu, C.; Kwok, H.S.; Zhan, X.; Liu, Y.; et al. Aggregation-induced emission of 1-methyl-1,2,3,4,5-pentaphenylsilole. *Chem. Commun.* **2001**, 1740–1741. [CrossRef] [PubMed]

74. Chen, Y.; Lam, J.W.Y.; Kwok, R.T.K.; Liu, B.; Tang, B.Z. Aggregation-induced emission: Fundamental understanding and future developments. *Mater. Horiz.* **2019**, *6*, 428–433. [CrossRef]

75. Saigusa, H.; Lim, E.C. Excited-State Dynamics of Aromatic Clusters: Correlation between Exciton Interactions and Excimer Formation Dynamics. *J. Phys. Chem.* **1995**, *99*, 15738–15747. [CrossRef]

76. Mei, J.; Leung, N.L.C.; Kwok, R.T.K.; Lam, J.W.Y.; Tang, B.Z. Aggregation-Induced Emission: Together We Shine, United We Soar! *Chem. Rev.* **2015**, *115*, 11718–11940. [CrossRef]

77. Jia, X.; Yang, X.; Li, J.; Li, D.; Wang, E. Stable Cu nanoclusters: From an aggregation-induced emission mechanism to biosensing and catalytic applications. *Chem. Commun.* **2014**, *50*, 237–239. [CrossRef]

78. Liu, Y.; Yao, D.; Zhang, H. Self-Assembly Driven Aggregation-Induced Emission of Copper Nanoclusters: A Novel Technology for Lighting. *ACS Appl. Mater. Interfaces* **2017**, *10*, 12071–12080. [CrossRef]

79. Jia, X.; Li, J.; Wang, E. Cu nanoclusters with aggregation induced emission enhancement. *Small* **2013**, *9*, 3873–3879. [CrossRef]

80. Wu, Z.; Dong, C.; Li, Y.; Hao, H.; Zhang, H.; Lu, Z.; Yang, B. Self-assembly of Au_{15} into single-cluster-thick sheets at the interface of two miscible high-boiling solvents. *Angew. Chem. Int. Ed.* **2013**, *52*, 9952–9955. [CrossRef]

81. Ai, L.; Liu, Z.; Zhou, D.; Liu, J.; Zou, H.; Wu, Z.; Liu, Y.; Zhang, H.; Yang, B. Copper inter-nanoclusters distance-modulated chromism of self-assembly induced emission. *Nanoscale* **2017**, *9*, 18845–18854. [CrossRef] [PubMed]

82. Liu, J.; Wu, Z.; Tian, Y.; Li, Y.; Ai, L.; Li, T.; Zou, H.; Liu, Y.; Zhang, X.; Zhang, H.; et al. Engineering the Self-Assembly Induced Emission of Cu Nanoclusters by Au(I) Doping. *ACS Appl. Mater. Interfaces* **2017**, *9*, 24899–24907. [CrossRef] [PubMed]

83. Wu, Z.; Liu, H.; Li, T.; Liu, J.; Yin, J.; Mohammed, O.F.; Bakr, O.M.; Liu, Y.; Yang, B.; Zhang, H. Contribution of Metal Defects in the Assembly Induced Emission of Cu Nanoclusters. *J. Am. Chem. Soc.* **2017**, *139*, 4318–4321. [CrossRef] [PubMed]

84. White-Morris, R.L.; Olmstead, M.M.; Balch, A.L. Orange Luminescence and Structural Properties of Three Isostructural Halocyclohexylisonitrilegold(I) Complexes. *Inorg. Chem.* **2003**, *42*, 6741–6748. [CrossRef]

85. Schmidbaur, H. The fascinating implications of new results in gold chemistry. *Gold Bull.* **1990**, *23*, 11–21. [CrossRef]

86. Chen, Y.; Yang, T.; Pan, H.; Yuan, Y.; Chen, L.; Liu, M.; Zhang, K.; Zhang, S.; Wu, P.; Xu, J. Photoemission mechanism of water-soluble silver nanoclusters: Ligand-to-metal-metal charge transfer vs. strong coupling between surface plasmon and emitters. *J. Am. Chem. Soc.* **2014**, *136*, 1686–1689. [CrossRef]

87. Yang, T.; Dai, S.; Tan, H.; Zong, Y.; Liu, Y.; Chen, J.; Zhang, K.; Wu, P.; Zhang, S.; Xu, J.; et al. Mechanism of Photoluminescence in Ag Nanoclusters: Metal-Centered Emission versus Synergistic Effect in Ligand-Centered Emission. *J. Phys. Chem. C* **2019**, *123*, 18638–18645. [CrossRef]

88. Yang, T.; Dai, S.; Yang, S.; Chen, L.; Liu, P.; Dong, K.; Zhou, J.; Chen, Y.; Pan, H.; Zhang, S.; et al. Interfacial Clustering-Triggered Fluorescence-Phosphorescence Dual Solvoluminescence of Metal Nanoclusters. *J. Phys. Chem. Lett.* **2017**, *8*, 3980–3985. [CrossRef]

89. Newberry, R.W.; Orke, S.J.; Raines, R.T. n→pi* Interactions Are Competitive with Hydrogen Bonds. *Org. Lett.* **2016**, *18*, 3614–3617. [CrossRef]

90. Kilgore, H.R.; Raines, R.T. n→π* Interactions Modulate the Properties of Cysteine Residues and Disulfide Bonds in Proteins. *J. Am. Chem. Soc.* **2018**, *140*, 17606–17611. [CrossRef]

91. León, I.; Alonso, E.R.; Cabezas, C.; Mata, S.; Alonso, J.L. Unveiling the n→π* interactions in dipeptides. *Commun. Chem.* **2019**, *2*, 3. [CrossRef]

92. Zhang, K.; Xu, L.L.; Jiang, J.G.; Calin, N.; Lam, K.F.; Zhang, S.J.; Wu, H.H.; Wu, G.D.; Albela, B.; Bonneviot, L.; et al. Facile large-scale synthesis of monodisperse mesoporous silica nanospheres with tunable pore structure. *J. Am. Chem. Soc.* **2013**, *135*, 2427–2430. [CrossRef] [PubMed]

93. Yu, Y.J.; Xing, J.L.; Pang, J.L.; Jiang, S.H.; Lam, K.F.; Yang, T.Q.; Xue, Q.S.; Zhang, K.; Wu, P. Facile synthesis of size controllable dendritic mesoporous silica nanoparticles. *ACS Appl. Mater. Interfaces* **2014**, *6*, 22655–22665. [CrossRef] [PubMed]

94. Liu, P.-C.; Yu, Y.-J.; Peng, B.; Ma, S.-Y.; Ning, T.-Y.; Shan, B.-Q.; Yang, T.-Q.; Xue, Q.-S.; Zhang, K.; Wu, P. A dual-templating strategy for the scale-up synthesis of dendritic mesoporous silica nanospheres. *Green Chem.* **2017**, *19*, 5575–5581. [CrossRef]

95. Zhang, K.; Yang, T.Q.; Shan, B.Q.; Liu, P.C.; Peng, B.; Xue, Q.S.; Yuan, E.H.; Wu, P.; Albela, B.; Bonneviot, L. Dendritic and core-shell-corona mesoporous sister nanospheres from polymer-surfactant-silica self-entanglement. *Chem. Eur. J.* **2018**, *24*, 478–486. [CrossRef]

96. Shan, B.-Q.; Xing, J.-L.; Yang, T.-Q.; Peng, B.; Hao, P.; Zong, Y.-X.; Chen, X.-Q.; Xue, Q.-S.; Zhang, K.; Wu, P. One-pot co-condensation strategy for dendritic mesoporous organosilica nanospheres with fine size and morphology control. *CrystEngComm* **2019**, *21*, 4030–4035. [CrossRef]

97. Peng, B.; Zong, Y.-X.; Nie, M.-Z.; Shan, B.-Q.; Yang, T.-Q.; Hao, P.; Ma, S.-Y.; Lam, K.-F.; Zhang, K. Interfacial charge shielding directs the synthesis of dendritic mesoporous silica nanospheres by a dual-templating approach. *New J. Chem.* **2019**, *43*, 15777–15784. [CrossRef]

98. Yang, T.-Q.; Ning, T.-Y.; Peng, B.; Shan, B.-Q.; Zong, Y.-X.; Hao, P.; Yuan, E.-H.; Chen, Q.-M.; Zhang, K. Interfacial electron transfer promotes photo-catalytic reduction of 4-nitrophenol by Au/Ag_2O nanoparticles confined in dendritic mesoporous silica nanospheres. *Catal. Sci. Technol.* **2019**, *9*, 5786–5792. [CrossRef]

99. Zhang, K.; Yuan, N.-H.; Xu, L.-L.; Xue, Q.-S.; Luo, C.; Albela, B.; Bonneviot, L. A New Method for Synthesis of High-Quality MCM-48 Mesoporous Silicas and the Mode of Action of the Template. *Eur. J. Inorg. Chem.* **2012**, 4183–4189. [CrossRef]

100. Zhang, K.; Albela, B.; He, M.-Y.; Wang, Y.; Bonneviot, L. Tetramethyl ammonium as masking agent for molecular stencil patterning in the confined space of the nano-channels of 2D hexagonal-templated porous silicas. *Phys. Chem. Chem. Phys.* **2009**, *11*, 2912–2921. [CrossRef]

101. Zhang, K.; Lam, K.F.; Albela, B.; Xue, T.; Khrouz, L.; Hou, Q.W.; Yuan, E.H.; He, M.Y.; Bonneviot, L. Mononuclear-dinuclear equilibrium of grafted copper complexes confined in the nanochannels of MCM-41 silica. *Chem. Eur. J.* **2011**, *17*, 14258–14266. [CrossRef] [PubMed]

102. Zhang, K.; Zhang, Y.; Hou, Q.-W.; Yuan, E.-H.; Jiang, J.-G.; Albela, B.; He, M.-Y.; Bonneviot, L. Novel synthesis and molecularly scaled surface hydrophobicity control of colloidal mesoporous silica. *Microporous Mesoporous Mater.* **2011**, *143*, 401–405. [CrossRef]

103. Campbell, E.K.; Holz, M.; Gerlich, D.; Maier, J.P. Laboratory confirmation of C_{60}^+ as the carrier of two diffuse interstellar bands. *Nature* **2015**, *523*, 322–323. [CrossRef] [PubMed]

104. Holmlid, L. The diffuse interstellar band carriers in interstellar space: All intense bands calculated from He doubly excited states embedded in Rydberg Matter. *Mon. Not. R. Astron. Soc.* **2008**, *384*, 764–774. [CrossRef]

105. Röhr, J.A.; Sá, J.; Konezny, S.J. The role of adsorbates in the green emission and conductivity of zinc oxide. *Commun. Chem.* **2019**, *2*, 52. [CrossRef]

106. Liu, L.; Wu, X.; Wang, L.; Xu, X.; Gan, L.; Si, Z.; Li, J.; Zhang, Q.; Liu, Y.; Zhao, Y.; et al. Atomic palladium on graphitic carbon nitride as a hydrogen evolution catalyst under visible light irradiation. *Commun. Chem.* **2019**, *2*, 18. [CrossRef]

107. Fisher, A.A.E.; Osborne, M.A.; Day, I.J.; Lucena Alcalde, G. Measurement of ligand coverage on cadmium selenide nanocrystals and its influence on dielectric dependent photoluminescence intermittency. *Commun. Chem.* **2019**, *2*, 63. [CrossRef]

108. Bubeck, C.; Widenmeyer, M.; Richter, G.; Coduri, M.; Goering, E.; Yoon, S.; Weidenkaff, A. Tailoring of an unusual oxidation state in a lanthanum tantalum(IV) oxynitride via precursor microstructure design. *Commun. Chem.* **2019**, *2*, 134. [CrossRef]

109. Wakasa, M.; Yago, T.; Sonoda, Y.; Katoh, R. Structure and dynamics of triplet-exciton pairs generated from singlet fission studied via magnetic field effects. *Commun. Chem.* **2018**, *1*, 9. [CrossRef]

110. Noblet, T.; Dreesen, L.; Boujday, S.; Méthivier, C.; Busson, B.; Tadjeddine, A.; Humbert, C. Semiconductor quantum dots reveal dipolar coupling from exciton to ligand vibration. *Commun. Chem.* **2018**, *1*, 76. [CrossRef]

111. Paris, C.; Floris, A.; Aeschlimann, S.; Neff, J.; Kling, F.; Kühnle, A.; Kantorovich, L. Kinetic control of molecular assembly on surfaces. *Commun. Chem.* **2018**, *1*, 66. [CrossRef]

112. Jin, C.; Ma, L.; Sun, W.; Han, P.; Tan, X.; Wu, H.; Liu, M.; Jin, H.; Wu, Z.; Wei, H.; et al. Single-atom nickel confined nanotube superstructure as support for catalytic wet air oxidation of acetic acid. *Commun. Chem.* **2019**, *2*, 135. [CrossRef]

113. Zhao, Q.; Fu, L.; Jiang, D.; Ouyang, J.; Hu, Y.; Yang, H.; Xi, Y. Nanoclay-modulated oxygen vacancies of metal oxide. *Commun. Chem.* **2019**, *2*, 11. [CrossRef]

114. Tabor, E.; Lemishka, M.; Sobalik, Z.; Mlekodaj, K.; Andrikopoulos, P.C.; Dedecek, J.; Sklenak, S. Low-temperature selective oxidation of methane over distant binuclear cationic centers in zeolites. *Commun. Chem.* **2019**, *2*, 71. [CrossRef]

 *nanomaterials*

Review

Heterogeneous Cross-Coupling over Gold Nanoclusters

Quanquan Shi [1,2], Zhaoxian Qin [2], Hui Xu [1] and Gao Li [2,*

[1] College of Science, Inner Mongolia Agricultural University, Hohhot 010018, China; qqshi@dicp.ac.cn (Q.S.); yqfxxuhui@163.com (H.X.)
[2] State Key Laboratory of Catalysis, Dalian Institute of Chemical Physics, Chinese Academy of Sciences, Dalian 116023, China; qinzhaoxian@dicp.ac.cn
* Correspondence: gaoli@dicp.ac.cn

Received: 23 April 2019; Accepted: 12 May 2019; Published: 1 June 2019

Abstract: Au clusters with the precise numbers of gold atoms, a novel nanogold material, have recently attracted increasing interest in the nanoscience because of very unique and unexpected properties. The unique interaction and electron transfer between gold clusters and reactants make the clusters promising catalysts during organic transformations. The Au_nL_m nanoclusters (where L represents organic ligands and n and m mean the number of gold atoms and ligands, respectively) have been well investigated and developed for selective oxidation, hydrogenation, photo-catalysis, and so on. These gold clusters possess unique frameworks, providing insights into the catalytic processes and an excellent arena to correlate the atomic frameworks with their intrinsic catalytic properties and to further investigate the tentative reaction mechanisms. This review comprehensively summarizes the very latest advances in the catalytic applications of the Au nanoclusters for the C−C cross-coupling reactions, e.g., Ullmann, Sonogashira, Suzuki cross-couplings, and A^3−coupling reactions. It is found that the proposed catalytically active sites are associated with the exposure of gold atoms on the surface of the metal core when partial capping organic ligands are selectively detached under the reaction conditions. Finally, the tentative catalytic mechanisms over the ligand-capped Au nanoclusters and the relationship of structure and catalytic performances at the atomic level using computational methods are explored in detail.

Keywords: gold nanocluster; cross-coupling; Ullmann hetero-coupling; Sonogashira coupling; Suzuki coupling; A^3−coupling; catalytic mechanism; ligand removal

---

1. Introduction

Since the work of Haruta's group in the late 1980s [1], supported gold nanoparticles with a particle size in the range of 3–20 nm have played a central role in a variety of reactions such as selective oxidation [2–5], hydrogenation [6–8], and photocatalysis [9,10]. These conventional gold catalysts have been realized via deposition–precipitation and co-precipitation impregnation with oxides by controlling the pH of the synthetic system. These obtained gold nanoparticles are usually polydisperse, which is a major issue in fundamental catalysis and investigations [11]. For example, the size-hierarchy of Au nanoparticles are often averaged out in polydispersion. It is difficult to correlate the relationship between the catalytic properties and the structure of the nanoparticles. Therefore, developing nanostructured catalysts with specific morphology (e.g., nanosheet, nanocube, and nanorod) is highly desirable to overcome this issue, further promoting the development of crystal-identified model catalysts [12,13].

On the other hand, the remarkable developments in the synthesis of atomically precise gold nanoclusters have been achieved in recent decades, which opens a new burgeoning area in nanoscience [14,15]. These gold nanoclusters are comprised of a few dozen to a few hundred gold atoms and protecting organic ligands (e.g., thiolate, phosphine, and alkyne), and the size is ultra-small, usually ca. 0.6–2 nm. Some gold nanoclusters are identifiable by X-ray crystallography technology. It offers a big opportunity for in-depth understanding of the relationship of the catalytic properties and the active-site structure at the atomic level [16].

It is observed that these gold nanoclusters exhibit good catalytic performance in heterogeneous catalysis (e.g., selective oxidation and hydrogenation) [16–18]. Sometime, the gold clusters show better catalytic behavior (e.g., activity and product selectivity) than the corresponding Au nanoparticles, because of their high surface-to-volume ratio (reaches up to ~100%), surface geometric effect (e.g., low-coordinated $Au^{\delta+}$ atoms, $0 < \delta < 1$), and the unique electronic properties and the quantum size effect. Furthermore, the protecting organic ligands can improve the product selectivity due to their electronic factors and steric hindrance and weak interaction (e.g., π-π interaction) between the reactants and cluster surface ligands during the catalysis process. In the recent decade, these gold nanoclusters exhibited good catalytic activity in the cross C–C couplings, e.g., Ullmann hetero-coupling, Suzuki and Sonogashira coupling, and A^3-coupling [19,20]. Traditionally, these catalyzed C–C coupling reactions are over the Cu, Pd, and Pt complexes and particles in the previous literatures [21–23].

In this review, we aim to provide an overview focused on the Au nanocluster-catalyzed coupling reactions, e.g., Ullmann hetero-coupling of Ar-I, Suzuki cross-coupling of $PhB(OH)_2$ and Ph-I (IB), Sonogashira cross-coupling of IB and Ph–C≡C–H (PA), and A^3-coupling (Scheme 1). The pathway of the selective detachment of the surface protecting ligands (under the reaction conditions), giving the catalytically active sites, is well discussed. Moreover, the proposed reaction pathway and mechanisms of these carbon–carbon coupling reactions were thoroughly summarized based on the precise framework of gold nanoclusters (e.g., two Au_{25}) as the mode theoretical calculations.

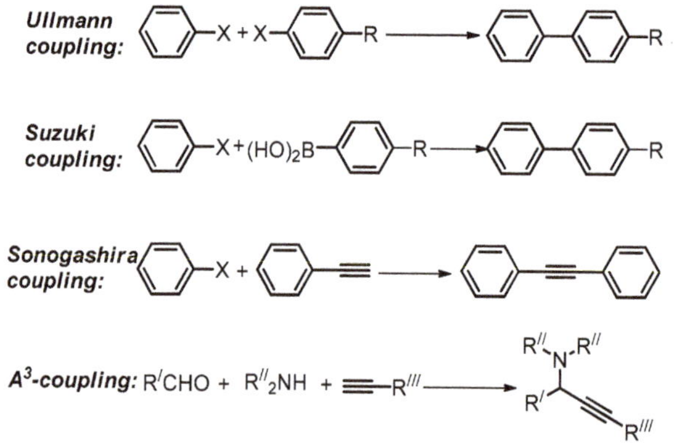

Scheme 1. Cross-coupling reactions over gold cluster in this review.

2. Activation of Ph–B(OH)₂, Ph–I, and C≡C–H Bonds over Au: Theoretical Simulation

The activation of the $Ph-B(OH)_2$, Ph–I, and C≡C–H bonds over the gold clusters plays an important role and step in the cross-coupling reactions, e.g., Ullmann hetero-coupling, Sonogashira coupling, Suzuki coupling, and A^3–coupling. The adsorption and activation process of the reactants over the different well-defined gold facets and sites are distinct [24–27]. The density functional theory (DFT) calculations should be a feasible and fast method to explore the activation of $Ph-B(OH)_2$, C–I, and C≡C–H bond by gold nanoclusters.

Firstly, the shape controlled Au nanoparticles (e.g., Au nanorod) with well-defined surfaces and morphologies are well investigated [28,29], providing a platform to establish site–activity relationships and to pursue the understanding of heterogeneous processes. Therefore, the gold nanorod is chosen for the theoretical simulation of the activation of Ph–I and C≡C–H bonds. DFT studies showed that the Ph-I reactant adsorbs onto the Au(100) and Au(111) with Au–I distance 2.95 and 2.86 Å. The C–I bond is more elongated on the Au(111) compared to Au(100) (2.17 Å vs. 2.13 Å) [26]; the C–I length of the free IB is 2.09 Å. Furthermore, the phenyl group is located on an Au atom. The iodine atom is strongly chemisorbed on the bridging and hollow sites. Regard to the activation of alkyne, the Ph–C≡C– fragment is also adsorbed on bridging and 3-fold hollow sites of Au(100) and Au(111). The coordination number of one Ph–C≡C– unit is 3 and 4 for Au(100) and Au(111) facets during the cross-coupling reaction, respectively, Figure 1, and the activation energy of cross-coupling is comparable [26].

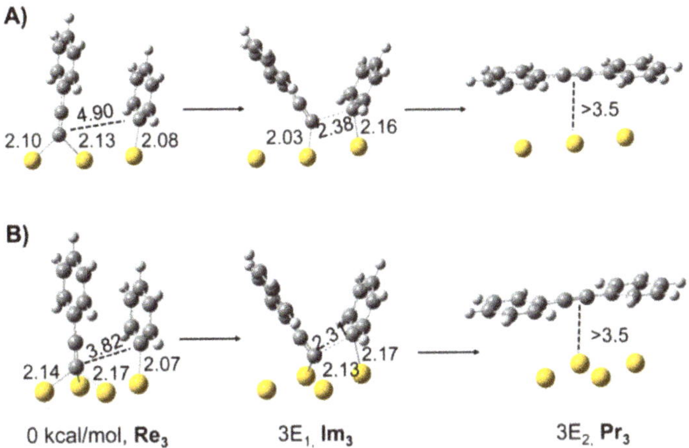

Figure 1. Proposed pathway for the cross-coupling on the (**A**) Au(100) and (**B**) Au(111). Reproduced with permission from [26]. Elsevier, 2015.

Next, the activation of Ph–B(OH)$_2$, Ph–I, and C≡C–H also was explored on a model cluster (Au$_{38}$ and partially oxidized Au$_{38}$O$_2$) [30–32]. It is worthy to note that the model clusters of the Au$_{38}$ and Au$_{38}$O$_2$ are somewhat different from the real catalyst in terms of the real structure. The Au$_{38}$O$_2$ cluster contains metallic Au^0 and cationic Au$^{\delta+}$ species, and each O atom bonds to three Au atoms. In the activation process of phenylboronate, DFT showed that the phenylboronate reactant is preferentially adsorbed on the Au^0 species rather than on the Au$^{\delta+}$ sites [30], leading to formation of the final product- biphenyl (Figure 2B). Meanwhile, DFT simulation also found that the interaction between the Au$^{\delta+}$ sites and Ph–I is weaker and the adsorption of C≡C–H on Au$^{\delta+}$ sites is relatively strong. Further, the proton of alkyne was detached with aid of O atom in the deprotonation step (Figure 2A). Therefore, a very low activation energy is required when the cross-coupling reactions occurred over the Au$_{38}$O$_2$ cluster. Of note, the cross-coupling step of the activated PA and –Ph on the Au$_{38}$O$_2$ should be the rate-determining step [32].

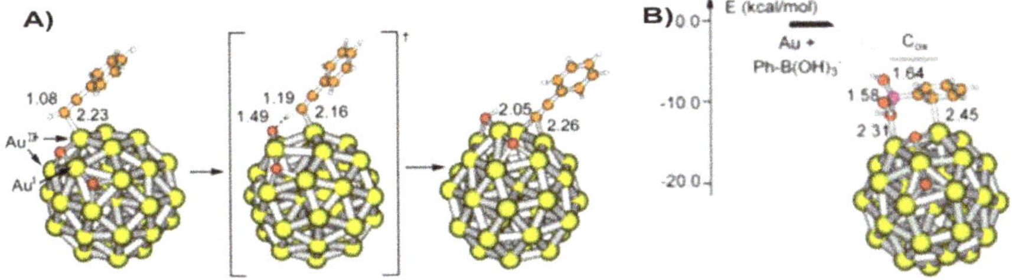

Figure 2. Calculated energy of the dissociation of PhB(OH)$_3^-$ anion (**A**) and deprotonation of PA (**B**) on the Au$_{38}$O$_2$ cluster. Reproduced with permission from [30]. American Chemical Society, 2012.

3. Physical Property of Au Nanoclusters

3.1. Framework

The gold clusters with crystal structure can be employed as the practical simulation models for mechanism study. In this Review, we only focus on the two cluster structures of the nanorod-shaped $[Au_{25}(PPh_3)_{10}L_5Cl_2]^{2+}$ (L = −SR and PA) and the nanosphere $[Au_{25}(SR)_{18}]^x$ (x = −1, 0, +1, etc.), used as the real model for DFT studies (*vide infra*). The $Au_{25}(SR)_{18}$ cluster comprises an Au_{13} core [33] and six staples of $Au_2(SR)_3$ [34]. $[Au_{25}(PPh_3)_{10}L_5Cl_2]^{2+}$ is composed of two Au_{13} cores by sharing one common vertex to form the waist sites [35], connected by thiolate or alkyne ligands [36] (Figure 3).

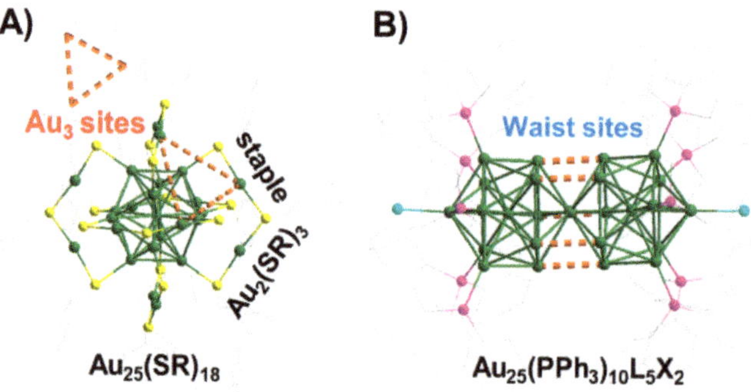

Figure 3. Framework of $Au_{25}(SR)_{18}$ (**A**) and $Au_{25}(PPh_3)_{10}L_5Cl_2$ (**B**). The orange areas in are the Au_3 and the waist active sites during the catalysis. Reproduced with permission from [36]. American Chemical Society, 2008 and Springer, 2019.

3.2. Redox Property of Au cluster

$Au_{38}S_2(SAdm)_{20}$ (SAdm = adamantanethiolate) nanoclusters exerted photosensitizing properties to give singlet oxygen ($^1O_2$) under visible light irradiation (e.g., 532 and 650 nm) [37]. The $Au_{38}S_2(SAdm)_{20}$ is intact during the whole photocatalysis process, evidenced by UV-vis tracing and mass spectroscopy analysis. The cyclic voltammetry analysis showed that the $Au_{38}S_2(SAdm)_{20}$ cluster had good charge transfer capacity to the redox $K_3Fe(CN)_6$ probe, Figure 4 [38]. However, the redox property of the cluster disappeared when β-cyclodextrins (β-CDs) was introduced in the THF solution. After detailed analysis, the huge β-CDs "umbrella" can trap the adamantane groups and then completely cover windows of the $Au_{38}S_2(SAdm)_{20}$ nanoclusters, thereby blocking direct interaction with foreign molecules and then quenching the charge transfer process (Figure 4B). It indicated that these gold nanoclusters have good redox properties and electron transfer (ET) capacity during the catalytic reactions [39].

Further, Kumar and coworkers studied the $Au_{25}(SG)_{18}$ catalyst in an electrochemical oxidation [40]. The $Au_{25}(SG)_{18}$ on the electrode gave good electro-activity during the oxidation of ascorbic acid and dopamine over a wide linear range from 0.71 to 44.4 μM. And pH dependent electrocatalytic activity was observed, attributed to the consequence of pH-dependent electrostatic attraction/repulsion between the charged $Au_{25}(SG)_{18}$ clusters and the charged analytes. Moreover, an amperometric sensing method for other compounds was developed. Next, Kauffman et al. investigated the electron transfer between CO_2 and $Au_{25}(PET)_{18}$ in solution [41]. Upon the DMF solution (containing $Au_{25}(SR)_{18}$) was saturated with CO_2 gas, the optical absorbance features showed the oxidized state of $Au_{25}(SR)_{18}$. Meanwhile, the photoluminescence increases and blue-shift. The CO_2^- induced optical changes can be simply reversed by purging the solution with N_2 gas to remove the CO_2, indicating an interaction between $Au_{25}(SR)_{18}$ and CO_2. DFT calculations revealed that the CO_2 molecule interacts with three S atoms of the Au_3 site of the $Au_{25}(SR)_{18}$ cluster, prompting the CO_2 electrochemical reduction. These observed unique interactions and electron transfers between gold clusters and reactants make the clusters promising catalysts during the organic transformations.

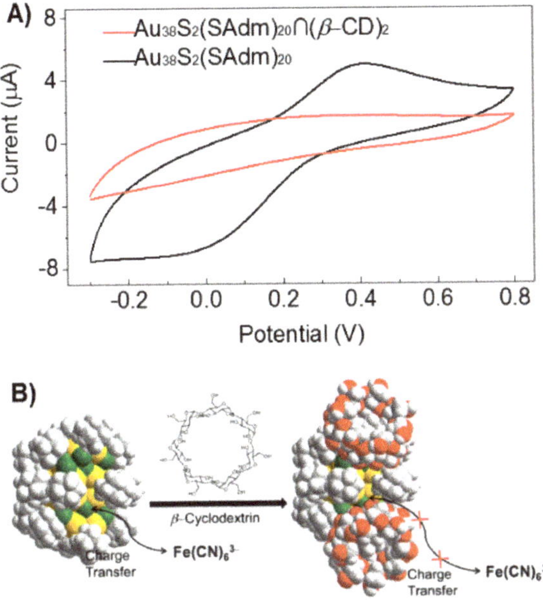

Figure 4. (**A**) Cyclic voltammograms (CV) for the redox reaction in potassium ferricyanide solution on $Au_{38}S_2(SAdm)_{20}$ clusters and $Au_{38}S_2(SAdm)_{20}$-$(\beta$-CD$)_2$ conjugates. (**B**) Schematic illustrations of the $Au_{38}S_2(SAdm)_{20}$-$(\beta$-CD$)_2$ formation and the charge transfer between $Fe(CN)_6^{3-}$ and the Au clusters and conjugates. Reproduced with permission from [38]. The Royal Society of Chemistry, 2016.

4. Catalytic Properties

4.1. Ullmann Coupling

At the beginning, the catalytic activity of the Au nanoclusters was examined in the Ullmann homo-coupling reactions of aryl iodides, which are generally catalyzed by palladium, nickel, and copper catalysts [42]. The supported gold cluster catalysts were simply prepared by a vortex-mixing of supports and a solution containing gold clusters at room temperature, and an annealing at 150 °C. The supported Au clusters were intact after the 150 °C annealing process (higher than the reaction temperatures), evidenced by UV−vis and scanning transmission electron microscopy (STEM) [43]. The X-ray photoelectron spectroscopy (XPS) analysis shows the chemical state of Au species in the oxide-supported cluster catalysts is positively charged (Au$^{\delta+}$) [44], where $0 < \delta < 1$, consistent with the free gold nanoclusters. The catalytic processes were carried out at 130 °C in the presence of base, which is similar with these catalyzed by Pd/Cu complexes or nanoparticles. The $Au_{25}(SR)_{18}/CeO_2$ showed the best catalytic activity, and the test was then expanded to a serial of substituents with functional side-groups (Table 1) [45]. Of note, the efficiency of gold nanoclusters was not as good as the palladium, nickel, and copper nanocomposites in the Ullmann homo-coupling reactions of aryl chlorides and aryl bromides.

Table 1. The catalytic results over the $Au_{25}(PET)_{18}/CeO_2$ catalysts in Ullmann homo-coupling. Reaction conditions: 0.2 mmol iodobenzene, 0.6 mmol K_2CO_3, 100 mg $Au_{25}(PET)_{18}/CeO_2$ (1 wt% cluster loading), 1 mL DMF, 130 °C, 2 day.

Entry	Substrate	Product	Conversion (%)
1	iodobenzene	biphenyl	99.8
2	MeO-C$_6$H$_4$-I	MeO-C$_6$H$_4$-C$_6$H$_4$-OMe	99.5
3	O$_2$N-C$_6$H$_4$-I	O$_2$N-C$_6$H$_4$-C$_6$H$_4$-NO$_2$	67.5
4	OHC-C$_6$H$_4$-I	OHC-C$_6$H$_4$-C$_6$H$_4$-CHO	78.2
5	1-iodonaphthalene	1,1'-binaphthalene	99.7

Later, these Au cluster were studied in the Ullmann hetero-coupling reactions. The catalytic conditions over the $Au_{25}(SR)_{18}/CeO_2$ catalysts were the same with the homo-coupling reactions (Table 1 vs. Table 2). The aromatic and aliphatic thiolate-capped Au_{25} nanoclusters (e.g., naphthalenethiolate (-SNap), benzenethiolate (-SPh), hexanethiolate (-SC$_6$H$_{13}$), and 2-phenylethanethiolate (PET)) were chosen for comparison and exploration in the Ullmann hetero-coupling of 4-MeC$_6$H$_4$I and 4-NO$_2$C$_6$H$_4$I [46]. Intriguingly, the aromatic thiolate ligated Au_{25} clusters gave much better catalytic performance (both the conversion of $NO_2C_6H_4I$ and selectivity for the hetero-coupling product (4-methyl-4′-nitrobiphenyl) than these protected by alkyl thiolate ligands. The $Au_{25}(SNap)_{18}$ cluster gave an 82% selectivity toward the hetero-coupling product, which was much higher than the Cu, Pd, and Au complexes (the selectivity: <30%, Table 2). Unfortunately, both of the conversion and selectivity decreased in the 2nd and 3rd cycles, which was due to the removal of the capping surface ligands and hence the decomposition of Au clusters, evidenced by the TEM images. These large gold nanoparticles jeopardized the catalytic performance in this coupling reaction. Thus, the protecting ligands on the clusters' surface play a key influence on their catalytic properties.

Table 2. The catalytic results over the $Au_{25}(SR)_{18}/CeO_2$ catalysts in Ullmann hetero-coupling. Reaction conditions: 0.06 mmoL 4-methyl iodobenzene, 0.05 mmoL 4-nitroiodobenzene, 0.3 mmol K_2CO_3, 100 mg $Au_{25}(SR)_{18}/CeO_2$ (~1 wt % cluster loading), 1 mL DMF, 130 °C, 24 h.

Catalyst	Conversion (%)	Selectivity (%)
$Au_{25}(SC_6H_{13})_{18}$	69	16
$Au_{25}(PET)_{18}$	72	19
$Au_{25}(SPh)_{18}$	80	50
$Au_{25}(SNap)_{18}$	91	82

DFT simulations were applied to explain the catalytic results. It is worthy to note that the reactants of both 4-MeC$_6$H$_4$I and 4-NO$_2$C$_6$H$_4$I cannot interact well with the intact Au$_{25}$(SR)$_{18}$ clusters, because of the steric effect of the protecting thiolate ligands on the clusters' surface. In the first step, one "-SR" unit on the Au$_{25}$(SR)$_{18}$ cluster was surmised to be detached under the reaction conditions in the presence of a K$_2$CO$_3$ base. Then the gold atoms on the motif were exposed to reactants and were associated with the catalytic sites [44]. Further, the activation energy for the homo- and hetero-couplings over the Au$_{25}$ protected by "-SCH$_3$" thiolate were compared by the nudged elastic band (NEB) approach (Figure 5). Intriguingly, the activation energy in the hetero-coupling was less than in the homo-coupling in the case of Au$_{25}$_SNap clusters, (Figure 4). It implied that the aromatic thiolate-capped gold cluster can not only improve the conversion rate but can also favor the hetero-coupled process [46].

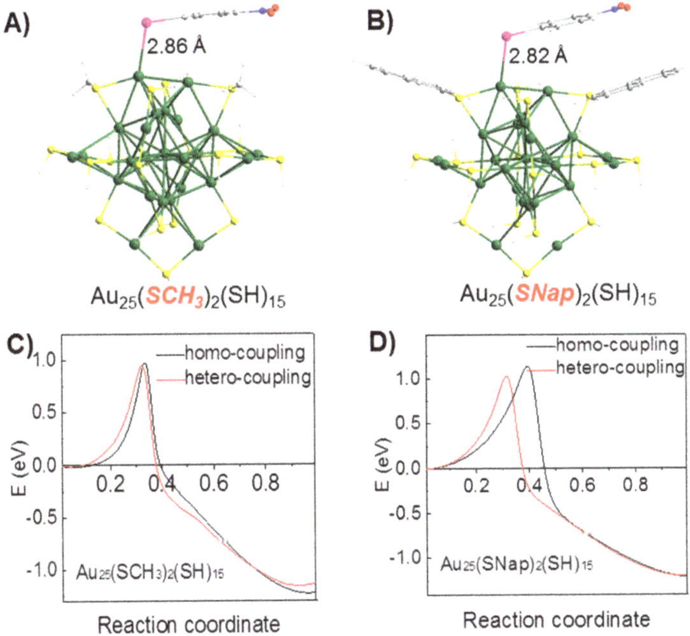

Figure 5. Interaction of 4-NO$_2$C$_6$H$_4$I with the exposed Au atoms on the Au$_{25}$(SCH$_3$)$_2$(SH)$_{15}$ (**A**) and Au$_{25}$(SNap)$_2$(SH)$_{15}$ (**B**). Energy vs. reaction coordinate with "-SCH$_3$" (**C**) and "-SNap" ligands (**D**) during the homo- and hetero-coupling reactions. Reproduced with permission from [46]. American Chemical Society, 2016.

DFT simulations were applied to explain the catalytic results. It is worthy to note that the reactants of both 4-MeC$_6$H$_4$I and 4-NO$_2$C$_6$H$_4$I cannot interact well with the intact Au$_{25}$(SR)$_{18}$ clusters, because of the steric effect of the protecting thiolate ligands on the clusters' surface. In the first step, one "-SR" unit on the Au$_{25}$(SR)$_{18}$ cluster was surmised to be detached under the reaction conditions in the presence of K$_2$CO$_3$ base. Then the gold atoms on the motif were exposed to reactants and were associated with the catalytic sites [44]. Further, the activation energy for the homo- and hetero-couplings over the Au$_{25}$ protected by "-SCH$_3$" thiolate are comparable by the nudged elastic band (NEB) approach (Figure 5). Intriguingly, the activation energy in the hetero-coupling less than in the homo-coupling in the case of Au$_{25}$_SNap clusters, (Figure 4). It implied that the aromatic thiolate-capped gold cluster not only can improve the conversion rate but also can favor the hetero-coupled process [46].

4.2. Suzuki Coupling

Further, the titania-supported Au_{25} clusters are studied in the Suzuki coupling in the presence of ionic liquids (ILs), which are catalyzed over palladium catalysts [42]. The Suzuki cross-coupling run at 90 °C using different solvents (e.g., ethanol, xylene, toluene, *N,N'*-dimethylformamide (DMF), ILs, etc.). The imidazolium-based ILs exerted a large effect on the $MeOC_6H_4I$ conversion to the desire products. A very low conversion (<5%) is observed when using the ethanol, toluene, o-xylene, and DMF as solvents in the Au_{25}/TiO_2 catalyzed coupling reactions (Table 3). Interestingly, the iodoanisole conversion over Au_{25}/TiO_2 drastically increased to 89%–99% when BMIM·X (BMIM: 1-butyl-3-methylimidazolium, X = Br or Cl or BF_4) solvents are introduced to the reaction system (Table 3). The catalytic results indicate that the imidazolium-based ILs acts as a promoter for the cross-coupling reactions [47]. Of note, only the BMIM cation (i.e., the acidic proton at position 2 of the imidazolium ions) play an important role during the reactions, as no activity is found in the presence of BDiMIM·BF_4 (BDiMIM: 1-butyl-2,3-dimethylimidazolium) solvent, which is further supported by the DFT calculations. It is worthy to note that the efficiency of gold nanoclusters was not as good as the palladium nanocomposites, however, the gold nanoclusters exhibited much better selectivity toward the target cross-coupling products.

Table 3. Catalytic results over the Au_{25}/TiO_2 catalysts in the Suzuki cross-coupling reactions of iodoanisole and phenylboronic acid using the imidazolium-based ionic liquids. Reaction conditions: 0.1 mmol iodoanisole, 0.12 mmol phenylboronic acid, 0.36 mmol K_2CO_3, 100 mg $Au_{25}(SR)_{18}/TiO_2$ (~1 wt % cluster loading), solvents (1 mL EtOH or $EtOH:H_2O$ (10:1, *v/v*), and 0.2 mL IL or other organic solvents), 90 °C, 18 h.

$$Ph\text{-}B(OH)_2 + I\text{-}\langle\bigcirc\rangle\text{-}OCH_3 \xrightarrow[\text{90 °C}]{Au_{25}/TiO_2} Ph\text{-}\langle\bigcirc\rangle\text{-}OCH_3$$

Solvent	Conversion (%)
o-xylene + $EtOH:H_2O$	<0.5
DMF + $EtOH:H_2O$	4
BMIM·Br + EtOH	89
BMIM·Cl + EtOH	90
BMIM·BF_4 + EtOH	94
BMIM·BF_4 + $EtOH:H_2O$ $EtOH:H_2O$	>99
BDiMIM·BF_4 + EtOH	<0.5

To explore the active species during the coupling reactions, the free $Au_{25}(PET)_{18}$ was mixed with the BMIM·BF_4 under the same reaction conditions [43]. Except the molecular peak of $Au_{25}(PET)_{18}$ cluster, four new mass peaks are clearly detected in the matrix-assisted laser desorption/ionization mass spectrometry (MALDI-MS). These new appeared mass peaks belonged to the $Au_{25-n}(SR)_{18-n}$ (where, n = 1–4) species (Figure 6). Of note, these new species are not the fragments caused by laser of the MALDI method. These species also were observed in the ESI-MS method [48]. The imidazolium-based ILs indeed assist the yield of $Au_{25-n}(SR)_{18-n}$ species under the reaction conditions, which may be the active sites for the cross-coupling reactions. The other explanation is that the Au-NHC complex (NHC: N-heterocyclic carbene) with the $Au_{25-n}(SR)_{18-n}$ species could be responsible for the active sites during the Suzuki cross-coupling reactions, although it needs further investigation. Of note, the Au-NHC complex was the product of the reaction of BMIM cations with the gold nanoclusters.

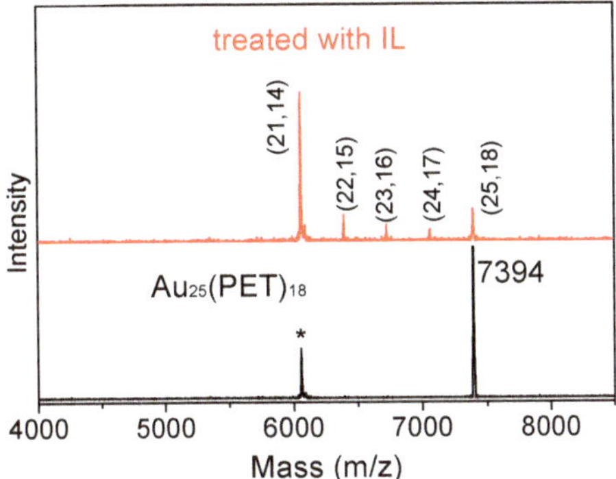

Figure 6. Matrix-assisted laser desorption/ionization mass spectrometry (MALDI-MS) of the fresh $Au_{25}(PET)_{18}$ clusters and samples treated with the ionic liquid (IL) of $BMIM \cdot BF_4$; the (n, m) presents the number of the $Au_n(PET)_m$ species. Reproduced with permission from [47]. American Chemical Society, 2015.

4.3. Sonogashira Coupling

As the IB and alkyne can be activated over gold clusters, hence, the catalytic performance of the gold nanoclusters may extend to Sonogashira cross-coupling reactions, often catalyzed over palladium catalysts [42]. The catalytic performance of the $Au_{25}(PET)_{18}$ cluster (supported on oxides) in the Sonogashira cross-coupling reaction was studied [49]. The supported catalyst was prepared by impregnating oxide powders (such as TiO_2, CeO_2, SiO_2, and MgO) in a CH_2Cl_2 solution of $Au_{25}(PET)_{18}$ (~1 wt % loading) with a 150 °C annealing. STEM and TG analyses showed that the protecting thiolate ligands were intact on the surface of gold clusters after thermal treatment. Then these Au_{25}/oxide catalysts were applied to the Sonogashira cross-coupling reaction. The optimized reaction conditions were using DMF as a solvent and K_2CO_3 as a base under an N_2 atmosphere at 160 °C, which is harsher than those for the above Suzuki and Ullmann couplings. The Au_{25}/CeO_2 catalyst showed the best activity (96.1% iodoanisole conversion with 88.1% selectivity toward the target product) (Table 4). The solvent and base can also influence the product selectivity. The size-dependent catalytic performance also was studied.

Table 4. The catalytic performance of $Au_{25}(SR)_{18}$/oxides in the Sonogashira cross-coupling reaction of p-iodoanisole and phenylacetylene. DMBP and MPEB stand for homo-coupling product of 4,4′-dimethoxy-1,1′-biphenyl and cross-coupling product of 1-methoxy-4-(2-phenylethynyl)benzene, respectively. Reaction conditions: 100 mg catalyst (1 wt % $Au_{25}(SR)_{18}$ or AuNC 2–3 nm (SC_6H_{13}) loading), 0.1 mmol p-iodoanisole, 0.15 mmol phenylacetylene, 0.3 mmol K_2CO_3, 1 mL DMF, 160 °C, 40 h. n.r. = no reaction. Conv. = conversion.

Entry	Catalyst	Conv. (%)	Selectivity (%)	
			MPEB	**DMBP**
1	$Au_{25}(SR)_{18}/CeO_2$	96.1	88.1	11.9
2	$Au_{25}(SR)_{18}/TiO_2$	92.8	82.9	17.1
3	$Au_{25}(SR)_{18}/SiO_2$	90.8	79.3	20.7
4	$Au_{25}(SR)_{18}/MgO$	93.3	80.6	19.4
5	AuNC 2–3 nm $(SC_6H_{13})/CeO_2$	65.5	57.2	42.8
6	CeO_2	n.r.		
7	TiO_2	n.r.		

The catalytic performance of small-sized $Au_{25}(PET)_{18}$ cluster catalysts was much better than large-sized of gold clusters of 2–3 nm and Au/CeO_2 (~20 nm). Support effects were studied in the cross-coupling, and no distinct effect of the oxide supports was observed (i.e., CeO_2, SiO_2, TiO_2, and MgO). The conversion was no obvious decrease, but the selectivity decreased from 88.1% to 64.5% after 5 cycles. It is noteworthy that TEM analysis shows that the gold clusters grow into larger nanoparticles (>3 nm), meaning that the gold clusters capped by organic ligands cannot stay intact under harsh reaction conditions (160 °C in the presence of a base). The gradual degradation of gold clusters leads to a decrease in selectivity, as the larger Au clusters showed a much lower selectivity. It is worthy to note that the efficiency of gold nanoclusters is much worse than the palladium-based catalysts, and the selectivity for the cross-coupling products over Au clusters is also worse.

DFT calculation found that the reactants (i.e., IB and PA) prefer to adsorb on the open facet (Au_3) of the Au_{25} cluster with the phenyl ring facing a surface Au atom (Figure 7). A total adsorption energy reaches −0.90 eV when the two reactants co-adsorb on the $Au_{25}(SR)_{18}$ catalyst. While, the IB/IB pair has an adsorption energy of −1.05 eV, indicating that the IB/IB pair interacts strongly with the cluster and the homocoupling of IBs is the dominant side-reaction competing with the cross-coupling between IB and PA. DFT results suggested that the catalytic active sites is associated with the $Au_{25}(SR)_{18}$ clusters, which is consistent with the experimental results.

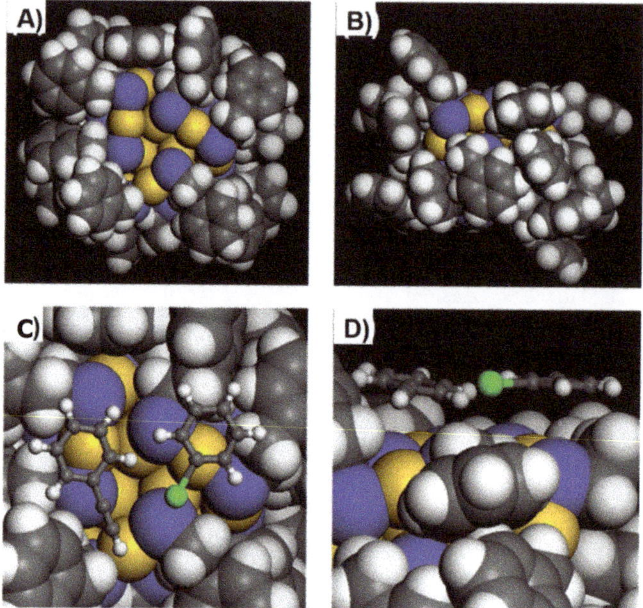

Figure 7. (**A**) Top view of one of the two open facets of $Au_{25}(PET)_{18}$ clusters where three external Au atoms are exposed; the other facet is on the back side. (**B**) Side view of the two facets which are at top and bottom. (**C**) Top view and (**D**) side view of the co-adsorption of PA and IB on the surface of the $Au_{25}(PET)_{18}$ clusters. Color code: Au, yellow; S, blue; C, gray; H, white; I, green. Reproduced with permission from [49]. Elsevier, 2013.

The structure of the 25-atom cluster is similar [50], but the electronic property and the catalytic activity of the bimetallic clusters can be largely regulated by the foreign dopants [50–55]. Recently, Li et al. [56] studied the doping effects of the $Au_{25}(SR)_{18}$ nanoclusters in the Sonogashira cross-coupling reaction base on the experiment and DFT simulations. The obtained results suggested that the Cu and Ag atoms are preferentially occupied at the cluster's kernel (Au_{13}) rather than the $Au_2(SR)_3$ staple motif, while a single Pt atom only can be doped individually and locates in the center of the cluster. The overall performance of $Ag_xAu_{25-x}(SR)_{18}$ was similar to that of $Au_{25}(SR)_{18}$ and $Pt_1Au_{24}(SR)_{18}$, which showed a decrease in catalytic activity (Table 5). The catalytic activity was from $Ag_xAu_{25-x}(SR)_{18}$ $\approx Au_{25}(SR)_{18} > Cu_xAu_{25-x}(SR)_{18} > Pt_1Au_{24}(SR)_{18}$. Interestingly, the $Cu_xAu_{25-x}(SR)_{18}$ produced a homo-coupling product base on the Ullmann homo-coupling pathway, which is contrary to the other three cluster catalysts. However, DFT calculations showed that the adsorption energy of one PA molecule on the $Pt_1Au_{24}(SR)_{18}$, $Cu_{1/2}Au_{24/23}(SR)_{18}$, and $Au_{25}(SR)_{18}$ nanoclusters was very similar (−0.50 to −0.52 eV, Table 6). The adsorption energy of one IB molecule onto the $Pt_1Au_{24}(SR)_{18}$, $Ag_{1/2}Au_{24/23}(SR)_{18}$ and $Au_{25}(SR)_{18}$ was also very similar (−0.59 to −0.61 eV, Table 6). These results suggested that the adsorption process of the PA and IB onto the alloy clusters is not the key step during the coupling reactions. Generally, the catalytic activity is largely affected by the electronic effect in the core of bimetallic clusters (i.e., Pt_1Au_{12}, Cu_xAu_{13-x}, Ag_xAu_{13-x}, and Au_{13}), and the selectivity of product is primarily turned by the atomic type on the shell of M_xAu_{12-x} [51,57].

Table 5. Catalytic performance of TiO_2-supported $M_xAu_{25-x}(SR)_{18}$ catalysts in the C–C coupling reaction between $MeOC_6H_4I$ and PA. Reaction conditions: 100 mg catalyst, 1 wt % $M_xAu_{25-x}(SR)_{18}$ loading, 0.1 mmol $MeOC_6H_4I$, 0.1 mmol PA, 0.3 mmol K_2CO_3, 1 mL DMF, 160 °C, 40 hr.

			Selectivity (%)	
Entry	Catalysts	Conversion (%)	MPEB	DMBP
1	$Au_{25}(SR)_{18}$	79.5	65.7	34.3
2	$Ag_xAu_{25-x}(SR)_{18}$	83.0	55.4	44.6
3	$Cu_xAu_{25-x}(SR)_{18}$	52.4	28.3	71.7
4	$Pt_1Au_{24}(SR)_{18}$	48.5	67.2	32.9

Table 6. Adsorption energy of the PA and IB on the Au_3 site of the $M_1Au_{24}(SR)_{18}$ (M: Pt, Ag, and Cu, and noted as M_1Au_{24}) and $M_2Au_{23}(SR)_{18}$ (M: Ag and Cu, M_2Au_{23}) cluster models, respectively.

Adsorption Energy (eV)	$M_xAu_{25-x}(SR)_{18}$ Cluster					
	Au_{25}	Pt_1Au_{24}	Ag_1Au_{24}	Ag_2Au_{23}	Cu_1Au_{24}	Cu_2Au_{23}
One PA	−0.51	−0.50	−0.61	−0.60	−0.52	−0.52
One IB	−0.60	−0.61	−0.59	−0.59	−0.54	−0.58
"PA + IB" pair	−1.11	−1.11	−1.20	−1.19	−1.06	−1.10
"IB + IB" pair	−1.20	−1.22	−1.18	−1.18	−1.08	−1.16

The structure of the 25-atom cluster was similar [50], but the electronic property and the catalytic activity of the bimetallic clusters could be largely regulated by the foreign dopants [50–55]. Recently, Li et al. [56] studied the doping effects of the $Au_{25}(SR)_{18}$ nanoclusters in a Sonogashira cross-coupling reaction based on an experiment and DFT simulations. The obtained results suggested that the Cu and Ag atoms were preferentially occupied at the cluster's kernel (Au_{13}) rather than the $Au_2(SR)_3$ staple motif, while a single Pt atom only can be doped individually and locates in the center of the cluster. The overall performance of $Ag_xAu_{25-x}(SR)_{18}$ was similar to that of $Au_{25}(SR)_{18}$ and $Pt_1Au_{24}(SR)_{18}$, which showed a decrease in catalytic activity (Table 5). The catalytic activity was $Ag_xAu_{25-x}(SR)_{18}$ $\approx Au_{25}(SR)_{18} > Cu_xAu_{25-x}(SR)_{18} > Pt_1Au_{24}(SR)_{18}$. Interestingly, the $Cu_xAu_{25-x}(SR)_{18}$ produced a homo-coupling product base on the Ullmann homo-coupling pathway, which was contrary to the other three cluster catalysts. However, DFT calculations showed that the adsorption energy of one PA molecule on the $Pt_1Au_{24}(SR)_{18}$, $Cu_{1/2}Au_{24/23}(SR)_{18}$, and $Au_{25}(SR)_{18}$ nanoclusters was very similar (−0.50 to −0.52 eV, Table 6). The adsorption energy of one IB molecule onto the $Pt_1Au_{24}(SR)_{18}$, $Ag_{1/2}Au_{24/23}(SR)_{18}$, and $Au_{25}(SR)_{18}$ was also very similar (−0.59 to −0.61 eV, Table 6). These results suggested that the adsorption process of the PA and IB onto the alloy clusters was not the key step during the coupling reactions. Generally, the catalytic activity was largely affected by the electronic effect in the core of bimetallic clusters (i.e., Pt_1Au_{12}, Cu_xAu_{13-x}, Ag_xAu_{13-x}, and Au_{13}), and the selectivity of the product is primarily turned by the atomic type on the shell of M_xAu_{12-x} [51,57].

4.4. A^3-Coupling

A^3-coupling reactions, where three reactants (aldehydes, amines, and alkynes) react each other in one-pot to yield only one product, have attracted overwhelming interest in the past decade. The A^3-coupling is favorable from environmental and economic perspectives: more efficient and less waste [58–61]. The A^3-coupling reactions involve new C–C and C–N bond formation in one procedure. The alkyne activation was deemed as the key step for the A^3-coupling; the aldehydes could react with amines spontaneously. Hence, the gold clusters should be active in this reaction, because the cluster catalyst has exhibited the capacity of the alkyne activation in the semi-hydrogenation of terminal alkynes [36].

The catalytic performance of the $Au_{25}(PPh_3)_{10}(PA)_5X_2$ was investigated in different solvents and reaction temperatures [62]. The most prominent feature was that the polarity of the solvent had a great influence on catalytic activity. The TiO_2-supported Au_{25} catalyst gave higher activity in the polar solvents. The gold clusters also showed good recyclability. Interestingly, the cluster catalyst showed no conversion using ketones as reactants (Figure 8), which was completely different from the catalytic behaviors of the gold complexes and bare gold nanoparticles. Therefore, the electronic factors and steric hindrance of the substituents had significant effects on the reaction conversion rate.

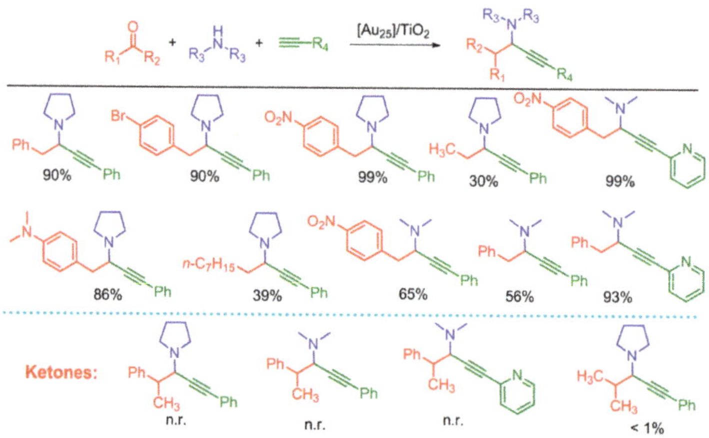

Figure 8. Catalytic results over the $Au_{25}(PPh_3)_{10}(PA)_5X_2$ ([Au_{25}]/TiO_2) catalysts in the A^3-coupling reaction. Reaction conditions: 100 mg catalyst (1 wt.% cluster loading), 1.0 mmol benzaldehyde, 1.2 mmol piperidine, 1.3 mmol phenylacetylene, 5 mL water, 100 °C, 18 h, under a N_2 atmosphere.

An induction period (0 to 3 h) appeared and the conversion slightly increased during the induction period. After the induction period, the reaction conversion of gold clusters significantly increased in the time evolution for A^3-coupling [62]. The results showed that some phosphine ligands are removed to generate catalytic active sites, which associates with the surface gold atoms. Further research found that the capped phosphine ligands can be selectively removed in the case of $Au_{11}(PPh_3)_7X_3$ with the aid of base (e.g., pyridine), evidenced by UV-vis and ESI-MS analyses [63].

Finally, the catalytic mechanism over the gold clusters was studied by DFT calculations. Firstly, the phosphine ligand is detached in the presence of the reaction system, and then the uncovered Au atoms are exposed to the reactants. Next, the PA molecules are adsorbed onto the M1 site via the interaction of Au···whole triple bond, Figure 9. Further, a terminal hydrogen deprotonation occurs in the presence of amine (e.g., $HN(CH_3)_2$). The iminium ion ($H_2C=N(CH_3)_2^+$) interacts with the $PhC\equiv C-$ on site M1 and finally give rise to the final product (i.e., propargylamine) [62].

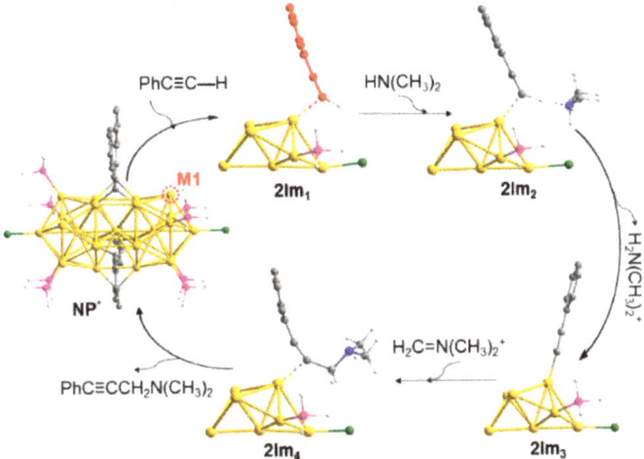

Figure 9. Proposed mechanism for the $[Au_{25}(PH_3)_{10}(PA)_5Cl_2]^{2+}$-catalyzed A^3-coupling reaction. Reproduced with permission from [62]. Elsevier, 2016.

Further, Jin et al. reported Au_{38} nanocluster-catalyzed the A^3-coupling reaction [64]. They argued that the synergistic effect of the partial positive charged Au surface ($Au^{\delta+}$, $0 < \delta < 1$) and the electron-rich Au_{23} kernel were responsible for the catalytic behaviors. Li et al. found that cadmium doped Au_{13} nanocluster also showed catalytic activity in the A^3-coupling [65]. The cooperation between the exerted cadmium atoms and the neighbor gold atoms on the surface of Au_{13} icosahedron are tentatively deemed as the active sites in the cross-coupling reactions.

5. Summary and Outlook

In the past decade years, remarkable advances have been made in developing catalytic applications of gold nanoclusters, which is a new perspective for gold nanocatalysis, especially in carbon–carbon cross-coupling reactions. These gold nanoclusters were more efficient in the cross C-C coupling reactions in this Review. The higher catalytic activity was mainly because of the distinctive frame structure and electronic properties of the gold nanoclusters and the protecting ligands. However, some of the gold nanoclusters with the protection of thiolate/phosphine could not stay intact under the harsh reaction conditions during the catalysis processes, leading to a decrease or even disappearance of catalytic activity due to the increasing size of formed particles. How to maintain the stability of gold nanoclusters during the catalysis has become a major challenge and subject for future research.

The present review demonstrated that CeO_2 and TiO_2 oxides have been observed to be excellent supports for gold nanoclusters [66]. Few studies have been done on other oxides. Developing new types of supports is another significant issue for substances such as zeolite [67,68], carbon materials (e.g., graphene or graphite oxide) [69,70], and MOFs [71]. For example, mesoporous materials, e.g., MOFs and zeolite, exhibited regular tunnels and cages and improved the stability of the Au clusters during the catalytic processes [72]. The tunnels and cages, by enriching concentration of the reactants, can also improve the catalytic activity. The steric effects of MOFs organic linker can improve the selectivity of products. In addition, the sites of Lewis acids and Bronstein acids in zeolites also can change the catalytic performance of the catalysts.

In a word, gold nanoclusters will be expected to be used into carbon–heteroatom cross-coupling reactions, which are not yet well investigated. The formed C–O and C–N bonds are very useful drug intermediates [73]. Recently, some attempts have been reported in these categories, such as the cross-aldol condensation and Michael addition (forming C=C bonds) and photo-oxidation of amines to imines (C=N bonds) [10,74]. Future research on gold nanoclusters will contribute to fundamental researches and provide clues for the new design of efficient catalysts for other specific chemical processes.

Author Contributions: Investigation, Q.S. and Z.Q. (resources, Q.S. and Z.Q.; writing—original draft preparation, Q.S. and Z.Q.; writing—review and editing, H.X. and G.L.; supervision, G.L.; project administration, G.L.

Funding: This research was funded by National Natural Science Foundation of China (21665020), the Natural Science Foundation of Inner Mongolia Autonomous Region (2018BS02004), and the Program of Higher-level Talents of Inner Mongolia Agricultural University (NDYB2016-03) for financial support.

Conflicts of Interest: The authors declare no conflict of interest.

References

1. Haruta, M.; Yamada, N.; Kobayashi, T.; Iijima, S. Gold catalysts prepared by coprecipitation for low-temperature oxidation of hydrogen and of carbon-monoxide. *J. Catal.* **1989**, *115*, 301–309. [CrossRef]
2. Tsukuda, T.; Tsunoyama, H.; Sakurai, H. Aerobic oxidations catalyzed by colloidal nanogold. *Chem. Asian J.* **2011**, *6*, 736–748. [CrossRef]
3. Della, P.C.; Falletta, E.; Rossi, M. Update on selective oxidation using gold. *Chem. Soc. Rev.* **2012**, *41*, 350–369.
4. Li, G.; Qian, H.F.; Jin, R.C. Gold nanocluster-catalyzed selective oxidation of sulfide to sulfoxide. *Nanoscale* **2012**, *4*, 6714–6717. [CrossRef]
5. Liu, C.; Yan, C.Y.; Lin, J.Z.; Yu, C.L.; Huang, J.H.; Li, G. One-pot synthesis of $Au_{144}(SCH_2Ph)_{60}$ nanoclusters and their catalytic application. *J. Mater. Chem. A* **2015**, *3*, 20167–20173. [CrossRef]
6. Hashmi, A.S.K.; Hutchings, G.J. Gold catalysis. *Angew. Chem. Int. Ed.* **2006**, *45*, 7896–7936. [CrossRef]
7. Li, G.; Zeng, C.J.; Jin, R.C. Thermally robust $Au_{99}(SPh)_{42}$ nanoclusters for chemoselective hydrogenation of nitrobenzaldehyde derivatives in water. *J. Am. Chem. Soc.* **2014**, *136*, 3673–3679. [CrossRef]
8. Li, G.; Jiang, D.E.; Kumar, S.; Chen, Y.X.; Jin, R.C. Size dependence of atomically precise gold nanoclusters in chemoselective hydrogenation and active site structure. *ACS Catal.* **2014**, *4*, 2463–2469. [CrossRef]
9. Naya, S.; Kimura, K.; Tada, H. One-step selective aerobic oxidation of amines to imines by gold nanoparticle-loaded Rutile titanium(IV) oxide plasmon photocatalyst. *ACS Catal.* **2013**, *3*, 10–13. [CrossRef]
10. Chen, H.J.; Liu, C.; Wang, M.; Zhang, C.F.; Luo, N.C.; Wang, Y.H.; Abroshan, H.; Li, G.; Wang, F. Visible light gold nanocluster photocatalyst: Selective aerobic oxidation of amines to imines. *ACS Catal.* **2017**, *7*, 3632–3638. [CrossRef]
11. Taketoshi, A.; Haruta, M. Size- and structure-specificity in catalysis by gold clusters. *Chem. Lett.* **2014**, *43*, 380–387. [CrossRef]
12. Lohse, S.E.; Murphy, C.J. The quest for shape control: A history of gold nanorod synthesis. *Chem. Mater.* **2013**, *25*, 1250–1261. [CrossRef]
13. Lu, C.L.; Prasad, K.S.; Wu, H.L.; Ho, J.A.A.; Huang, M.H. Au nanocube-directed fabrication of Au-Pd core-shell nanocrystals with tetrahexahedral, concave octahedral, and octahedral structures and their electrocatalytic activity. *J. Am. Chem. Soc.* **2010**, *132*, 14546–14553. [CrossRef]
14. Zhang, J.; Li, Z.; Zheng, K.; Li, G. Synthesis and characterization of size-controlled atomically-precise gold clusters. *Phys. Sci. Rev.* **2018**, *3*. [CrossRef]
15. Jin, R.C. Atomically precise metal nanoclusters: Stable sizes and optical properties. *Nanoscale* **2015**, *7*, 1549–1565. [CrossRef]
16. Li, G.; Jin, R.C. Atomically precise gold nanoclusters as new model catalysts. *Acc. Chem. Res.* **2013**, *46*, 1749–1758. [CrossRef]
17. Yamazoe, S.; Koyasu, K.; Tsukuda, T. Nonscalable oxidation catalysis of gold clusters. *Acc. Chem. Res.* **2014**, *47*, 816–824. [CrossRef] [PubMed]
18. Li, Z.M.; Abroshan, H.; Liu, C.; Li, G. A Critical review on the catalytic applications of non-metallic gold nanoclusters: Selective oxidation, hydrogenation, and coupling reactions. *Curr. Org. Chem.* **2017**, *21*, 476–488. [CrossRef]
19. Li, G.; Jin, R.C. Catalysis by gold NPs: Carbon-carbon coupling reactions. *Nanotechnol. Rev.* **2013**, *5*, 529–545.

20. Zhou, Y.; Li, G. A Critical review on carbon-carbon coupling over ultra-small gold nanoclusters. *Acta Phys.-Chim. Sin.* **2017**, *33*, 1297–1309.

21. Alonso, F.; Beletskaya, I.P.; Yus, M. Non-conventional methodologies for transition-metal catalysed carbon-carbon coupling: A critical overview. Part 2: The Suzuki reaction. *Tetrahedron* **2008**, *64*, 3047–3101. [CrossRef]

22. Seechurn, C.; Kitching, M.O.; Colacot, T.J.; Snieckus, V. Palladium-catalyzed cross-coupling: A historical contextual perspective to the 2010 Nobel Prize. *Angew. Chem. Int. Ed.* **2012**, *51*, 5062–5085. [CrossRef] [PubMed]

23. Fihri, A.; Bouhrara, M.; Nekoueishahraki, B.; Polshettiwar, V. Nanocatalysts for Suzuki cross-coupling reactions. *Chem. Soc. Rev.* **2011**, *40*, 5181–5203. [CrossRef]

24. Kanuru, V.K.; Kyriakou, G.; Beaumont, S.K.; Papageorgiou, A.C.; Watson, D.J.; Lambert, R.M. Sonogashira coupling on an extended gold surface in vacuo: Reaction of phenylacetylene with iodobenzene on Au(111). *J. Am. Chem. Soc.* **2010**, *132*, 8081–8086. [CrossRef]

25. González-Arellano, C.; Abad, A.; Corma, A.; García, H.; Iglesias, M.; Sánchez, F. Catalysis by gold(I) and gold(III): A parallelism between homo- and heterogeneous catalysts for copper-free Sonogashira cross-coupling reactions. *Angew. Chem. Int. Ed.* **2007**, *46*, 1536–1538. [CrossRef] [PubMed]

26. Lin, J.Z.; Abroshan, H.; Liu, C.; Zhu, M.Z.; Li, G.; Haruta, M. Sonogashira cross-coupling on the Au(111) and Au(100) facets of gold nanorod catalysts: Experimental and computational investigation. *J. Catal.* **2015**, *330*, 354–361. [CrossRef]

27. Li, G.; Zeng, C.J.; Jin, R.C. Chemoselective hydrogenation of nitrobenzaldehyde to nitrobenzyl alcohol with unsupported au nanorod catalysts in water. *J. Phys. Chem. C* **2015**, *119*, 11143–11147. [CrossRef]

28. Murphy, C.J.; Gole, A.M.; Hunyadi, S.E.; Stone, J.W.; Sisco, P.N.; Alkilany, A.; Kinard, B.E.; Hankins, P. Chemical sensing and imaging with metallic nanorods. *Chem. Commun.* **2008**, *5*, 544–547. [CrossRef]

29. Bai, X.T.; Gao, Y.N.; Liu, H.G.; Zheng, L.Q. Synthesis of amphiphilic ionic liquids terminated gold nanorods and their superior catalytic activity for the reduction of nitro compounds. *J. Phys. Chem. C* **2009**, *113*, 17730–17736. [CrossRef]

30. Boronat, M.; Combita, D.; Concepción, P.; Corma, A.; García, H.; Juárez, R.; Laursen, S.; López-Castro, J.D. Making C-C bonds with gold: Identification of selective gold sites for homo- and cross-coupling reactions between iodobenzene and alkynes. *J. Phys. Chem. C* **2012**, *116*, 24855–24867. [CrossRef]

31. Boronat, M.; Corma, A. Molecular approaches to catalysis naked gold NPs as quasi-molecular catalysts for green processes. *J. Catal.* **2011**, *284*, 138–147. [CrossRef]

32. Corma, A.; Juárez, R.; Boronat, M.; Sánchez, F.; Iglesiasc, M.; García, H. Gold catalyzes the Sonogashira coupling reaction without the requirement of palladium impurities. *Chem. Commun.* **2011**, *47*, 1446–1448. [CrossRef] [PubMed]

33. Zhang, J.; Zhou, Y.; Zheng, K.; Abroshan, H.; Kauffman, D.R.; Sun, J.; Li, G. Diphosphine-induced chiral propeller arrangement of gold nanoclusters for singlet oxygen photogeneration. *Nano Res.* **2018**, *11*, 5787–5798. [CrossRef]

34. Zhu, M.; Aikens, C.M.; Hollander, F.J.; Schatz, G.C.; Jin, R.C. Correlating the crystal structure of a thiol-protected Au25 cluster and optical properties. *J. Am. Chem. Soc.* **2008**, *130*, 5883–5885. [CrossRef]

35. Zheng, K.; Zhang, J.W.; Zhao, D.; Yang, Y.; Li, Z.M.; Li, G. Motif Mediated $Au_{25}(SPh)_5(PPh_3)_{10}X_2$ Nanorod of Conjugated Electron Delocalization. *Nano Res.* **2019**, *12*, 501–507. [CrossRef]

36. Li, G.; Jin, R.C. Gold nanocluster-catalyzed semihydrogenation: A unique activation pathway for terminal alkynes. *J. Am. Chem. Soc.* **2014**, *136*, 11347–11354. [CrossRef]

37. Li, Z.M.; Liu, C.; Abroshan, H.; Kauffman, D.R.; Li, G. $Au_{38}S_2(SAdm)_{20}$ Photocatalyst for One-Step Selective Aerobic Oxidations. *ACS Catal.* **2017**, *7*, 3368–3374. [CrossRef]

38. Yan, C.Y.; Liu, C.; Abroshan, H.; Li, Z.M.; Qiu, R.; Li, G. Surface modification of adamantane-terminated gold nanoclusters using cyclodextrins. *Phys. Chem. Chem. Phys.* **2016**, *18*, 23358–23364. [CrossRef] [PubMed]

39. Zhang, C.; Chen, Y.; Wang, H.; Li, Z.; Zheng, K.; Li, S.; Li, G. Transition-metal-mediated catalytic properties of CeO_2-supported gold clusters in aerobic alcohol oxidation. *Nano Res.* **2018**, *11*, 2139–2148. [CrossRef]

40. Kumar, S.S.; Kwak, K.; Lee, D. Amperometric sensing based on glutathione protected Au_{25} nanoparticles and their pH dependent electrocatalytic activity. *Electroanalysis* **2011**, *23*, 2116–2124. [CrossRef]

41. Kauffman, D.R.; Alfonso, D.; Matranga, C.; Qian, H.; Jin, R. Experimental and computational investigation of Au_{25} clusters and CO_2: A unique interaction and enhanced electrocatalytic activity. *J. Am. Chem. Soc.* **2012**, *134*, 10237–10243. [CrossRef] [PubMed]

42. Monnier, F.; Taillefer, M. Catalytic C-C, C-N, and C-O Ullmann-type coupling reactions. *Angew. Chem. Int. Ed.* **2009**, *48*, 6954–6971. [CrossRef] [PubMed]

43. Yu, C.L.; Li, G.; Kumar, K.S.; Kawasaki, H.; Jin, R.C. Stable $Au_{25}(SR)_{18}/TiO_2$ Composite nanostructure with enhanced visible light photocatalytic activity. *J. Phys. Chem. Lett.* **2013**, *4*, 2847–2852. [CrossRef]

44. Chen, H.J.; Liu, C.; Wang, M.; Zhang, C.F.; Li, G.; Wang, F. Thermally robust silica-enclosed Au_{25} nanocluster and its catalysis. *Chin. J. Catal.* **2016**, *37*, 1787–1793. [CrossRef]

45. Li, G.; Liu, C.; Lei, Y.; Jin, R.C. Au_{25} nanocluster-catalyzed Ullmann-type homocoupling reaction of aryl iodides. *Chem. Commun.* **2012**, *48*, 12005–12007. [CrossRef] [PubMed]

46. Li, G.; Abroshan, H.; Liu, C.; Zhuo, S.; Li, Z.M.; Xie, Y.; Kim, H.J.; Rosi, N.L.; Jin, R.C. Tailoring the electronic and catalytic properties of Au_{25} nanoclusters via ligand engineering. *ACS Nano* **2016**, *10*, 7998–8005. [CrossRef] [PubMed]

47. Abroshan, H.; Li, G.; Lin, J.Z.; Kim, H.J.; Jin, R.C. Molecular mechanism for the activation of $Au_{25}(SCH_2CH_2Ph)_{18}$ nanoclusters by imidazolium-based ionic liquids for catalysis. *J. Catal.* **2016**, *337*, 72–79. [CrossRef]

48. Li, G.; Abroshan, H.; Chen, Y.X.; Jin, R.C.; Kim, H.J. Experimental and mechanistic understanding of aldehyde hydrogenation using Au_{25} nanoclusters with Lewis acids: Unique sites for catalytic reactions. *J. Am. Chem. Soc.* **2015**, *137*, 14295–14304. [CrossRef]

49. Li, G.; Jiang, D.E.; Liu, C.; Yu, C.L.; Jin, R.C. Oxide-supported atomically precise gold nanocluster for catalyzing Sonogashira cross-coupling. *J. Catal.* **2013**, *306*, 177–183. [CrossRef]

50. Kumara, C.; Aikens, C.M.; Dass, A. X-ray crystal structure and theoretical analysis of $Au_{25-x}Ag_x(SCH_2CH_2Ph)_{18}$- alloy. *J. Phys. Chem. Lett.* **2014**, *5*, 461–466. [CrossRef]

51. Li, W.L.; Liu, C.; Abroshan, H.; Ge, Q.J.; Yang, X.J.; Xu, H.Y.; Li, G. Catalytic CO oxidation using bimetallic MxAu25-x clusters: A combined experimental and computational study on doping effects. *J. Phys. Chem. C* **2016**, *120*, 10261–10267. [CrossRef]

52. Jiang, D.E.; Whetten, R.L. Magnetic doping of a thiolated-gold superatom: First-principles density functional theory calculations. *Phys. Rev. B* **2009**, *80*, 115402. [CrossRef]

53. Qian, H.F.; Jiang, D.E.; Li, G.; Gayathri, C.; Das, A.; Gil, R.R.; Jin, R.C. Monoplatinum Doping of Gold Nanoclusters and Catalytic Application. *J. Am. Chem. Soc.* **2012**, *134*, 16159–16162. [CrossRef]

54. Xie, S.H.; Tsunoyama, H.; Kurashige, W.; Negishi, Y.; Tsukuda, T. Enhancement in aerobic alcohol oxidation catalysis of Au25 clusters by single Pd atom doping. *ACS Catal.* **2012**, *2*, 1519–1523. [CrossRef]

55. Li, G.; Jin, R.C. Atomic level tuning of the catalytic properties: Doping effects of 25-atom bimetallic nanoclusters on styrene oxidation. *Catal. Today* **2016**, *278*, 187–191. [CrossRef]

56. Li, Z.M.; Yang, X.J.; Liu, C.; Wang, J.; Li, G. Effects of doping in 25-atom bimetallic nanocluster catalysts for carbon-carbon coupling reaction of iodoanisole and phenylacetylene. *Proc. Nat. Sci. Mater. Int.* **2016**, *26*, 477–482. [CrossRef]

57. Qin, Z.; Zhao, D.; Zhao, L.; Xiao, Q.; Wu, T.; Zhang, J.; Wan, C.-Q.; Li, G. Tailoring the Stability, Photocatalysis and Photoluminescence Properties of Au_{11} Nanocluster via Doping Engineering. *Nanoscale Adv.* **2019**, *2*. [CrossRef]

58. Dömling, A. Recent developments in isocyanide based multicomponent reactions in applied chemistry. *Chem. Rev.* **2006**, *106*, 17–89. [CrossRef]

59. Ganem, B. Strategies for innovation in multicomponent reaction design. *Acc. Chem. Res.* **2009**, *42*, 463–472. [CrossRef]

60. Climent, M.J.; Corma, A.; Iborra, S. Homogeneous and heterogeneous catalysts for multicomponent reactions. *RSC Adv.* **2012**, *2*, 16–58. [CrossRef]

61. Wei, C.M.; Li, C.J. A highly efficient three-component coupling of aldehyde, alkyne, and amines via C-H activation catalyzed by gold in water. *J. Am. Chem. Soc.* **2003**, *125*, 9584–9585. [CrossRef] [PubMed]

62. Chen, Y.D.; Liu, C.; Abroshan, H.; Li, Z.M.; Wang, J.; Li, G.; Haruta, M. Phosphine/phenylacetylide-ligated Au clusters for multicomponent coupling reactions. *J. Catal.* **2016**, *337*, 287–294. [CrossRef]

63. Liu, C.; Abroshan, H.; Yan, C.Y.; Li, G.; Haruta, M. One-pot synthesis of $Au_{11}(PPh_2Py)_7Br_3$ for the highly chemoselective hydrogenation of nitrobenzaldehyde. *ACS Catal.* **2016**, *6*, 92–99. [CrossRef]

64. Li, Q.; Das, A.; Wang, S.X.; Chen, Y.X.; Jin, R.C. Highly efficient three-component coupling reaction catalysed by atomically precise ligand-protected $Au_{38}(SC_2H_4Ph)_{24}$ nanoclusters. *Chem. Commun.* **2016**, *52*, 14298–14301. [CrossRef]

65. Li, M.; Tian, S.; Wu, Z. Improving the catalytic activity of Au_{25} nanocluster by peeling and doping. *Chin. J. Chem.* **2017**, *35*, 567–571. [CrossRef]

66. Wen, Z.Y.; Li, Z.M.; Ge, Q.J.; Zhou, Y.; Sun, J.; Li, G. Robust nickel cluster@Mes-HZSM-5 composite nanostructure with enhanced catalytic activity in the DTG reaction. *J. Catal.* **2018**, *363*, 26–33. [CrossRef]

67. Li, Z.M.; Li, W.L.; Abroshan, H.; Ge, Q.J.; Zhou, Y.; Zhang, C.L.; Li, G.; Jin, R.C. Dual Effects of Water Vapor over Ceria-Supported Gold Clusters. *Nanoscale* **2018**, *10*, 6558–6565. [CrossRef]

68. Wang, F.; Wen, Z.; Fang, Q.; Ge, Q.; Sun, J.; Li, G. Manganese cluster induce the control synthesis of RHO- and CHA-type silicoaluminaphosphates for dimethylether to light olefin conversion. *Fuel* **2019**, *244*, 104–109. [CrossRef]

69. Guo, S.; Zhang, S.H.; Fang, Q.H.; Abroshan, H.; Kim, H.; Haruta, M.; Li, G. Gold-palladium nanoalloys supported by graphene oxide and lamellar TiO_2 for direct synthesis of hydrogen peroxide. *ACS Appl. Mater. Interfaces* **2018**, *10*, 40599–40607. [CrossRef]

70. Liu, C.; Zhang, J.Y.; Huang, J.H.; Zhang, C.L.; Hong, F.; Zhou, Y.; Li, G.; Haruta, M. Efficient aerobic oxidation of glucose to gluconic acid over activated carbon-supported gold clusters. *ChemSuSChem* **2017**, *10*, 1976–1980. [CrossRef]

71. Vilhelmsen, L.B.; Walton, K.S.; Sholl, D.S. Structure and mobility of metal clusters in MOFs: Au, Pd, and AuPd clusters in MOF-74. *J. Am. Chem. Soc.* **2012**, *134*, 12807–12816. [CrossRef] [PubMed]

72. Chen, H.; Li, Z.; Qin, Z.; Kim, H.J.; Abroshan, H.; Li, G. Silica-encapsulated gold nanoclusters for efficient acetylene hydrogenation to ethylene. *ACS Appl. Nano Mater.* **2019**. [CrossRef]

73. Ciobanu, M.; Cojocaru, B.; Teodorescu, C.; Vasiliu, F.; Coman, S.M.; Leitner, W.; Parvulescu, V.I. Heterogeneous amination of bromobenzene over titania-supported gold catalysts. *J. Catal.* **2012**, *296*, 43–54. [CrossRef]

74. Fang, Q.; Qin, Z.; Shi, Y.; Liu, F.; Barkaoui, S.; Abroshan, H.; Li, G. Au/NiO composite: A catalyst for one-pot cascade conversion of furfural. *ACS Appl. Energy Mater.* **2019**, *2*, 2654–2661. [CrossRef]

 *nanomaterials*

Review

Gold Nanoclusters as Electrocatalysts for Energy Conversion

Tokuhisa Kawawaki [1,2] **and Yuichi Negishi** [1,2,*]

[1] Department of Applied Chemistry, Faculty of Science, Tokyo University of Science, 1–3 Kagurazaka, Shinjuku-ku, Tokyo 162–8601, Japan; kawawaki@rs.tus.ac.jp

[2] Photocatalysis International Research Center, Tokyo University of Science, 2641 Yamazaki, Noda, Chiba 278–8510, Japan

* Correspondence: negishi@rs.kagu.tus.ac.jp

Received: 11 January 2020; Accepted: 27 January 2020; Published: 29 January 2020

Abstract: Gold nanoclusters (Au$_n$ NCs) exhibit a size-specific electronic structure unlike bulk gold and can therefore be used as catalysts in various reactions. Ligand-protected Au$_n$ NCs can be synthesized with atomic precision, and the geometric structures of many Au$_n$ NCs have been determined by single-crystal X-ray diffraction analysis. In addition, Au$_n$ NCs can be doped with various types of elements. Clarification of the effects of changes to the chemical composition, geometric structure, and associated electronic state on catalytic activity would enable a deep understanding of the active sites and mechanisms in catalytic reactions as well as key factors for high activation. Furthermore, it may be possible to synthesize Au$_n$ NCs with properties that surpass those of conventional catalysts using the obtained design guidelines. With these expectations, catalyst research using Au$_n$ NCs as a model catalyst has been actively conducted in recent years. This review focuses on the application of Au$_n$ NCs as an electrocatalyst and outlines recent research progress.

Keywords: gold; cluster; catalyst; hydrogen evolution reaction; oxygen evolution reaction; oxygen reduction reaction; water splitting; fuel cells; alloy; ligand-protected

1. Introduction

Gold nanoclusters (Au$_n$ NCs) have physical/chemical properties that differ from those of bulk Au owing to their size-specific electrical/geometrical structure [1–22]. Therefore, Au$_n$ NCs have been actively studied since the 1960s from the viewpoints of both basic science and application. Since Brust et al., discovered a method for synthesizing Au$_n$ NCs protected by thiolate (Au$_n$(SR)$_m$) in 1994 [1], researches on Au$_n$ NCs in particular have grown [6]. Au$_n$(SR)$_m$ NCs exhibit high stability both in solution and in the solid state because Au forms a strong bond with SR. In addition, Au$_n$(SR)$_m$ NCs can be synthesized by simply mixing reagents under the ambient atmosphere. Au$_n$(SR)$_m$ NCs with these unique characteristics have a low handling threshold even for researchers unfamiliar with the chemical synthesis of metal clusters. Au$_n$(SR)$_m$ NCs are thus currently one of the most studied metal NCs [1–18]. For these Au$_n$(SR)$_m$ NCs, it became possible to synthesize a series of Au$_n$(SR)$_m$ NCs with atomic precision in 2005 [19]. In addition, since 2007, the geometric structures of many Au$_n$(SR)$_m$ NCs have been determined through single-crystal X-ray diffraction (SC-XRD) analysis [20]. Since 2009, partial replacement of the Au atoms of Au$_n$(SR)$_m$ NCs with other elements such as silver (Ag), copper (Cu), platinum (Pt), palladium (Pd), cadmium (Cd), and mercury (Hg) has also been realized [3–5,23–44].

In parallel to these synthesis and structural analysis studies, studies on the functions of Au$_n$ NCs have also been actively conducted. Au$_n$ NCs have been observed to possess catalytic activity for several reactions, including carbon monoxide oxidation [45–55], alcohol oxidation [56–65], styrene oxidation [66–70], aromatic compound oxidation [71,72], sulfide oxidation [73–75], and carbon dioxide reduction [76–83]. One of the reasons for these active studies on the catalysis of Au_n NCs is that their electronic and geometric structures are well understood. Thus, if the obtained catalytic properties are compared with the electronic/geometrical structures of $Au_n(SR)_m$ NCs, information on active sites, mechanisms, and key factors for high activation in catalytic reactions can be obtained. With these expectations, $Au_n(SR)_m$ NCs have received great attention as model catalysts [45–83].

In addition, several studies on $Au_n(SR)_m$ NCs as electrocatalysts have also been performed recently. To prevent serious environmental issues including the depletion of fossil fuels and global warming, the establishment of a system in which hydrogen (H_2) is generated from water and solar energy using a photocatalyst is desired, with the generated H_2 used for the generation of electricity using fuel cells [84,85]. Once such an energy conversion system is established, it will be possible to circulate an energy medium (H_2) in addition to obtaining electricity only from solar energy and abundant water resources. However, realization of such an ultimate energy conversion system requires further improvement of the reaction efficiency of each half reaction of water splitting and fuel cells, including the hydrogen evolution reaction (HER), oxygen evolution reaction (OER), hydrogen oxidation reaction (HOR), and oxygen reduction reaction (ORR; Figure 1A).

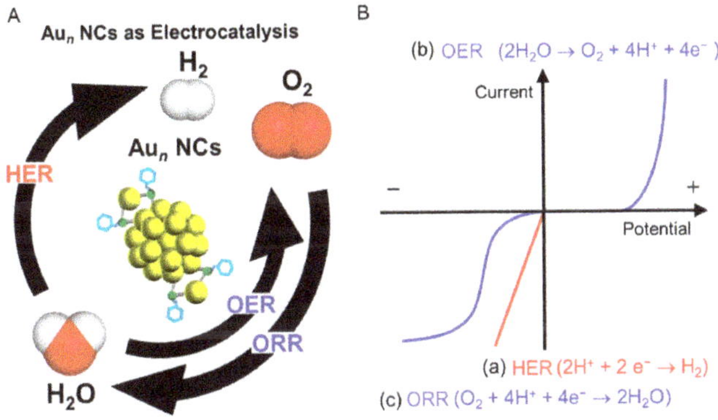

Figure 1. (**A**) Schematic illustration of gold nanoclusters (Au_n NCs) for an electrocatalytic reaction in water splitting (hydrogen evolution reaction (HER) and oxygen evolution reaction (OER)) and fuel cells (oxygen reduction reaction (ORR)). (**B**) Current–potential characteristics for (a) HER, (b) OER, and (c) ORR.

To improve the reactivity per unit volume, it is necessary to increase the specific surface area of the active sites and increase the reaction rate at the active sites. For the former, size reduction of the catalyst is one effective method. However, the latter is strongly related to the adsorption energy of reactive molecules on the catalyst surface. The activity of the chemical reaction on the catalyst surface is the highest when the Gibbs energy of adsorption between the catalyst and reactant is moderate according to the Sabatier principle [86]. This is because the reaction does not occur without the adsorption of reactants but is inhibited by the strong adsorption of reactants. Therefore, the relationship between the reaction efficiency and the Gibbs energy for the adsorption of reactants follows a curved line called an activity volcano plot [87]. Fine nanoparticle catalysts suitable for the HER [88–92], OER [93–95], and ORR [96–101] have been developed based on theoretical predictions of activity volcano plots using various metals and alloy nanoparticles (NPs). Au_n NCs have recently been observed to possess catalytic activity for the HER, OER, and ORR [77,102–116] (Figure 1). Therefore, Au_n NCs are expected to become a model catalyst even in such an energy conversion system. A better understanding of the correlation between electronic/geometrical structures and the catalytic activity of the HER, OER, and ORR in Au_n NCs might lead to the discovery of new key factors for achieving high activation. Furthermore, because Au_n NCs are composed of several tens of atoms or less, the use of fine Au_n NCs as a catalyst is also effective in reducing the consumption of expensive noble metals. Thus, it may be possible to create HER, OER, and ORR catalysts with properties that surpass those of conventional catalysts using these unique characteristics of Au_n NCs. With these expectations, several groups are conducting research on the application of Au_n NCs as electrocatalysts. This article reviews the basic theory of electrocatalysts and recent research on HER, OER, and ORR catalysts using Au_n NCs and their alloy NCs.

2. Electrocatalytic Reaction in Water Splitting

H_2 is expected to be an important energy source to support a sustainable energy society. Currently, H_2 is generated as a by-product during steam reforming or coke production. However, if a water-splitting reaction using an electrocatalyst can be applied for hydrogen production, the large-scale facility of the current system would not be required. In addition, it would be possible to produce H_2 only with water and electricity using the surplus power from a power plant. Therefore, water electrolysis is considered one of the cleanest energy production reactions for a sustainable energy society.

The water-splitting reaction consists of two half reactions, the HER and OER. When a voltage is applied to the metal electrode, a reduction reaction proceeds at the cathode and an oxidation reaction proceeds at the anode, resulting in the decomposition of water molecules into H_2 and O_2 at each electrode. However, the reactions do not proceed even if a potential equal to or higher than both the oxidation and reduction potentials in each reaction (HER: 0 V vs. SHE, OER: 1.23 V vs. SHE; SHE = standard hydrogen electrode) is applied to the electrode. This is because the activation energy of each reaction is too high. Therefore, noble metal NPs are used as a catalyst to reduce the activation energy of the reaction.

2.1. Hydrogen Evolution Reaction

In the HER, metal surface atoms of the catalyst form bonding orbitals with protons (H^+) through the Volmer–Heyrovsky or Volmer–Tafel mechanism, producing molecular hydrogen [117].

Under acidic conditions, the following reactions occur:

$$\text{Volmer reaction: } M + H^+ + e^- \rightarrow \text{M-H} \tag{1}$$

$$\text{Heyrovsky reaction: } \text{M-H} + H^+ + e^- \rightarrow \text{M-H}_2 \tag{2}$$

$$\text{Tafel reaction: } 2\text{M-H} \rightarrow 2M + H_2 \tag{3}$$

However, under alkaline conditions, the following reactions occur:

$$\text{Volmer reaction: } 2M + 2H_2O + 2e^- \rightarrow 2\text{M-H} + 2OH^- \tag{4}$$

$$\text{Heyrovsky reaction: } \text{M-H} + H_2O + e^- \rightarrow \text{M-H}_2 + OH^- \tag{5}$$

$$\text{Tafel reaction: } 2\text{M-H} \rightarrow 2M + H_2 \tag{6}$$

Bulk Au possesses almost no HER activity, whereas $Au_n(SR)_m$ NCs possess HER activity. In addition, their activity can be further improved by doping $Au_n(SR)_m$ NCs with appropriate heterogeneous elements. These effects were reported by Lee and Jiang et al., in 2017 [102]. They evaluated the HER activity using linear sweep voltammetry (LSV) in tetrahydrofuran (THF) solution with 1.0 M trifluoroacetic acid (TFA) and 0.1 M tetrabutylammonium hexafluorophosphate (Bu_4NPF_6) in the absence (black) and presence of $Au_{25}(SC_6H_{13})_{18}$ or $Au_{24}Pt(SC_6H_{13})_{18}$ (SC_6H_{13} = 1-hexanethiolate) on a glassy carbon electrode (GCE). The onset potential of the HER (Figure 1B(a)) occurred at −1.25 V for the GCE blank (Figure 2A, black line), whereas it occurred at −1.1 V for the GCE with $Au_{25}(SC_6H_{13})_{18}$ (Figure 2A, red line). In addition, for the GCE with $Au_{24}Pt(SC_6H_{13})_{18}$, the onset potential of the HER was further reduced to −0.89 V (Figure 2A, blue line). These findings indicated that $Au_n(SR)_m$ NCs has catalytic activity for the HER and that the HER activity can be further improved by substituting one Au atom of the $Au_n(SR)_m$ NCs with a Pt atom (Table 1). They estimated the HER energies of $Au_{25}(SCH_3)_{18}$ and $Au_{24}Pt(SCH_3)_{18}$ (SCH_3 = methanethiolate) using density functional theory (DFT) calculations to elucidate the reasons for this behavior (Figure 2C). In these DFT calculations, H^+ solvated by two THF molecules was used as H^+. The resulting energy change in the Volmer step was 0.539 eV for $[Au_{25}(SCH_3)_{18}]^-$, indicating that this reaction is endothermic. However, the energy change in the Volmer step was -0.059 eV for $[Au_{24}Pt(SCH_3)_{18}]^{2-}$, indicating that there is almost no energy change (Figure 2C, step 1). The higher HER activity of $Au_{24}Pt(SC_6H_{13})_{18}$ was explained by these differences in the energy barriers in the reaction. In addition, $Au_{24}Pt(SC_6H_{13})_{18}$ possessed higher HER activity even compared with Pt NPs, which are highly active materials for the HER (Figure 2B).

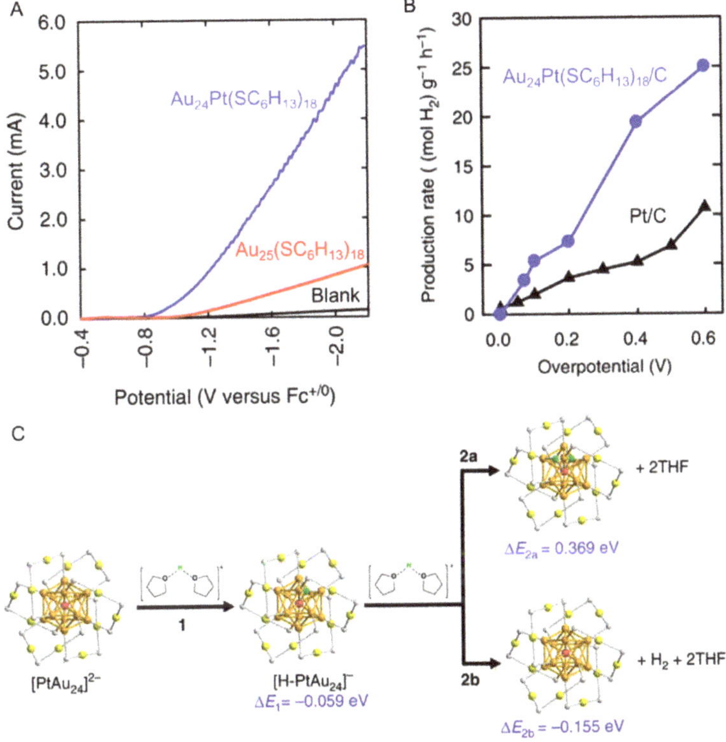

Figure 2. (**A**) HER polarization curves of $Au_{25}(SC_6H_{13})_{18}$- or $Au_{24}Pt(SC_6H_{13})_{18}$-adsorbed glassy carbon electrode (GCE), or GCE. (**B**) H_2 production rates per mass of metals in the catalyst of $Au_{24}Pt(SC_6H_{13})_{18}$/C (blue circles) and Pt/C (black triangles) electrodes. (**C**) DFT calculation results for $Au_{24}Pt(SCH_3)_{18}$. Color code: golden = Au core; olive = Au shell; purple = Pt; green = adsorbed H from the liquid medium; grey = S. Panels (**A–C**) are reproduced with permission from reference [102]. Copyright Springer Nature, 2017.

Lee and Jiang et al., observed that a high HER activity and a high catalyst turnover frequency (TOF) can be achieved by doping $Au_{25}(SC_6H_{13})_{18}$ with not only Pt but also Pd ($Au_{24}Pt(SC_6H_{13})_{18}$ > $Au_{24}Pd(SC_6H_{13})_{18}$ > $Au_{25}(SC_6H_{13})_{18}$) [103]. They reported that TOF values of $Au_{25}(SC_6H_{13})_{18}$, $Au_{24}Pd(SC_6H_{13})_{18}$, and $Au_{24}Pt(SC_6H_{13})_{18}$ were 8.2, 13.0, and 33.3 mol H_2 (mol catalyst)$^{-1}$ s^{-1} at -0.60 V vs. the reversible hydrogen electrode (RHE), respectively. In addition, it was revealed that the doping of $Au_{38}(SR)_{24}$ with different elements results in a similar activity enhancement effect with $Au_{25}(SC_6H_{13})_{18}$ ($Au_{36}Pt_2(SC_6H_{13})_{24}$ > $Au_{36}Pd_2(SC_6H_{13})_{24}$ > $Au_{38}(SC_6H_{13})_{24}$) [103]. These results are in good agreement with the DFT calculation results. In addition to these studies, Jiang et al., also investigated the doping effects of various elements (Pt, Pd, Ag, Cu, Hg, and Cd) in $Au_{25}(SCH_3)_{18}$ using DFT calculations [105]. The results predicted that $Au_{24}Pt(SCH_3)_{18}$, $Au_{24}Pd(SCH_3)_{18}$, and $Au_{24}Cu(SCH_3)_{18}$, in which the heteroatom (Pt, Pd, or Cu) is located at the center of the metal core, have a higher HER activity than $Au_{25}(SCH_3)_{18}$. Zhu et al., reported that another fine alloy NC, $Au_2Pd_6(S_4(PPh_3)_4(PhF_2S)_6)$ (PPh_3 = triphenylphosphine, PhF_2S = 3,4-difluorobenzenethiolate), also exhibits HER activity (Table 1) [106]. These studies revealed that $Au_n(SR)_m$ and their alloy NCs have HER activity and it can be improved by controlling the electronic structure of Au_n NCs through heteroatom doping.

Table 1. Representative references on HER activity of Au_n NCs and related alloy NCs.

Ligand	Support	Experimental Condition	Activity	Reference
SC_6H_{13}	–	1.0 M TFA and 0.1 M Bu_4NPF_6 in THF [c]	$Au_{24}Pt(SC_6H_{13})_{18}$ > $Au_{25}(SC_6H_{13})_{18}$	[102]
SC_6H_{13}	carbon black	1 M Britton–Robinson buffer solution in 2 M KCl aq (pH 3) [c,d]	$Au_{24}Pt(SC_6H_{13})_{18}$ > $Au_{24}Pd(SC_6H_{13})_{18}$ > $Au_{25}(SC_6H_{13})_{18}$	[103]
SC_6H_{13}	carbon black	1 M Britton–Robinson buffer solution in 2 M KCl aq (pH 3) [c,d]	$Au_{36}Pt_2(SC_6H_{13})_{24}$ > $Au_{36}Pd_2(SC_6H_{13})_{24}$ > $Au_{38}(SC_6H_{13})_{24}$	[103]
PPh_3 PPh_2 [a] Cl [b] PhF_2S	MoS_2	0.5 M phosphate buffer solution (pH 6.7) [c,d]	$Au_2Pd_6(S_4(PPh_3)_4(PhF_2S)_6)/MoS_2$ > Mixture of $Au_2Cl_2C(PPh_2)_2$ and $Pd_3(Cl(PPh_2)_2(PPh_3)_3)/MoS_2$ > $Pd_3(Cl(PPh_2)_2(PPh_3)_3)/MoS_2$ > $Au_2Cl_2C(PPh_2)_2/MoS_2$ > MoS_2	[104]
porphyrin SC_1P porphyrin SC_2P PET	–	0.5 M H_2SO_4 aq [e]	Au(1.3 nm)(porphyrin SC_1P) > Au(1.3 nm)(porphyrin SC_2P) > Au(1.3 nm)(PET)	[107]
PET SePh	MoS_2	0.5 M H_2SO_4 aq [c,d]	$Au_{25}(PET)_{18}/MoS_2$ > $Au_{25}(SePh)_{18}/MoS_2$ > MoS_2	[108]
SC_6H_{13} MPA MPS	–	0.1 M KCl aq [c]	$Au_{24}Pt(MPS)_{18}$ > $Au_{25}(MPS)_{18}$ > $Au_{25}(MPA)_{18}$ > $Au_{25}(SC_6H_{13})_{18}$	[109]

[a] Diphenylphosphine. [b] Chlorine. [c] WE: Working electrode; GCE. [d] WE: Containing Nafion. [e] WE: Carbon tape.

The HER activity varies depending not only on the chemical composition of the metal core but also on the properties of the ligand. In 2018, Teranishi and Sakamoto et al., used Au_n NCs coordinated with SR-containing porphyrin (porphyrin SC_xP). They investigated the effects of the ligand structure on the HER activity of $Au_n(SR)_m$ NCs [107]. In these clusters, the porphyrin ring coordinates horizontally to the gold core. Then, the distance between the porphyrin ring and the Au surface was controlled by changing the length of the alkyl chain between the porphyrin ring and the acetylthio group (Figure 3A,C) [118,119]. The alkyl chain is a methylene chain for porphyrin SC_1P and an ethylene chain for porphyrin SC_2P. The distance between the porphyrin ring and the acetylthio group was determined to be 3.4 Å for porphyrin SC_1P and 4.9 Å for porphyrin SC_2P by SC-XRD analysis. The researchers synthesized three sizes of Au_n NCs with a core size of approximately 1.3, 2.2, or 3.8 nm using porphyrin SC_1P, porphyrin SC_2P, or a common protective ligand, 2-phenylethanethiolate (PET). Transmission electron microscope (TEM) images of the synthesized $Au_n(SR)_m$ NCs (SR = porphyrin SC_1P, porphyrin SC_2P, or PET) with a core size of approximately 1.3 nm are presented in Figure 3B,D,F, respectively. Among these products, matrix-assisted laser desorption/ionization mass spectrometry indicated that $Au_n(\text{porphyrin } SC_1P)_m$ NCs consisted of 77 Au atoms and 8 porphyrin SC_1P molecules and $Au_n(\text{porphyrin } SC_2P)_m$ NCs consisted of 75 Au atoms and 11 porphyrin SC_2P molecules. The effects of the ligand structure and Au core size on the HER activity of $Au_n(SR)_m$ NCs were investigated using the obtained nine types of $Au_n(SR)_m$ NCs. As a result, in $Au_n(SR)_m$ NCs with a core size of approximately 1.3 nm, $Au_n(\text{porphyrin } SC_1P)_m$ and $Au_n(\text{porphyrin } SC_2P)_m$ NCs exhibited higher current densities of the HER than $Au_n(PET)_m$ NCs (Table 1). For instance, $Au_n(\text{porphyrin } SC_1P)_m$ NCs resulted in a 4.6 times higher current density of the HER than $Au_n(PET)_m$ NCs at −0.4 V vs. RHE. In addition, using $Au_n(\text{porphyrin } SC_1P)_m$ NCs, the HER occurred at a smaller overvoltage than using $Au_n(\text{porphyrin } SC_2P)_m$ NCs. These results indicate that the HER activity of Au_n NCs depends on the type of ligand and the distance between the ligand and the metal core in Au_n NCs [107]. In this work, the $Au_n(SR)_m$ NCs with a core size of approximately 2.2 nm showed higher catalytic activity than those with a core size of approximately 1.3 nm (Figure 3G,H). This size dependence of the catalytic activity is a little strange considering the surface area of the metal core because a reduction of a core size of $Au_n(SR)_m$ NCs typically leads to the increase in the surface area of Au metal core, which are active sites in HER. The authors have not discussed the details on this point in this paper probably due to the difficulty in precisely estimating the surface area of each $Au_n(SR)_m$ NCs.

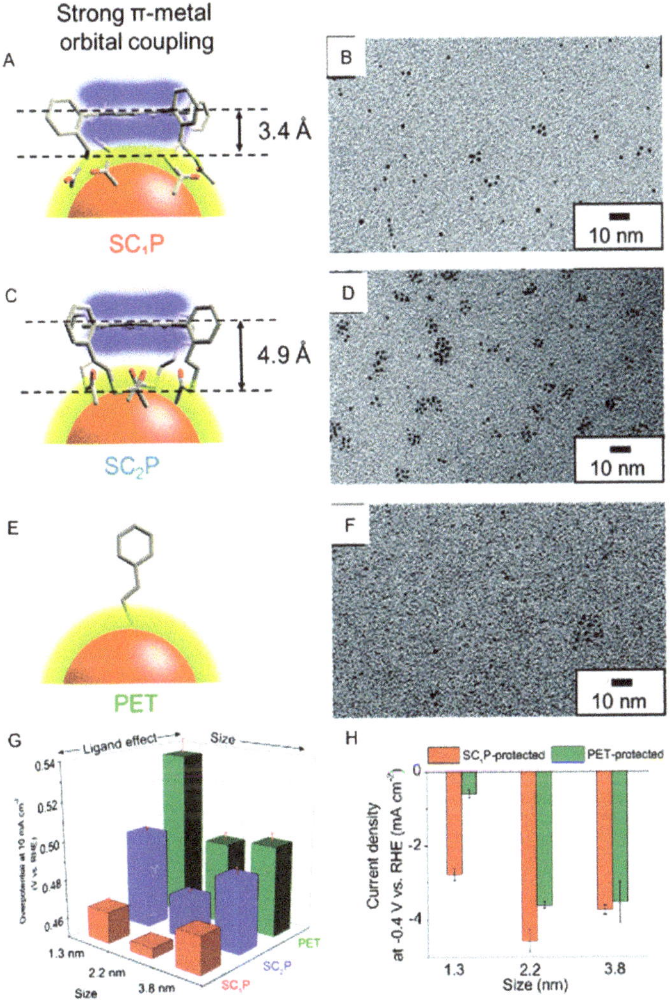

Figure 3. (**A,C,E**) Schematic illustration of coordination of ligands: (**A**) porphyrin SC_1P, (**C**) porphyrin SC_2P, and (**E**) PET. (**B,D,F**) TEM images of Au NCs with a core size of approximately 1.3 nm protected by porphyrin SC_1P, porphyrin SC_2P, or PET, respectively. (**G**) Comparison of overpotential at -10 mA cm^{-2} and (**H**) current density at -0.4 V of each size of Au NCs protected with each ligand. Panels (**A–H**) are reproduced with permission from reference [107]. Copyright Royal Society of Chemistry, 2018.

The property of the ligand also strongly affects the interaction between $Au_n(SR)_m$ NCs and the electrode as well as the affinity between $Au_n(SR)_m$ NCs and water molecules. Lee and Jiang et al., synthesized $Au_n(SR)_m$ NCs with SC_6H_{13}, 3-mercaptopropionic acid (MPA), or 3-mercapto-1-propanesulfonic acid (MPS; Figure 4B) as a ligand ($Au_{25}(SC_6H_{13})_{18}$, $Au_{25}(MPA)_{18}$, and $Au_{25}(MPS)_{18}$) and used them to investigate the effect of ligand properties on the HER activity [109]. In the experiment, $Au_{25}(SC_6H_{13})_{18}$, $Au_{25}(MPA)_{18}$, or $Au_{25}(MPS)_{18}$ was dissolved at a concentration of 1 mM in 0.1 M KCl aqueous solution, and LSV measurements were performed using a GCE (50 mV s^{-1}).

Although the blank current was 0.01 mA at −0.7 V vs. RHE (Figure 4C, black line), the HER current of the sample including $Au_{25}(MPA)_{18}$ increased up to 0.13 mA at −0.7 V vs. RHE (Figure 4C, red line). When $Au_{25}(MPS)_{18}$ was used, a higher HER current of 1.0 mA was observed at −0.7 V vs. RHE (Figure 4C, blue line). MPS and MPA have a hydrophilic functional group (sulfonic acid or carboxylic acid group, respectively) unlike SC_6H_{13}. These hydrophilic functional groups have the property of releasing H^+ in an aqueous solution. In addition, the sulfonic acid group of MPS (pKa < 1) is expected to have higher H^+ releasing ability than the carboxylic acid group of MPA (pKa = 3.7). For these reasons, it was interpreted that the difference in the HER activity described above is largely related to the difference in the H^+ releasing ability of these ligands (Table 1). It was speculated that the energy barrier associated with the intermolecular and intramolecular H^+ transfer steps is lowered by H^+ relay in Au_n NCs with high HER activity (Figure 4A). In this paper, they also reported that the use of $Au_{24}Pt(MPS)_{18}$, in which $Au_{25}(MPS)_{18}$ is replaced with Pt, results in even higher HER activity than $Au_{25}(MPS)_{18}$ (Figure 4D and Table 1). They descried that the TOF value of $Au_{24}Pt(MPS)_{18}$ was 127 mol H_2 (mol catalyst)$^{-1}$ s^{-1}, which was 4 times higher than that of $Au_{25}(MPS)_{18}$ at −0.7 V vs. RHE.

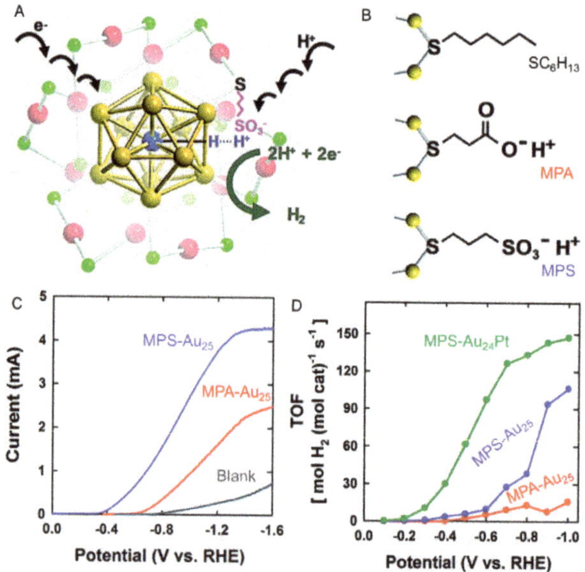

Figure 4. (**A**) Schematic illustration of proton relay mechanism of $Au_{24}Pt(SR)_{18}$ nanocluster for formation of H_2 and (**B**) ligand structures: SC_6H_{13}, MPA, and MPS. Color codes: blue = Pt; golden = core Au; red = shell Au; and green = S. (**C**) HER polarization curves in 0.1 M KCl aqueous solution containing 180 mM acetic acid for MPA-Au_{25} (red) or MPS-Au_{25} (blue). (**D**) turnover frequencies (TOFs) obtained at various potentials in water (3.0 M KCl) containing 180 mM HOAc for MPA-Au_{25} (red), MPS-Au_{25} (blue), or MPS-$Au_{24}Pt$ (green). Panels (**A**–**D**) are reproduced with permission from reference [109]. Copyright Royal Society of Chemistry, 2018.

An electronic interaction also occurs between the $Au_n(SR)_m$ NCs and a catalytic support. This phenomenon was revealed by Jin et al., by measuring the HER activity of MoS_2 nanosheets (catalytic support) carrying $Au_{25}(PET)_{18}$ ($Au_{25}(PET)_{18}/MoS_2$) [108]. In this experiment, $Au_{25}(PET)_{18}/MoS_2$ was prepared by mixing the MoS_2 nanosheets synthesized by the hydrothermal method and $Au_{25}(PET)_{18}$ in dichloromethane for 1 h and drying the obtained products under nitrogen atmosphere. High-angle annular dark-field scanning TEM (HAADF-STEM) images confirmed that $Au_{25}(PET)_{18}$ was uniformly supported on MoS_2 (Figure 5A). $Au_{25}(PET)_{18}/MoS_2$ was then loaded on a GCE, and the HER polarization curve of $Au_{25}(PET)_{18}/MoS_2$ was obtained by scanning the potential in a 0.5 M H_2SO_4 aqueous solution using the rotating disk electrode (RDE) method (Figure 5B,D). MoS_2 without $Au_{25}(PET)_{18}$ exhibited a HER overvoltage of 0.33 V at a current density of 10 mA cm^{-2}, whereas $Au_{25}(PET)_{18}/MoS_2$ exhibited a smaller HER overvoltage of approximately −0.28 V at the same current density. In addition, $Au_{25}(PET)_{18}/MoS_2$ (59.3 mA cm^{-2}) exhibited a 1.79 times higher current density than that of MoS_2 (33.2 mA cm^{-2}) at an applied voltage of −0.4 V vs. RHE. Thus, the HER activity of the MoS_2 nanosheets was greatly improved by carrying $Au_{25}(PET)_{18}$ (Table 1). This improvement of the HER activity was interpreted to be greatly related to the electronic interaction between $Au_{25}(PET)_{18}$ and MoS_2. In fact, X-ray photoelectron spectroscopy (XPS) analysis confirmed that the binding energy of MoS_2 in the Mo 3 d orbit was negatively shifted by 0.4 eV after $Au_{25}(PET)_{18}$ was loaded (Figure 5C). It was assumed that the charge transfer from $Au_{25}(PET)_{18}$ to MoS_2 occurred in $Au_{25}(PET)_{18}/MoS_2$, causing a high HER activity of $Au_{25}(PET)_{18}/MoS_2$. In this study, the HER activity of MoS_2 nanosheets carrying $Au_{25}(SePh)_{18}$ (SePh = phenylselenolate) ($Au_{25}(SePh)_{18}/MoS_2$) was also investigated. $Au_{25}(SePh)_{18}/MoS_2$ was shown to also exhibit higher HER activity than MoS_2 nanosheets (Table 1). However, the improvement of the activity was smaller than that when carrying $Au_{25}(PET)_{18}$ (Figure 5D). This difference was attributed to the difference in the electron interaction and electron relay between Au cores of Au_n NCs and the MoS_2 nanosheet depending on the ligands. In this way, the HER activity of the Au_n NCs-loaded catalyst was shown to depend on the electronic interaction between the Au_n NCs and the catalytic support.

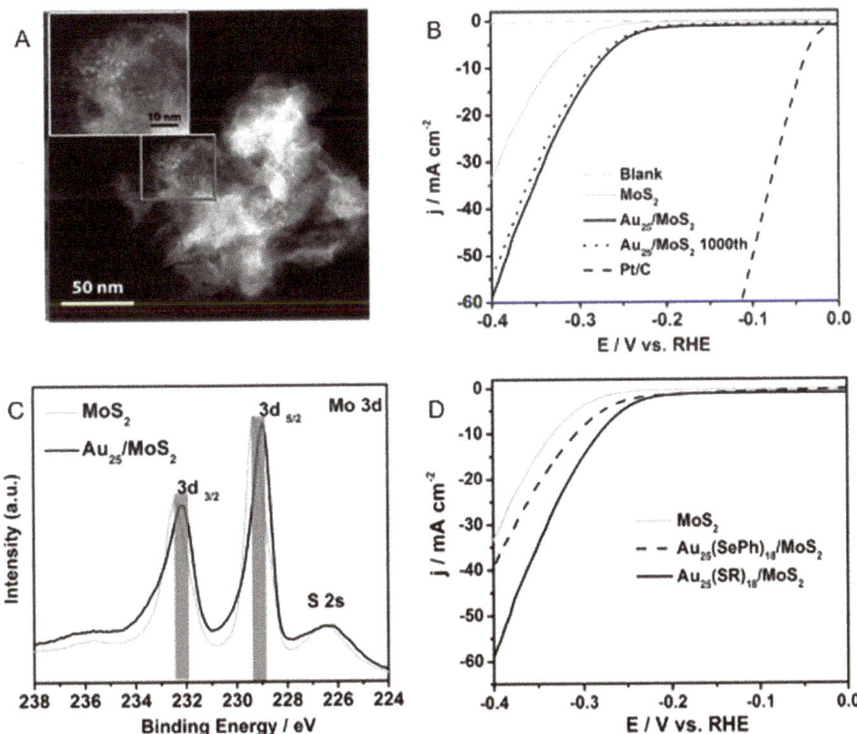

Figure 5. (**A**) High-angle annular dark-field scanning TEM (HAADF-STEM) images, (**B**) HER polarization curves, and (**C**) Mo 3d X-ray photoelectron spectroscopy (XPS) spectra of $Au_{25}(PET)_{18}/MoS_2$. (**D**) HER polarization curves of $Au_{25}(SePh)_{18}/MoS_2$. Panels (**A–D**) are reproduced with permission from reference [108]. Copyright Wiley-VCH, 2017.

2.2. Oxygen Evolution Reaction

The OER is a multi-step four-electron reaction in which the reaction proceeds along different reaction paths depending on the binding energy between the metal and the OER intermediate (O, OH, and OOH).

Under acidic conditions, the following reactions occur:

$$M + H_2O \rightarrow M{-}OH + H^+ + e^- \tag{7}$$

$$M{-}OH \rightarrow M{-}O + H^+ + e^- \tag{8}$$

$$2(M{-}O) \rightarrow 2M + O_2 \tag{9}$$

$$M{-}O + H_2O \rightarrow M{-}OOH + H^+ + e^- \tag{10}$$

$$M{-}OOH \rightarrow M + O_2 + H^+ + e^- \tag{11}$$

However, under alkaline conditions, the following reactions occur:

$$M + OH^- \rightarrow M{-}OH + e^- \tag{12}$$

$$M{-}OH + OH^- \rightarrow M{-}O + H_2O + e^- \tag{13}$$

$$2(M{-}O) \rightarrow 2M + O_2 \tag{14}$$

$$M{-}O + OH^- \rightarrow M{-}OOH + e^- \tag{15}$$

$$M{-}OOH + OH^- \rightarrow M + O_2 + H_2O + e^- \tag{16}$$

As described above, because the reaction route of OER depends on the intermediates (O, OH, and OOH) on the surface of catalyst, the OER activity of the catalyst also depends on these intermediates. Catalysts that have neither too high nor too low binding energy with oxygen species are suitable for the OER. Previous studies have demonstrated that iridium oxide and ruthenium oxide have such desirable properties. Therefore, miniaturization of these metal oxides and prediction of their physical properties by theoretical calculation have been actively performed [120–123]. However, because these precious metals are expensive and have the problem of depletion, a search for low-cost catalysts is also being conducted. Related studies have shown that cobalt (Co)-based materials (oxides, hydroxides, selenides, and phosphides) can be used as good OER catalysts. Furthermore, it has been reported that when Au NPs are composited with such Co materials, the OER performance is greatly enhanced as a result of the improved electron conductivity and preferential formation of OOH intermediates on the surface of the catalyst [124–126].

Jin et al., have shown that these mixing effects also occur when Au_n NCs are used instead of Au NPs [110]. In this study, the $Au_{25}(PET)_{18}$-loaded $CoSe_2$ nanosheet ($Au_{25}(PET)_{18}/CoSe_2$) was prepared by stirring $Au_{25}(PET)_{18}$ and $CoSe_2$ nanosheets in dichloromethane for 1 h. HAADF-STEM analysis confirmed that $Au_{25}(PET)_{18}$ was uniformly supported on the $CoSe_2$ nanosheets (Figure 6A,B). $Au_{25}(PET)_{18}/CoSe_2$ was loaded on the GCE, and their OER polarization curves were obtained by scanning the applied potential (5 mV s^{-1}) in 0.1 M KOH aqueous solution. The $CoSe_2$ nanosheets without $Au_{25}(PET)_{18}$ exhibited an OER overvoltage of 0.52 V at a current density of 10 mA cm^{-2} (Figure 1B(b)), whereas $Au_{25}(PET)_{18}/CoSe_2$ exhibited a smaller OER overvoltage of 0.43 V at the same current density (Figure 6C). XPS (Figure 6E) and Raman spectroscopy (Figure 6F) analyses revealed that the electronic interaction occurred between the $Au_{25}(PET)_{18}$ and $CoSe_2$ nanosheet even in such a composite catalyst. Furthermore, DFT calculation revealed that the formation of the intermediate via OH$^-$ is more advantageous by 0.21 eV mol^{-1} at the interface of Co–Au than at the surface of Co. It was thus interpreted that $Au_{25}(PET)_{18}/CoSe_2$ exhibited higher OER activity than the $CoSe_2$ nanosheets because $Au_{25}(PET)_{18}/CoSe_2$ stabilized the generation of an OOH intermediate compared with only the

CoSe$_2$ nanosheet (Table 2). This study also revealed that the OER activity increases with the core size of Au$_n$(SR)$_m$ NCs (Figure 6D).

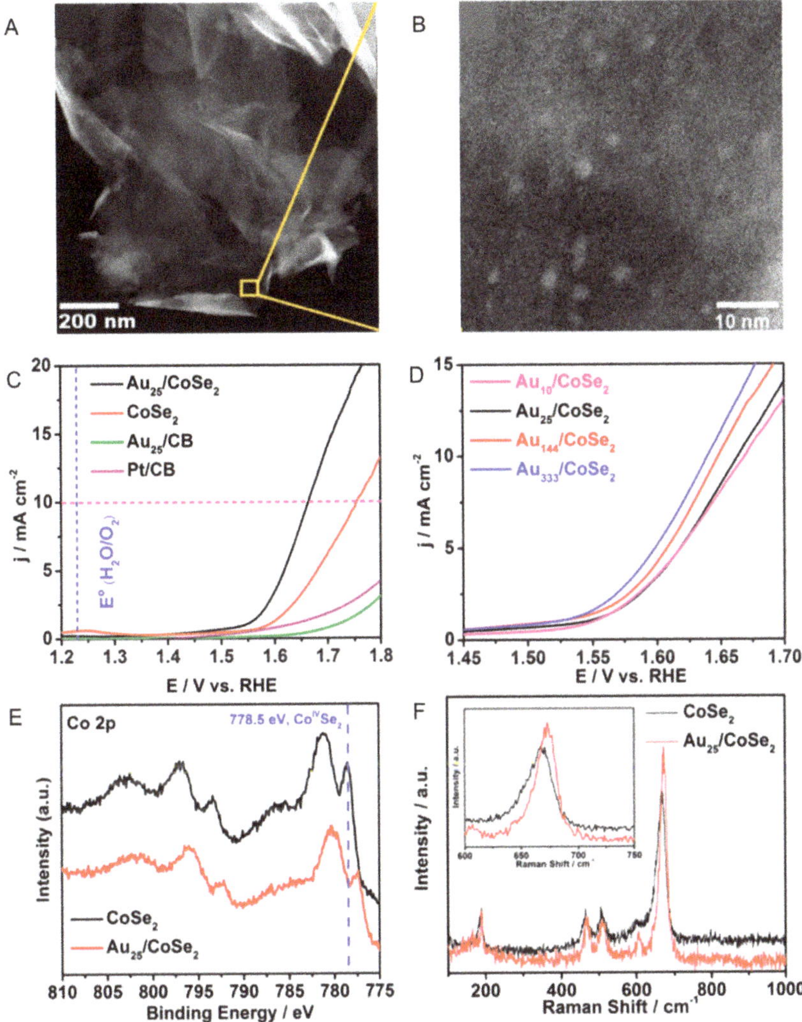

Figure 6. (**A,B**) HAADF-STEM images of Au$_{25}$(PET)$_{18}$/CoSe$_2$ composite at different magnifications. (**C,D**) OER polarization curves of CoSe$_2$, Au$_{10}$(SPh-^{t}Bu)$_{10}$/CoSe$_2$, Au$_{25}$(PET)$_{18}$/CoSe$_2$, Au$_{144}$(PET)$_{60}$/CoSe$_2$, Au$_{333}$(PET)$_{79}$/CoSe$_2$, and Pt$_{NP}$/CB (CB = carbon black). (**E**) Co 2p XPS spectra and (**F**) Raman spectra of CoSe$_2$ and Au$_{25}$(PET)$_{18}$/CoSe$_2$ composite. Panels (**A–F**) are reproduced with permission from reference [110]. Copyright American Chemical Society, 2017.

Table 2. Representative reference on OER activity of Au$_n$(SR)$_m$ NCs.

Ligand	Support	Experimental Condition	Activity	Reference
PET	CoSe$_2$	0.1 M KOH aq [a, b]	Au$_{25}$(PET)$_{18}$/CoSe$_2$ > CoSe$_2$	[110]

[a] WE: GCE. [b] WE: Containing Nafion.

3. Electrocatalytic Reactions in Fuel Cells

To establish a circulating energy system that does not use fossil fuels and only produces water and a small amount of carbon dioxide as waste, it is essential to further improve the functions of fuel cells. Fuel cells can be roughly classified into those using hydrogen and those using alcohol as a fuel. In fuel cells using hydrogen as a fuel, the HOR and ORR are involved in the system. The HOR is a one-electron reaction, and generally an HER-active catalyst is also useful for the HOR. However, the ORR is a four-electron reaction, and the reaction process is complicated. In addition, the OER is a reaction under oxidizing conditions, whereas the ORR is a reaction under reducing conditions. The surface state of the catalyst and the accompanying binding to the reactants also differ greatly between the OER and ORR. Therefore, catalysts that are active for OER are not necessarily useful for the ORR. Because the ORR is rate-limiting step in a fuel cell, controlling the ORR is important for further development of fuel cells. The ORR pathways under acidic and alkaline conditions are as follows [94].

Under acidic conditions:

$$O_2 + 4H^+ + 4e^- \rightarrow 2H_2O \tag{17}$$

$$O_2 + 2H^+ + 2e^- \rightarrow H_2O_2 \tag{18}$$

$$H_2O_2 + 2H^+ + 2e^- \rightarrow 2H_2O \tag{19}$$

Under alkaline conditions:

$$O_2 + 2H_2O + 4e^- \rightarrow 4OH^- \tag{20}$$

$$O_2 + H_2O + 2e^- \rightarrow OOH^- + OH^- \tag{21}$$

$$OOH^- + H_2O + 2e^- \rightarrow 3OH^- \tag{22}$$

Equations (17) and (20) are four-electron reactions, and Equations (18), (19), (21), and (22) are two-electron reactions. For both sets of reactions, the reactions start with the breaking of the O–O bond. The theoretical redox potential is 1.23 V vs. SHE in the direct four-electron path and 0.68 V vs. SHE in the indirect two-electron path. Therefore, a higher energy conversion efficiency can be achieved using the direct four-electron path, and this reaction path is thus more desirable for fuel cells [81]. Although Pt is a useful catalyst for such a reaction pathway, it is expected to be replaced with another metal element because of the high cost of Pt and the resource depletion issue. In addition, synthesis methods of Pt_n NCs in ambient atmosphere with atomic precision are limited, and therefore, it is difficult to study the ORR mechanism using Pt_n NCs as model catalysts. However, for Au_n NCs, there are many examples of synthesis with atomic precision, and these catalysts are stable in ambient atmosphere. In addition, theoretical calculations [127,128] and experimental results [65,129] have predicted that O_2 molecules can be highly activated on the surface of Au_n NCs. For these reasons, several studies have also been performed on the application of Au_n NCs as ORR catalysts.

In 2009, Chen et al., evaluated the ORR catalytic activity of $Au_{11}(PPh_3)_8Cl_3$, $Au_{25}(PET)_{18}$, $Au_{55}(PPh_3)_{12}Cl_6$, and $Au_{140}(SC_6H_{13})_{53}$ (Cl = chlorine) [111]. In this experiment, after a series of Au_n NCs were loaded on the GCE, the ORR activity was measured by scanning the potential using the RDE method in a 0.1 M KOH aqueous solution filled with O_2. When $Au_{11}(PPh_3)_8Cl_3$ was used as the Au_n NCs, the onset potential of the ORR (Figure 1B(c)) was about −0.08 V, and the peak current density was 2.4 mA cm^{-2} (Figure 7A). However, when $Au_{140}(SC_6H_{13})_{53}$ was used as the Au_n NCs, the onset potential shifted to the more cathodic −0.22 V and the reduction peak current decreased to less than 1.0 mA cm^{-2}. These results and those for the other two Au_n NCs indicated that the ORR activity increased with decreasing Au core size ($Au_{11}(PPh_3)_8Cl_3$ > $Au_{25}(PET)_{18}$ > $Au_{55}(PPh_3)_{12}Cl_6$ > $Au_{140}(SC_6H_{13})_{53}$) (Figure 7A,B and Table 3). From estimation of the number of electrons for the ORR from a Koutecky–Levich plot [85], it was observed that the relatively small size of Au_n NCs ($Au_{11}(PPh_3)_8Cl_3$, $Au_{25}(PET)_{18}$, and $Au_{55}(PPh_3)_{12}Cl_6$) resulted in the occurrence of the four-electron reaction, whereas $Au_{140}(SC_6H_{13})_{53}$ tended to follow the two-electron reaction pathway (Figure 7C,D). Later, these researchers also synthesized a series of $Au_n(SR)_m$ NCs ($Au_{25}(PET)_{18}$, $Au_{38}(PET)_{24}$, and

$Au_{144}(PET)_{60}$) with PET ligands and measured their ORR activities. The results revealed that a smaller core size was associated with higher ORR activity: $Au_{25}(PET)_{18} > Au_{38}(PET)_{24} > Au_{144}(PET)_{60}$ (Table 3) [112]. As the core size decreased, the ratio of low-coordinated surface atoms increased and the d-band center of the Fermi level changed. It was interpreted that smaller $Au_n(SR)_m$ NCs exhibited higher ORR activity because the promotion of oxygen adsorption on the gold core surface was accelerated by miniaturization of the metal core.

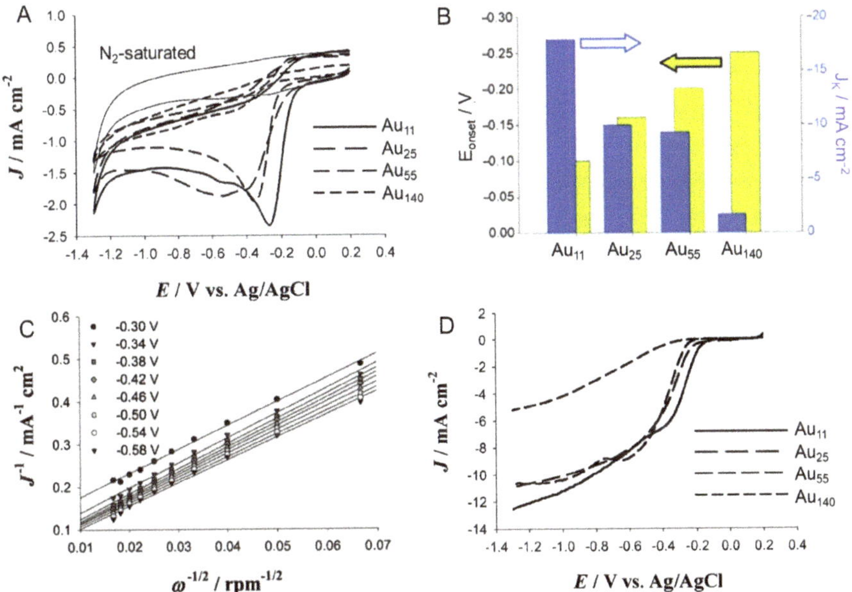

Figure 7. (**A**) Cyclic voltammograms of $Au_n(SR)_m$/GCE (n = 11, 25, 55, and 140) saturated with O_2 and $Au_{11}(PPh_3)_8Cl_3$/GCE saturated with N_2 (thin solid curve). (**B**) Current density and overpotential of ORR activity with each size of Au_n NCs. (**C**) Koutecky–Levich plots at different applied potentials of a GCE modified with $Au_{11}(PPh_3)_8Cl_3$. (**D**) Rotating-disk voltammograms (rotation rate: 3600 rpm) of various $Au_n(SR)_m$/GCE (n = 11, 25, 55, and 140). Panels (**A–D**) are reproduced with permission from reference [111]. Copyright Wiley-VCH, 2009.

Table 3. Representative references on ORR activity of Au_n NCs.

Ligand	Support	Experimental Condition	Activity	Reference
PET SC$_6$H$_{13}$ Cl PPh$_3$	–	0.1 M KOH aq ^{a}	$Au_{11}(PPh_3)_8Cl_3 > Au_{25}(PET)_{18} > Au_{55}(PPh_3)_{12}Cl_6 > Au_{140}(SC_6H_{13})_{53}$	[101]
PET	Reduced graphene oxide	0.1 M KOH aq $^{a,\,b}$	$Au_{25}(PET)_{18} > Au_{38}(PET)_{24} > Au_{144}(PET)_{60}$	[112]
TBBT	SWNTs	0.1 M KOH aq $^{a,\,b}$	$Au_{36}(TBBT)_{24} > Au_{133}(TBBT)_{52} > Au_{279}(TBBT)_{84} > Au_{28}(TBBT)_{20}$	[113]
S-^{t}Bu	SWNTs	0.1 M KOH aq $^{a,\,b}$	$Au_{65}(S\text{-}^tBu)_{29} > Au_{46}(S\text{-}^tBu)_{24} > Au_{30}(S\text{-}^tBu)_{18} > Au_{23}(S\text{-}^tBu)_{16}$	[114]
SC$_{12}$H$_{25}$	–	0.1 M KOH aq $^{a,\,b}$	$[Au_{25}(SC_{12}H_{25})_{18}]^- > [Au_{25}(SC_{12}H_{25})_{18}]^0 > [Au_{25}(SC_{12}H_{25})_{18}]^{+\,c}$	[115]

^{a} WE: GCE. ^{b} WE; Containing Nafion. ^{c} Tow-electron reduction.

On the other hand, Dass et al., studied the dependence of the ORR activity on the core size using Au_n NCs protected by 4-*tert*-butylbenzenethiolate (TBBT), whose structure differs significantly from that of PET [113]. In this experiment, single-walled carbon nanotubes (SWNTs) carrying $Au_n(TBBT)_m$ NCs (n = 28, 36, 133, and 279; Figure 8A; $Au_n(TBBT)_m$ NCs/SWNTs) were loaded onto the GCE. The ORR actives were measured by scanning the potential using the RDE method in a 0.1 M KOH aqueous solution filled with O_2 (Figure 8B). The overvoltage of the ORR was smaller in the order of $Au_{36}(TBBT)_{24} > Au_{133}(TBBT)_{52} > Au_{279}(TBBT)_{84} > Au_{28}(TBBT)_{20}$. However, the selectivity of the four-electron reduction reaction was superior in the order of $Au_{36}(TBBT)_{24} \approx Au_{133}(TBBT)_{52} >$

Au$_{279}$(TBBT)$_{84}$ > Au$_{28}$(TBBT)$_{20}$ [113] (Figure 8C). Notably, this trend was similar to that of the size dependence of the stability of Au$_n$(TBBT)$_m$ NCs itself. The same group performed similar studies using *tert*-butylthiolate (S-^{t}Bu) instead of TBBT as a ligand [114]. S-^{t}Bu has a bulky framework and when this ligand is used in the synthesis of Au$_n$(SR)$_m$ NCs, the ratio of the metal atom and the ligand in the generated Au$_n$(SR)$_m$ NCs is different from that in Au$_n$(SR)$_m$ NCs synthesized using another ligand. Such Au$_n$(S-^{t}Bu)$_m$ NCs exhibit a unique size dependency for ORR activity (Au$_{65}$(S-^{t}Bu)$_{29}$ > Au$_{46}$(S-^{t}Bu)$_{24}$ > Au$_{30}$(S-^{t}Bu)$_{18}$ > Au$_{23}$(S-^{t}Bu)$_{16}$) [114].

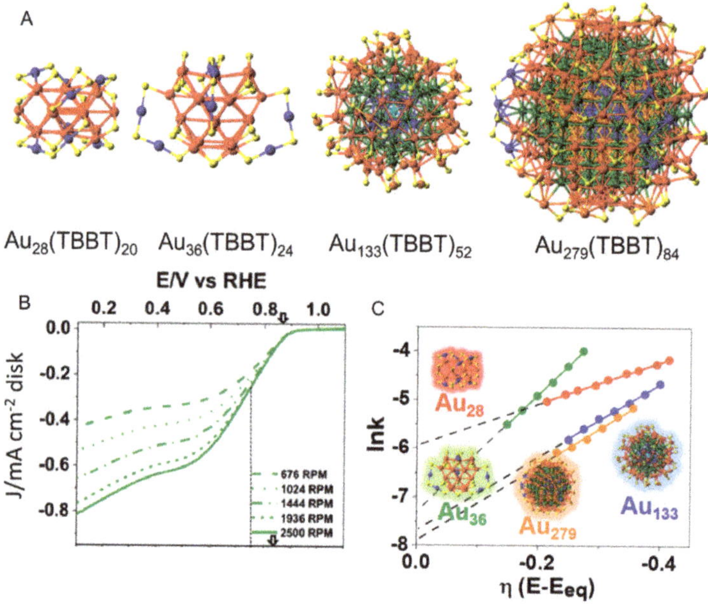

Figure 8. (**A**) X-ray crystal structures of Au$_n$(TBBT)$_m$ NCs (n = 28, 36, 133, and 279). (**B**) Rotating-disk voltammograms recorded for the ORR activity of Au$_{36}$(TBBT)$_{24}$/GCE at different rotation rates. (**C**) Reaction rate constant ln(k) vs. overpotential E plots with each size of Au$_n$(TBBT)$_m$ (n = 28, 36, 133, and 279). Panels (**A–C**) are reproduced with permission from reference [113]. Copyright American Chemical Society, 2018.

In addition to these effects of core sizes and ligands, the ORR activity also depended on the charge state of Au$_n$(SR)$_m$ NCs. Chen et al., carried [Au$_{25}$(SC$_{12}$H$_{25}$)$_{18}$]$^-$, [Au$_{25}$(SC$_{12}$H$_{25}$)$_{18}$]^0, and [Au$_{25}$(SC$_{12}$H$_{25}$)$_{18}$]$^+$ (SC$_{12}$H$_{25}$ = 1-dodecanethiolate) on the GCE, and their ORR activities were evaluated by scanning the potential in a 0.1 M KOH aqueous solution using a rotating ring-disk electrode (RRDE) filled with O$_2$ [115]. In addition, the generation of H$_2$O$_2$ was evaluated from the RRDE current at a fixed ring potential (0.5 V vs. saturated calomel electrode (SCE)). When [Au$_{25}$(SC$_{12}$H$_{25}$)$_{18}$]$^-$, [Au$_{25}$(SC$_{12}$H$_{25}$)$_{18}$]^0, and [Au$_{25}$(SC$_{12}$H$_{25}$)$_{18}$]$^+$ were used, the efficiencies of H$_2$O$_2$ were 86%, 82%, and 72%, respectively. In addition, the number of electrons for the ORR was estimated to be 2.28 ([Au$_{25}$(SC$_{12}$H$_{25}$)$_{18}$]$^-$), 2.35 ([Au$_{25}$(SC$_{12}$H$_{25}$)$_{18}$]^0), and 2.56 ([Au$_{25}$(SC$_{12}$H$_{25}$)$_{18}$]$^+$; Figure 9A–C). For [Au$_{25}$(SC$_{12}$H$_{25}$)$_{18}$]$^-$, which showed the highest production rate of H$_2$O$_2$, the activity decreased only 9% even after 1000 cycles (Figure 9D). These results indicate that [Au$_{25}$(SC$_{12}$H$_{25}$)$_{18}$]$^-$ has high H$_2$O$_2$ generating ability (Table 3) [115]. Since H$_2$O$_2$ is a useful raw material for chemical products, the development of their highly selective production reactions is important. Jin et al., also studied the dependence of the ORR activity on the charge state of Au$_n$(SR)$_m$ NCs using [Au$_{25}$(PET)$_{18}$]$^-$, [Au$_{25}$(PET)$_{18}$]^0, and [Au$_{25}$(PET)$_{18}$]$^+$. They reported that too strong of an OH$^-$ adsorbing ability of

[Au$_{25}$(PET)$_{18}$]$^+$ reduces the ORR activity [77]. Thus, it has been clarified that the charge state of Au$_n$(SR)$_m$ NCs also has a significant effect on the ORR activity of Au$_n$(SR)$_m$ NCs.

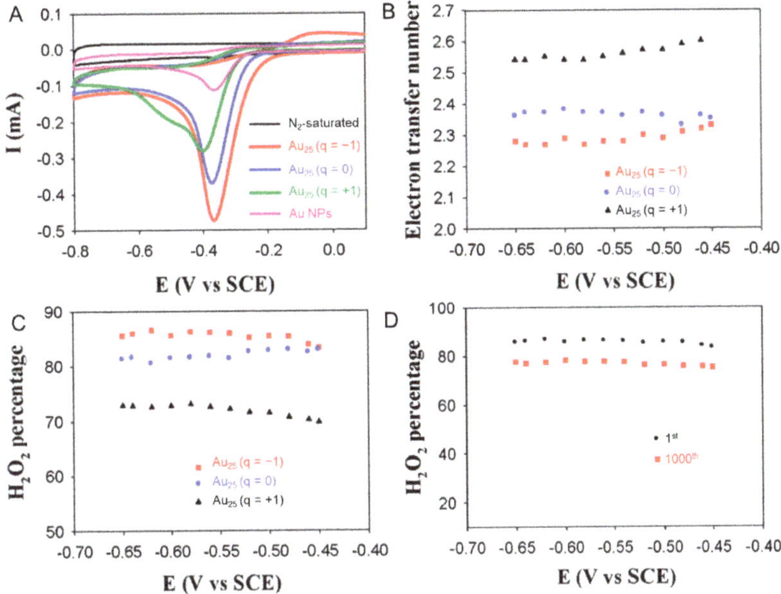

Figure 9. (**A**) Cyclic voltammograms, (**B**) electron transfer number (*n*), and (**C**) percentage of H$_2$O$_2$ of the ORR on Au$_{25}$(SC$_{12}$H$_{25}$)$_{18}$ with different charge states ([Au$_{25}$(SC$_{12}$H$_{25}$)$_{18}$]$^-$, [Au$_{25}$(SC$_{12}$H$_{25}$)$_{18}$]^0, and [Au$_{25}$(SC$_{12}$H$_{25}$)$_{18}$]$^+$) in 0.1 M KOH aq saturated with O$_2$. (**D**) Accelerated durability tests of [Au$_{25}$(SC$_{12}$H$_{25}$)$_{18}$]$^-$ performed for 1000 cycles. Panels (**A–D**) are reproduced with permission from reference [115]. Copyright Royal Society of Chemistry, 2014.

4. Conclusions

A system for the generation of a fuel such as hydrogen or methanol using natural energy (e.g., solar cells or photocatalytic water splitting) and the production of electricity by fuel cells using these fuels would be one of the ultimate energy conversion systems for our society. To realize such a system, high activation of the HER, OER, HOR, and ORR is indispensable. Recently, Au$_n$ NCs have attracted considerable attention as model catalysts for these reactions. In this review, recent works on these materials were summarized. The overall characteristics of the HER, OER, and ORR can be summarized as follows.

1) Since the core size, doping metal, ligand structure, and charge state affect the electronic and geometrical structures of Au$_n$ NCs, these parameters also have a great effect on the catalytic activity of Au$_n$ NCs.

2) Although these three reactions proceed via different mechanisms, reducing the core size of Au$_n$ NCs and improving the ligand conductivity tend to improve the activities.

3) When Au$_n$ NCs are carried on a conventional catalytic support, their electronic structure changes and thus their catalytic activity also changes. Therefore, Au$_n$ NCs are also useful for improving the catalytic activity of conventional catalytic materials.

5. Perspectives

Until recently, the materials with relatively high activity for all of HER, OER, and ORR are considered to be limited to Ir, Rh, Ru, and Pt [84,85]. However, the recent studies demonstrated that these properties could also be caused in Au by the discretization of the band structure (e.g., shift of d-band center [107,111]). For Au_n NCs, it is possible to precisely control the electronic/geometrical structures and thereby to elucidate the correlation between catalytic activity and electronic/geometrical structure. In addition, the use of fine Au_n NCs as a catalyst is effective in reducing the consumption of expensive noble metals. It is expected that the studies on the catalytic activities of Au_n NCs lead to solve the mechanism in catalytic reactions on the metal surface and create the amazing catalysts we have never seen.

However, to create such HER, OER, and ORR catalysts using Au_n NCs and their alloy NCs, further studies are required. Previous studies have shown that doping with Group 10 elements (Pt and Pd) induces high activation. Thus, a method for increasing the doping concentration of these elements is expected to be developed in the future. In addition, regarding the HER and OER, in spite of decomposing water, most studies thus far have used hydrophobic ligands that are not compatible with water. This may be related to the fact that the synthesis of hydrophobic Au_n NCs is easier than that of hydrophilic Au_n NCs. In particular, it is difficult to selectively synthesize a group-10-element-doped cluster using a hydrophilic ligand using the conventional synthesis method. However, as shown in this review, it is more appropriate to use hydrophilic Au_n NCs as HER and OER catalysts. Therefore, in the future, additional research on hydrophilic Au_n NCs is expected to increase the types of ligands and core sizes of hydrophilic Au_n NCs. Such studies are expected to lead to the creation of highly active HER, OER, and ORR catalysts and eventually to the development of design guidelines for establishing ultimate energy conversion systems.

Author Contributions: T.K. and Y.N. developed the idea for this review article and wrote the paper. All authors have read and agreed to the published version of the manuscript.

Funding: This work was supported by the Japan Society for the Promotion of Science (JSPS) KAKENHI (grant number JP18K14074 and JP18H01953), Scientific Research on Innovative Areas "Coordination Asymmetry" (grant number 17H05385) and Scientific Research on Innovative Areas "Innovations for Light-Energy Conversion" (grant number 18H05178).

Conflicts of Interest: There are no conflicts to declare.

References

1. Brust, M.; Walker, M.; Bethell, D.; Schiffrin, D.J.; Whyman, R. Synthesis of Thiol-Derivatised Gold Nanoparticles in a Two-Phase Liquid–Liquid System. *J. Chem. Soc. Chem. Commun.* **1994**, 801–802. [CrossRef]

2. Jin, R.; Zeng, C.; Zhou, M.; Chen, Y. Atomically Precise Colloidal Metal Nanoclusters and Nanoparticles: Fundamentals and Opportunities. *Chem. Rev.* **2016**, *116*, 10346–10413. [CrossRef] [PubMed]

3. Kurashige, W.; Niihori, Y.; Sharma, S.; Negishi, Y. Precise Synthesis, Functionalization and Application of Thiolate-Protected Gold Clusters. *Coord. Chem. Rev.* **2016**, *320*, 238–250. [CrossRef]

4. Hossain, S.; Niihori, Y.; Nair, L.V.; Kumar, B.; Kurashige, W.; Negishi, Y. Alloy Clusters: Precise Synthesis and Mixing Effects. *Acc. Chem. Res.* **2018**, *51*, 3114–3124. [CrossRef] [PubMed]

5. Gan, Z.; Xia, N.; Wu, Z. Discovery, Mechanism; Application of Antigalvanic Reaction. *Acc. Chem. Res.* **2018**, *51*, 2774–2783. [CrossRef] [PubMed]

6. Chakraborty, I.; Pradeep, T. Atomically Precise Clusters of Noble Metals: Emerging Link between Atoms and Nanoparticles. *Chem. Rev.* **2017**, *117*, 8208–8271. [CrossRef]

7. Yao, Q.; Chen, T.; Yuan, X.; Xie, J. Toward Total Synthesis of Thiolate-Protected Metal Nanoclusters. *Acc. Chem. Res.* **2018**, *51*, 1338–1348. [CrossRef]

8. Qian, H.; Zhu, M.; Wu, Z.; Jin, R. Quantum Sized Gold Nanoclusters with Atomic Precision. *Acc. Chem. Res.* **2012**, *45*, 1470–1479. [CrossRef]

9. Whetten, R.L.; Weissker, H.-C.; Pelayo, J.J.; Mullins, S.M.; López-Lozano, X.; Garzón, I.L. Chiral-Icosahedral (I) Symmetry in Ubiquitous Metallic Cluster Compounds (145A,60X): Structure and Bonding Principles. *Acc. Chem. Res.* **2019**, *52*, 34–43. [CrossRef]

10. Aikens, C.M. Electronic and Geometric Structure, Optical Properties, and Excited State Behavior in Atomically Precise Thiolate-Stabilized Noble Metal Nanoclusters. *Acc. Chem. Res.* **2018**, *51*, 3065–3073. [CrossRef]

11. Nieto-Ortega, B.; Bürgi, T. Vibrational Properties of Thiolate-Protected Gold Nanoclusters. *Acc. Chem. Res.* **2018**, *51*, 2811–2819. [CrossRef] [PubMed]

12. Agrachev, M.; Ruzzi, M.; Venzo, A.; Maran, F. Nuclear and Electron Magnetic Resonance Spectroscopies of Atomically Precise Gold Nanoclusters. *Acc. Chem. Res.* **2019**, *52*, 44–52. [CrossRef] [PubMed]

13. Pei, Y.; Wang, P.; Ma, Z.; Xiong, L. Growth-Rule-Guided Structural Exploration of Thiolate-Protected Gold Nanoclusters. *Acc. Chem. Res.* **2019**, *52*, 23–33. [CrossRef] [PubMed]

14. Ghosh, A.; Mohammed, O.F.; Bakr, O.M. Atomic-Level Doping of Metal Clusters. *Acc. Chem. Res.* **2018**, *51*, 3094–3103. [CrossRef]

15. Bigioni, T.P.; Whetten, R.L.; Dag, Ö. Near-Infrared Luminescence from Small Gold Nanocrystals. *J. Phys. Chem. B* **2000**, *104*, 6983–6986. [CrossRef]

16. Yan, J.; Teo, B.K.; Zheng, N. Surface Chemistry of Atomically Precise Coinage–Metal Nanoclusters: From Structural Control to Surface Reactivity and Catalysis. *Acc. Chem. Res.* **2018**, *51*, 3084–3093. [CrossRef]

17. Sakthivel, N.A.; Dass, A. Aromatic Thiolate-Protected Series of Gold Nanomolecules and a Contrary Structural Trend in Size Evolution. *Acc. Chem. Res.* **2018**, *51*, 1774–1783. [CrossRef]

18. Tang, Q.; Hu, G.; Fung, V.; Jiang, D.-E. Insights into Interfaces, Stability, Electronic Properties, and Catalytic Activities of Atomically Precise Metal Nanoclusters from First Principles. *Acc. Chem. Res.* **2018**, *51*, 2793–2802. [CrossRef]

19. Negishi, Y.; Nobusada, K.; Tsukuda, T. Glutathione-Protected Gold Clusters Revisited: Bridging the Gap between Gold(I)–Thiolate Complexes and Thiolate-Protected Gold Nanocrystals. *J. Am. Chem. Soc.* **2005**, *127*, 5261–5270. [CrossRef]

20. Jadzinsky, P.D.; Calero, G.; Ackerson, C.J.; Bushnell, D.A.; Kornberg, R.D. Structure of a Thiol Monolayer-Protected Gold Nanoparticle at 1.1 Å Resolution. *Science* **2007**, *318*, 430–433. [CrossRef]

21. Wang, S.; Li, Q.; Kang, X.; Zhu, M. Customizing the Structure, Composition, and Properties of Alloy Nanoclusters by Metal Exchange. *Acc. Chem. Res.* **2018**, *51*, 2784–2792. [CrossRef] [PubMed]

22. Higaki, T.; Li, Q.; Zhou, M.; Zhao, S.; Li, Y.; Li, S.; Jin, R. Toward the Tailoring Chemistry of Metal Nanoclusters for Enhancing Functionalities. *Acc. Chem. Res.* **2018**, *51*, 2764–2773. [CrossRef] [PubMed]

23. Negishi, Y.; Kurashige, W.; Niihori, Y.; Iwasa, T.; Nobusada, K. Isolation, Structure, and Stability of a Dodecanethiolate-Protected Pd_1Au_{24} Cluster. *Phys. Chem. Chem. Phys.* **2010**, *12*, 6219–6225. [CrossRef]

24. Negishi, Y.; Iwai, T.; Ide, M. Continuous Modulation of Electronic Structure of Stable Thiolate-Protected Au_{25} Cluster by Ag Doping. *Chem. Commun.* **2010**, *46*, 4713–4715. [CrossRef] [PubMed]

25. Negishi, Y.; Igarashi, K.; Munakata, K.; Ohgake, W.; Nobusada, K. Palladium Doping of Magic Gold Cluster $Au_{38}(SC_2H_4Ph)_{24}$: Formation of $Pd_2Au_{36}(SC_2H_4Ph)_{24}$ with Higher Stability than $Au_{38}(SC_2H_4Ph)_{24}$. *Chem. Commun.* **2012**, *48*, 660–662. [CrossRef] [PubMed]

26. Negishi, Y.; Munakata, K.; Ohgake, W.; Nobusada, K. Effect of Copper Doping on Electronic Structure, Geometric Structure, and Stability of Thiolate-Protected Au_{25} Nanoclusters. *J. Phys. Chem. Lett.* **2012**, *3*, 2209–2214. [CrossRef]

27. Niihori, Y.; Kurashige, W.; Matsuzaki, M.; Negishi, Y. Remarkable Enhancement in Ligand-Exchange Reactivity of Thiolate-Protected Au_{25} Nanoclusters by Single Pd Atom Doping. *Nanoscale* **2013**, *5*, 508–512. [CrossRef]

28. Negishi, Y.; Kurashige, W.; Kobayashi, Y.; Yamazoe, S.; Kojima, N.; Seto, M.; Tsukuda, T. Formation of a $Pd@Au_{12}$ Superatomic Core in $Au_{24}Pd_1(SC_{12}H_{25})_{18}$ Probed by ^{197}Au Mössbauer and Pd K-edge EXAFS Spectroscopy. *J. Phys. Chem. Lett.* **2013**, *4*, 3579–3583. [CrossRef]

29. Negishi, Y.; Kurashige, W.; Niihori, Y.; Nobusada, K. Toward the Creation of Stable, Functionalized Metal Clusters. *Phys. Chem. Chem. Phys.* **2013**, *15*, 18736–18751. [CrossRef]

30. Niihori, Y.; Matsuzaki, M.; Uchida, C.; Negishi, Y. Advanced Use of High-Performance Liquid Chromatography for Synthesis of Controlled Metal Clusters. *Nanoscale* **2014**, *6*, 7889–7896. [CrossRef]

31. Yamazoe, S.; Kurashige, W.; Nobusada, K.; Negishi, Y.; Tsukuda, T. Preferential Location of Coinage Metal Dopants (M = Ag or Cu) in $[Au_{25-x}M_x(SC_2H_4Ph)_{18}]^-$ ($x \sim 1$) as Determined by Extended X-ray Absorption Fine Structure and Density Functional Theory Calculations. *J. Phys. Chem. C* **2014**, *118*, 25284–25290. [CrossRef]

32. Sharma, S.; Kurashige, W.; Nobusada, K.; Negishi, Y. Effect of Trimetallization in Thiolate-Protected $Au_{24-n}Cu_nPd$ Clusters. *Nanoscale* **2015**, *7*, 10606–10612. [CrossRef] [PubMed]

33. Niihori, Y.; Eguro, M.; Kato, A.; Sharma, S.; Kumar, B.; Kurashige, W.; Nobusada, K.; Negishi, Y. Improvements in the Ligand-Exchange Reactivity of Phenylethanethiolate-Protected Au_{25} Nanocluster by Ag or Cu Incorporation. *J. Phys. Chem. C* **2016**, *120*, 14301–14309. [CrossRef]

34. Niihori, Y.; Hossain, S.; Kumar, B.; Nair, L.V.; Kurashige, W.; Negishi, Y. Perspective: Exchange Reactions in Thiolate-Protected Metal Clusters. *APL Mater.* **2017**, *5*, 053201. [CrossRef]

35. Niihori, Y.; Hossain, S.; Sharma, S.; Kumar, B.; Kurashige, W.; Negishi, Y. Understanding and Practical Use of Ligand and Metal Exchange Reactions in Thiolate-Protected Metal Clusters to Synthesize Controlled Metal Clusters. *Chem. Rec.* **2017**, *17*, 473–484. [CrossRef] [PubMed]

36. Niihori, Y.; Shima, D.; Yoshida, K.; Hamada, K.; Nair, L.V.; Hossain, S.; Kurashige, W.; Negishi, Y. High-Performance Liquid Chromatography Mass Spectrometry of Gold and Alloy Clusters Protected by Hydrophilic Thiolates. *Nanoscale* **2018**, *10*, 1641–1649. [CrossRef]

37. Hossain, S.; Ono, T.; Yoshioka, M.; Hu, G.; Hosoi, M.; Chen, Z.; Nair, L.V.; Niihori, Y.; Kurashige, W.; Jiang, D.-E.; et al. Thiolate-Protected Trimetallic $Au_{\sim 20}Ag_{\sim 4}Pd$ and $Au_{\sim 20}Ag_{\sim 4}Pt$ Alloy Clusters with Controlled Chemical Composition and Metal Positions. *J. Phys. Chem. Lett.* **2018**, *9*, 2590–2594. [CrossRef]

38. Yokoyama, T.; Hirata, N.; Tsunoyama, H.; Negishi, Y.; Nakajima, A. Characterization of Floating-gate Memory Device with Thiolate-Protected Gold and Gold-Palladium Nanoclusters. *AIP Adv.* **2018**, *8*, 065002. [CrossRef]

39. Niihori, Y.; Koyama, Y.; Watanabe, S.; Hashimoto, S.; Hossain, S.; Nair, L.V.; Kumar, B.; Kurashige, W.; Negishi, Y. Atomic and Isomeric Separation of Thiolate-Protected Alloy Clusters. *J. Phys. Chem. Lett.* **2018**, *9*, 4930–4934. [CrossRef]

40. Niihori, Y.; Hashimoto, S.; Koyama, Y.; Hossain, S.; Kurashige, W.; Negishi, Y. Dynamic Behavior of Thiolate-Protected Gold–Silver 38-Atom Alloy Clusters in Solution. *J. Phys. Chem. C* **2019**, *123*, 13324–13329. [CrossRef]

41. Kurashige, W.; Hayashi, R.; Wakamatsu, K.; Kataoka, Y.; Hossain, S.; Iwase, A.; Kudo, A.; Yamazoe, S.; Negishi, Y. Atomic-Level Understanding of the Effect of Heteroatom Doping of the Cocatalyst on Water-Splitting Activity in AuPd or AuPt Alloy Cluster-Loaded $BaLa_4Ti_4O_{15}$. *ACS Appl. Energy Mater.* **2019**, *2*, 4175–4187. [CrossRef]

42. Hossain, S.; Imai, Y.; Suzuki, D.; Choi, W.; Chen, Z.; Suzuki, T.; Yoshioka, M.; Kawawaki, T.; Lee, D.; Negishi, Y. Elucidating Ligand Effects in Thiolate-Protected Metal Clusters Using $Au_{24}Pt(TBBT)_{18}$ as a Model Cluster. *Nanoscale* **2019**, *11*, 22089–22098. [CrossRef] [PubMed]

43. Kawawaki, T.; Negishi, Y.; Kawasaki, H. Photo/electrocatalysis and Photosensitization Using Metal Nanoclusters for Green Energy and Medical Applications. *Nanoscale Adv.* **2020**, *2*, 17–36. [CrossRef]

44. Hossain, S.; Imai, Y.; Motohashi, Y.; Chen, Z.; Suzuki, D.; Suzuki, T.; Kataoka, Y.; Hirata, M.; Ono, T.; Kurashige, W.; et al. Understanding and Designing One-Dimensional Assemblies of Ligand-Protected Metal Nanoclusters. *Mater. Horiz.*. in press. [CrossRef]

45. Haruta, M. Spiers Memorial Lecture Role of Perimeter Interfaces in Catalysis by Gold Nanoparticles. *Faraday Discuss.* **2011**, *152*, 11–32. [CrossRef] [PubMed]

46. Nie, X.; Qian, H.; Ge, Q.; Xu, H.; Jin, R. CO Oxidation Catalyzed by Oxide-Supported $Au_{25}(SR)_{18}$ Nanoclusters and Identification of Perimeter Sites as Active Centers. *ACS Nano* **2012**, *6*, 6014–6022. [CrossRef]

47. Nie, X.; Zeng, C.; Ma, X.; Qian, H.; Ge, Q.; Xu, H.; Jin, R. CeO_2-Supported $Au_{38}(SR)_{24}$ Nanocluster Catalysts for CO Oxidation: A Comparison of Ligand-on and -off Catalysts. *Nanoscale* **2013**, *5*, 5912–5918. [CrossRef]

48. Wu, Z.; Jiang, D.-e.; Mann, A.K.P.; Mullins, D.R.; Qiao, Z.-A.; Allard, L.F.; Zeng, C.; Jin, R.; Overbury, S.H. Thiolate Ligands as a Double-Edged Sword for CO Oxidation on CeO_2 Supported $Au_{25}(SCH_2CH_2Ph)_{18}$ Nanoclusters. *J. Am. Chem. Soc.* **2014**, *136*, 6111–6122. [CrossRef]

49. Li, W.; Ge, Q.; Ma, X.; Chen, Y.; Zhu, M.; Xu, H.; Jin, R. Mild Activation of CeO_2-Supported Gold Nanoclusters and Insight into the Catalytic Behavior in CO Oxidation. *Nanoscale* **2016**, *8*, 2378–2385. [CrossRef]

50. Gaur, S.; Miller, J.T.; Stellwagen, D.; Sanampudi, A.; Kumar, C.S.S.R.; Spivey, J.J. Synthesis, Characterization, and Testing of Supported Au Catalysts Prepared from Atomically-Tailored $Au_{38}(SC_{12}H_{25})_{24}$ Clusters. *Phys. Chem. Chem. Phys.* **2012**, *14*, 1627–1634. [CrossRef]

51. Wu, Z.; Hu, G.; Jiang, D.-e.; Mullins, D.R.; Zhang, Q.-F.; Allard, L.F.; Wang, L.-S.; Overbury, S.H. Diphosphine-Protected Au_{22} Nanoclusters on Oxide Supports Are Active for Gas-Phase Catalysis without Ligand Removal. *Nano Lett.* **2016**, *16*, 6560–6567. [CrossRef]

52. Wu, Z.; Mullins, D.R.; Allard, L.F.; Zhang, Q.; Wang, L. CO Oxidation over Ceria Supported Au$_{22}$ Nanoclusters: Shape Effect of the Support. *Chin. Chem. Lett.* **2018**, *29*, 795–799. [CrossRef]

53. Lin, J.; Li, W.; Liu, C.; Huang, P.; Zhu, M.; Ge, Q.; Li, G. One-Phase Controlled Synthesis of Au$_{25}$ Nanospheres and Nanorods from 1.3 nm Au : PPh$_3$ Nanoparticles: The Ligand Effects. *Nanoscale* **2015**, *7*, 13663–13670. [CrossRef] [PubMed]

54. Li, W.; Liu, C.; Abroshan, H.; Ge, Q.; Yang, X.; Xu, H.; Li, G. Catalytic CO Oxidation Using Bimetallic M$_X$Au$_{25-X}$ Clusters: A Combined Experimental and Computational Study on Doping Effects. *J. Phys. Chem. C* **2016**, *120*, 10261–10267. [CrossRef]

55. Good, J.; Duchesne, P.N.; Zhang, P.; Koshut, W.; Zhou, M.; Jin, R. On the Functional Role of the Cerium Oxide Support in the Au$_{38}$(SR)$_{24}$/CeO$_2$ Catalyst for CO Oxidation. *Catal. Today* **2017**, *280*, 239–245. [CrossRef]

56. Du, Y.; Sheng, H.; Astruc, D.; Zhu, M. Atomically Precise Noble Metal Nanoclusters as Efficient Catalysts: A Bridge between Structure and Properties. *Chem. Rev.* **2020**, *120*, 526–622. [CrossRef]

57. Xie, S.; Tsunoyama, H.; Kurashige, W.; Negishi, Y.; Tsukuda, T. Enhancement in Aerobic Alcohol Oxidation Catalysis of Au$_{25}$ Clusters by Single Pd Atom Doping. *ACS Catal.* **2012**, *2*, 1519–1523. [CrossRef]

58. Yoskamtorn, T.; Yamazoe, S.; Takahata, R.; Nishigaki, J.-i.; Thivasasith, A.; Limtrakul, J.; Tsukuda, T. Thiolate-Mediated Selectivity Control in Aerobic Alcohol Oxidation by Porous Carbon-Supported Au$_{25}$ Clusters. *ACS Catal.* **2014**, *4*, 3696–3700. [CrossRef]

59. Lavenn, C.; Demessence, A.; Tuel, A. Au$_{25}$(SPh-*p*NH$_2$)$_{17}$ Nanoclusters Deposited on SBA-15 as Catalysts for Aerobic Benzyl Alcohol Oxidation. *J. Catal.* **2015**, *322*, 130–138. [CrossRef]

60. Deng, H.; Wang, S.; Jin, S.; Yang, S.; Xu, Y.; Liu, L.; Xiang, J.; Hu, D.; Zhu, M. Active Metal (Cadmium) Doping Enhanced the Stability of Inert Metal (Gold) Nanocluster under O$_2$ Atmosphere and the Catalysis Activity of Benzyl Alcohol Oxidation. *Gold Bull.* **2015**, *48*, 161–167. [CrossRef]

61. Li, L.; Dou, L.; Zhang, H. Layered Double Hydroxide Supported Gold Nanoclusters by Glutathione-Capped Au Nanoclusters Precursor Method for Highly Efficient Aerobic Oxidation of Alcohols. *Nanoscale* **2014**, *6*, 3753–3763. [CrossRef] [PubMed]

62. Wang, S.; Yin, S.; Chen, G.; Li, L.; Zhang, H. Nearly Atomic Precise Gold Nanoclusters on Nickel-Based Layered Double Hydroxides for Extraordinarily Efficient Aerobic Oxidation of Alcohols. *Catal. Sci. Technol.* **2016**, *6*, 4090–4104. [CrossRef]

63. Yin, S.; Li, J.; Zhang, H. Hierarchical Hollow Nanostructured Core@Shell Recyclable Catalysts γ-Fe$_2$O$_3$@LDH@Au$_{25-X}$ for Highly Efficient Alcohol Oxidation. *Green Chem.* **2016**, *18*, 5900–5914. [CrossRef]

64. Lee, K.E.; Shivhare, A.; Hu, Y.; Scott, R.W.J. Supported Bimetallic AuPd Clusters Using Activated Au$_{25}$ Clusters. *Catal. Today* **2017**, *280*, 259–265. [CrossRef]

65. Tsunoyama, H.; Ichikuni, N.; Sakurai, H.; Tsukuda, T. Effect of Electronic Structures of Au Clusters Stabilized by Poly(*N*-vinyl-2-pyrrolidone) on Aerobic Oxidation Catalysis. *J. Am. Chem. Soc.* **2009**, *131*, 7086–7093. [CrossRef]

66. Zhu, Y.; Qian, H.; Zhu, M.; Jin, R. Thiolate-Protected Au$_n$ Nanoclusters as Catalysts for Selective Oxidation and Hydrogenation Processes. *Adv. Mater.* **2010**, *22*, 1915–1920. [CrossRef]

67. Zhu, Y.; Qian, H.; Jin, R. An Atomic-Level Strategy for Unraveling Gold Nanocatalysis from the Perspective of Au$_n$(SR)$_m$ Nanoclusters. *Chem. Eur. J.* **2010**, *16*, 11455–11462. [CrossRef]

68. Wang, S.; Jin, S.; Yang, S.; Chen, S.; Song, Y.; Zhang, J.; Zhu, M. Total Structure Determination of Surface Doping [Ag$_{46}$Au$_{24}$(SR)$_{32}$](BPh$_4$)$_2$ Nanocluster and Its Structure-Related Catalytic Property. *Science Adv.* **2015**, *1*, e1500441. [CrossRef]

69. Chai, J.; Chong, H.; Wang, S.; Yang, S.; Wu, M.; Zhu, M. Controlling the Selectivity of Catalytic Oxidation of Styrene over Nanocluster Catalysts. *RSC Adv.* **2016**, *6*, 111399–111405. [CrossRef]

70. Turner, M.; Golovko, V.B.; Vaughan, O.P.H.; Abdulkin, P.; Berenguer-Murcia, A.; Tikhov, M.S.; Johnson, B.F.G.; Lambert, R.M. Selective Oxidation with Dioxygen by Gold Nanoparticle Catalysts Derived from 55-Atom Clusters. *Nature* **2008**, *454*, 981–983. [CrossRef]

71. Zhang, B.; Kaziz, S.; Li, H.; Hevia, M.G.; Wodka, D.; Mazet, C.; Bürgi, T.; Barrabés, N. Modulation of Active Sites in Supported Au$_{38}$(SC$_2$H$_4$Ph)$_{24}$ Cluster Catalysts: Effect of Atmosphere and Support Material. *J. Phys. Chem. C* **2015**, *119*, 11193–11199. [CrossRef]

72. Liu, Y.; Tsunoyama, H.; Akita, T.; Xie, S.; Tsukuda, T. Aerobic Oxidation of Cyclohexane Catalyzed by Size-Controlled Au Clusters on Hydroxyapatite: Size Effect in the Sub-2 nm Regime. *ACS Catal.* **2011**, *1*, 2–6. [CrossRef]

73. Liu, C.; Yan, C.; Lin, J.; Yu, C.; Huang, J.; Li, G. One-Pot Synthesis of Au$_{144}$(SCH$_2$Ph)$_{60}$ Nanoclusters and Their Catalytic Application. *J. Mater. Chem. A* **2015**, *3*, 20167–20173. [CrossRef]

74. Li, G.; Qian, H.; Jin, R. Gold Nanocluster-Catalyzed Selective Oxidation of Sulfide to Sulfoxide. *Nanoscale* **2012**, *4*, 6714–6717. [CrossRef]

75. Chen, Y.; Wang, J.; Liu, C.; Li, Z.; Li, G. Kinetically Controlled Synthesis of Au$_{102}$(SPh)$_{44}$ Nanoclusters and Catalytic Application. *Nanoscale* **2016**, *8*, 10059–10065. [CrossRef] [PubMed]

76. Kauffman, D.R.; Alfonso, D.; Matranga, C.; Ohodnicki, P.; Deng, X.; Siva, R.C.; Zeng, C.; Jin, R. Probing Active Site Chemistry with Differently Charged Au$_{25}{}^q$ Nanoclusters ($q = -1, 0, +1$). *Chem. Sci.* **2014**, *5*, 3151–3157. [CrossRef]

77. Kauffman, D.R.; Alfonso, D.; Matranga, C.; Qian, H.; Jin, R. Experimental and Computational Investigation of Au$_{25}$ Clusters and CO$_2$: A Unique Interaction and Enhanced Electrocatalytic Activity. *J. Am. Chem. Soc.* **2012**, *134*, 10237–10243. [CrossRef]

78. Zhao, S.; Jin, R.; Jin, R. Opportunities and Challenges in CO$_2$ Reduction by Gold- and Silver-Based Electrocatalysts: From Bulk Metals to Nanoparticles and Atomically Precise Nanoclusters. *ACS Energy Lett.* **2018**, *3*, 452–462. [CrossRef]

79. Andrews, E.; Katla, S.; Kumar, C.; Patterson, M.; Sprunger, P.; Flake, J. Electrocatalytic Reduction of CO$_2$ at Au Nanoparticle Electrodes: Effects of Interfacial Chemistry on Reduction Behavior. *J. Electrochem. Soc.* **2015**, *162*, F1373–F1378. [CrossRef]

80. Kauffman, D.R.; Thakkar, J.; Siva, R.; Matranga, C.; Ohodnicki, P.R.; Zeng, C.; Jin, R. Efficient Electrochemical CO$_2$ Conversion Powered by Renewable Energy. *ACS Appl. Mater. Inter.* **2015**, *7*, 15626–15632. [CrossRef]

81. Zhao, S.; Austin, N.; Li, M.; Song, Y.; House, S.D.; Bernhard, S.; Yang, J.C.; Mpourmpakis, G.; Jin, R. Influence of Atomic-Level Morphology on Catalysis: The Case of Sphere and Rod-Like Gold Nanoclusters for CO$_2$ Electroreduction. *ACS Catal.* **2018**, *8*, 4996–5001. [CrossRef]

82. Jupally, V.R.; Dharmaratne, A.C.; Crasto, D.; Huckaba, A.J.; Kumara, C.; Nimmala, P.R.; Kothalawala, N.; Delcamp, J.H.; Dass, A. Au$_{137}$(SR)$_{56}$ Nanomolecules: Composition, Optical Spectroscopy, Electrochemistry and Electrocatalytic Reduction of CO$_2$. *Chem. Commun.* **2014**, *50*, 9895–9898. [CrossRef] [PubMed]

83. Alfonso, D.R.; Kauffman, D.; Matranga, C. Active Sites of Ligand-Protected Au$_{25}$ Nanoparticle Catalysts for CO$_2$ Electroreduction to CO. *J. Chem. Phys.* **2016**, *144*, 184705. [CrossRef] [PubMed]

84. Seh, Z.W.; Kibsgaard, J.; Dickens, C.F.; Chorkendorff, I.; Nørskov, J.K.; Jaramillo, T.F. Combining Theory and Experiment in Electrocatalysis: Insights into Materials Design. *Science* **2017**, *355*, eaad4998. [CrossRef] [PubMed]

85. Jiao, Y.; Zheng, Y.; Jaroniec, M.; Qiao, S.Z. Design of Electrocatalysts for Oxygen- and Hydrogen-Involving Energy Conversion Reactions. *Chem. Soc. Rev.* **2015**, *44*, 2060–2086. [CrossRef]

86. Sabatier, P. Hydrogénations Et Déshydrogénations Par Catalyse. *Ber. Dtsch. Chem. Ges.* **1911**, *44*, 1984–2001. [CrossRef]

87. Parsons, R. The Rate of Electrolytic Hydrogen Evolution and the Heat of Adsorption of Hydrogen. *Trans. Faraday Soc.* **1958**, *54*, 1053–1063. [CrossRef]

88. Benck, J.D.; Hellstern, T.R.; Kibsgaard, J.; Chakthranont, P.; Jaramillo, T.F. Catalyzing the Hydrogen Evolution Reaction (HER) with Molybdenum Sulfide Nanomaterials. *ACS Catal.* **2014**, *4*, 3957–3971. [CrossRef]

89. Jaramillo, T.F.; Jørgensen, K.P.; Bonde, J.; Nielsen, J.H.; Horch, S.; Chorkendorff, I. Identification of Active Edge Sites for Electrochemical H$_2$ Evolution from MoS$_2$ Nanocatalysts. *Science* **2007**, *317*, 100–102. [CrossRef]

90. Cao, B.; Veith, G.M.; Neuefeind, J.C.; Adzic, R.R.; Khalifah, P.G. Mixed Close-Packed Cobalt Molybdenum Nitrides as Non-Noble Metal Electrocatalysts for the Hydrogen Evolution Reaction. *J. Am. Chem. Soc.* **2013**, *135*, 19186–19192. [CrossRef]

91. Popczun, E.J.; McKone, J.R.; Read, C.G.; Biacchi, A.J.; Wiltrout, A.M.; Lewis, N.S.; Schaak, R.E. Nanostructured Nickel Phosphide as an Electrocatalyst for the Hydrogen Evolution Reaction. *J. Am. Chem. Soc.* **2013**, *135*, 9267–9270. [CrossRef] [PubMed]

92. Kibler, L.A.; El-Aziz, A.M.; Hoyer, R.; Kolb, D.M. Tuning Reaction Rates by Lateral Strain in a Palladium Monolayer. *Angew. Chem. Int. Ed.* **2005**, *44*, 2080–2084. [CrossRef] [PubMed]

93. McCrory, C.C.L.; Jung, S.; Peters, J.C.; Jaramillo, T.F. Benchmarking Heterogeneous Electrocatalysts for the Oxygen Evolution Reaction. *J. Am. Chem. Soc.* **2013**, *135*, 16977–16987. [CrossRef] [PubMed]

94. Stoerzinger, K.A.; Qiao, L.; Biegalski, M.D.; Shao-Horn, Y. Orientation-Dependent Oxygen Evolution Activities of Rutile IrO$_2$ and RuO$_2$. *J. Phys. Chem. Lett.* **2014**, *5*, 1636–1641. [CrossRef] [PubMed]

95. Zhang, B.; Zheng, X.; Voznyy, O.; Comin, R.; Bajdich, M.; García-Melchor, M.; Han, L.; Xu, J.; Liu, M.; Zheng, L.; et al. Homogeneously Dispersed Multimetal Oxygen-Evolving Catalysts. *Science* **2016**, *352*, 333–337. [CrossRef] [PubMed]

96. Gasteiger, H.A.; Kocha, S.S.; Sompalli, B.; Wagner, F.T. Activity Benchmarks and Requirements for Pt, Pt-Alloy, and Non-Pt Oxygen Reduction Catalysts for PEMFCs. *Appl. Catal. B* **2005**, *56*, 9–35. [CrossRef]

97. Gasteiger, H.A.; Marković, N.M. Just a Dream–or Future Reality? *Science* **2009**, *324*, 48–49. [CrossRef]

98. Siahrostami, S.; Verdaguer-Casadevall, A.; Karamad, M.; Deiana, D.; Malacrida, P.; Wickman, B.; Escudero-Escribano, M.; Paoli, E.A.; Frydendal, R.; Hansen, T.W.; et al. Enabling Direct H_2O_2 Production through Rational Electrocatalyst Design. *Nat. Mater.* **2013**, *12*, 1137–1143. [CrossRef]

99. Nørskov, J.K.; Rossmeisl, J.; Logadottir, A.; Lindqvist, L.; Kitchin, J.R.; Bligaard, T.; Jónsson, H. Origin of the Overpotential for Oxygen Reduction at a Fuel-Cell Cathode. *J. Phys. Chem. B* **2004**, *108*, 17886–17892. [CrossRef]

100. Huang, X.; Zhao, Z.; Cao, L.; Chen, Y.; Zhu, E.; Lin, Z.; Li, M.; Yan, A.; Zettl, A.; Wang, Y.M.; et al. High-Performance Transition Metal–Doped Pt_3Ni Octahedra for Oxygen Reduction Reaction. *Science* **2015**, *348*, 1230–1234. [CrossRef]

101. Peng, Z.; Yang, H. Synthesis and Oxygen Reduction Electrocatalytic Property of Pt-on-Pd Bimetallic Heteronanostructures. *J. Am. Chem. Soc.* **2009**, *131*, 7542–7543. [CrossRef] [PubMed]

102. Kwak, K.; Choi, W.; Tang, Q.; Kim, M.; Lee, Y.; Jiang, D.-e.; Lee, D. A Molecule-Like $PtAu_{24}(SC_6H_{13})_{18}$ Nanocluster as an Electrocatalyst for Hydrogen Production. *Nat. Commun.* **2017**, *8*, 14723. [CrossRef] [PubMed]

103. Choi, W.; Hu, G.; Kwak, K.; Kim, M.; Jiang, D.-e.; Choi, J.-P.; Lee, D. Effects of Metal-Doping on Hydrogen Evolution Reaction Catalyzed by MAu_{24} and M_2Au_{36} Nanoclusters (M = Pt, Pd). *ACS Appl. Mater. Inter.* **2018**, *10*, 44645–44653. [CrossRef] [PubMed]

104. Kwak, K.; Lee, D. Electrochemistry of Atomically Precise Metal Nanoclusters. *Acc. Chem. Res.* **2019**, *52*, 12–22. [CrossRef] [PubMed]

105. Hu, G.; Tang, Q.; Lee, D.; Wu, Z.; Jiang, D.-e. Metallic Hydrogen in Atomically Precise Gold Nanoclusters. *Chem. Mater.* **2017**, *29*, 4840–4847. [CrossRef]

106. Du, Y.; Xiang, J.; Ni, K.; Yun, Y.; Sun, G.; Yuan, X.; Sheng, H.; Zhu, Y.; Zhu, M. Design of Atomically Precise Au_2Pd_6 Nanoclusters for Boosting Electrocatalytic Hydrogen Evolution on MoS_2. *Inorg. Chem. Front.* **2018**, *5*, 2948–2954. [CrossRef]

107. Eguchi, D.; Sakamoto, M.; Teranishi, T. Ligand Effect on the Catalytic Activity of Porphyrin Protected Gold Clusters in the Electrochemical Hydrogen Evolution Reaction. *Chem. Sci.* **2018**, *9*, 261–265. [CrossRef]

108. Zhao, S.; Jin, R.; Song, Y.; Zhang, H.; House, S.D.; Yang, J.C.; Jin, R. Atomically Precise Gold Nanoclusters Accelerate Hydrogen Evolution over MoS_2 Nanosheets: The Dual Interfacial Effect. *Small* **2017**, *13*, 1701519. [CrossRef]

109. Kwak, K.; Choi, W.; Tang, Q.; Jiang, D.-e.; Lee, D. Rationally Designed Metal Nanocluster for Electrocatalytic Hydrogen Production from Water. *J. Mater. Chem. A* **2018**, *6*, 19495–19501. [CrossRef]

110. Zhao, S.; Jin, R.; Abroshan, H.; Zeng, C.; Zhang, H.; House, S.D.; Gottlieb, E.; Kim, H.J.; Yang, J.C.; Jin, R. Gold Nanoclusters Promote Electrocatalytic Water Oxidation at the Nanocluster/$CoSe_2$ Interface. *J. Am. Chem. Soc.* **2017**, *139*, 1077–1080. [CrossRef]

111. Chen, W.; Chen, S. Oxygen Electroreduction Catalyzed by Gold Nanoclusters: Strong Core Size Effects. *Angew. Chem. Int. Ed.* **2009**, *48*, 4386–4389. [CrossRef] [PubMed]

112. Wang, L.; Tang, Z.; Yan, W.; Yang, H.; Wang, Q.; Chen, S. Porous Carbon-Supported Gold Nanoparticles for Oxygen Reduction Reaction: Effects of Nanoparticle Size. *ACS Appl. Mater. Inter.* **2016**, *8*, 20635–20641. [CrossRef] [PubMed]

113. Sumner, L.; Sakthivel, N.A.; Schrock, H.; Artyushkova, K.; Dass, A.; Chakraborty, S. Electrocatalytic Oxygen Reduction Activities of Thiol-Protected Nanomolecules Ranging in Size from $Au_{28}(SR)_{20}$ to $Au_{279}(SR)_{84}$. *J. Phys. Chem. C* **2018**, *122*, 24809–24817. [CrossRef]

114. Jones, T.C.; Sumner, L.; Ramakrishna, G.; Hatshan, M.b.; Abuhagr, A.; Chakraborty, S.; Dass, A. Bulky *t*-Butyl Thiolated Gold Nanomolecular Series: Synthesis, Characterization, Optical Properties, and Electrocatalysis. *J. Phys. Chem. C* **2018**, *122*, 17726–17737. [CrossRef]

115. Lu, Y.; Jiang, Y.; Gao, X.; Chen, W. Charge State-Dependent Catalytic Activity of $[Au_{25}(SC_{12}H_{25})_{18}]$ Nanoclusters for the Two-Electron Reduction of Dioxygen to Hydrogen Peroxide. *Chem. Commun.* **2014**, *50*, 8464–8467. [CrossRef]

116. Kwak, K.; Azad, U.P.; Choi, W.; Pyo, K.; Jang, M.; Lee, D. Efficient Oxygen Reduction Electrocatalysts Based on Gold Nanocluster–Graphene Composites. *ChemElectroChem* **2016**, *3*, 1253–1260. [CrossRef]

117. Skúlason, E.; Tripkovic, V.; Björketun, M.E.; Gudmundsdóttir, S.; Karlberg, G.; Rossmeisl, J.; Bligaard, T.; Jónsson, H.; Nørskov, J.K. Modeling the Electrochemical Hydrogen Oxidation and Evolution Reactions on the Basis of Density Functional Theory Calculations. *J. Phys. Chem. C* **2010**, *114*, 18182–18197. [CrossRef]

118. Sakamoto, M.; Tanaka, D.; Tsunoyama, H.; Tsukuda, T.; Minagawa, Y.; Majima, Y.; Teranishi, T. Platonic Hexahedron Composed of Six Organic Faces with an Inscribed Au Cluster. *J. Am. Chem. Soc.* **2012**, *134*, 816–819. [CrossRef]

119. Tanaka, D.; Inuta, Y.; Sakamoto, M.; Furube, A.; Haruta, M.; So, Y.-G.; Kimoto, K.; Hamada, I.; Teranishi, T. Strongest Π–Metal Orbital Coupling in a Porphyrin/Gold Cluster System. *Chem. Sci.* **2014**, *5*, 2007–2010. [CrossRef]

120. Li, P.; Wang, M.; Duan, X.; Zheng, L.; Cheng, X.; Zhang, Y.; Kuang, Y.; Li, Y.; Ma, Q.; Feng, Z.; et al. Boosting Oxygen Evolution of Single-Atomic Ruthenium through Electronic Coupling with Cobalt-Iron Layered Double Hydroxides. *Nat. Commun.* **2019**, *10*, 1711. [CrossRef]

121. Lee, Y.; Suntivich, J.; May, K.J.; Perry, E.E.; Shao-Horn, Y. Synthesis and Activities of Rutile IrO_2 and RuO_2 Nanoparticles for Oxygen Evolution in Acid and Alkaline Solutions. *J. Phys. Chem. Lett.* **2012**, *3*, 399–404. [CrossRef] [PubMed]

122. Mattioli, G.; Giannozzi, P.; Amore Bonapasta, A.; Guidoni, L. Reaction Pathways for Oxygen Evolution Promoted by Cobalt Catalyst. *J. Am. Chem. Soc.* **2013**, *135*, 15353–15363. [CrossRef] [PubMed]

123. Frydendal, R.; Paoli, E.A.; Knudsen, B.P.; Wickman, B.; Malacrida, P.; Stephens, I.E.L.; Chorkendorff, I. Benchmarking the Stability of Oxygen Evolution Reaction Catalysts: The Importance of Monitoring Mass Losses. *ChemElectroChem* **2014**, *1*, 2075–2081. [CrossRef]

124. Zhuang, Z.; Sheng, W.; Yan, Y. Synthesis of Monodispere $Au@Co_3O_4$ Core-Shell Nanocrystals and Their Enhanced Catalytic Activity for Oxygen Evolution Reaction. *Adv. Mater.* **2014**, *26*, 3950–3955. [CrossRef]

125. Zhao, X.; Gao, P.; Yan, Y.; Li, X.; Xing, Y.; Li, H.; Peng, Z.; Yang, J.; Zeng, J. Gold Atom-Decorated $CoSe_2$ Nanobelts with Engineered Active Sites for Enhanced Oxygen Evolution. *J. Mater. Chem. A* **2017**, *5*, 20202–20207. [CrossRef]

126. Li, Z.-y.; Ye, K.-h.; Zhong, Q.-s.; Zhang, C.-j.; Shi, S.-t.; Xu, C.-w. $Au–Co_3O_4/C$ as an Efficient Electrocatalyst for the Oxygen Evolution Reaction. *ChemPlusChem* **2014**, *79*, 1569–1572. [CrossRef]

127. Mills, G.; Gordon, M.S.; Metiu, H. Oxygen Adsorption on Au Clusters and a Rough Au(111) Surface: The Role of Surface Flatness, Electron Confinement, Excess Electrons, and Band Gap. *J. Chem. Phys.* **2003**, *118*, 4198–4205. [CrossRef]

128. Okumura, M.; Kitagawa, Y.; Kawakami, T.; Haruta, M. Theoretical Investigation of the Hetero-Junction Effect in PVP-Stabilized Au_{13} Clusters. The Role of PVP in Their Catalytic Activities. *Chem. Phys. Lett.* **2008**, *459*, 133–136. [CrossRef]

129. Yin, H.; Tang, H.; Wang, D.; Gao, Y.; Tang, Z. Facile Synthesis of Surfactant-Free Au Cluster/Graphene Hybrids for High-Performance Oxygen Reduction Reaction. *ACS Nano* **2012**, *6*, 8288–8297. [CrossRef]

Communication

Intramolecular Metal Exchange Reaction Promoted by Thiol Ligands

Yangfeng Li, Man Chen, Shuxin Wang * and Manzhou Zhu *

Department of Chemistry and Centre for Atomic Engineering of Advanced Materials, Anhui Province Key Laboratory of Chemistry for Inorganic/Organic Hybrid Functionalized Materials, Anhui University, Hefei 230601, Anhui, China; 18255127378@163.com (Y.L.); 13101@ahu.edu.cn (M.C.)
* Correspondence: lxing@ahu.edu.cn (S.W.); zmz@ahu.edu.cn (M.Z.); Tel.: +86-551-63861487 (S.W. & M.Z.)

Received: 29 November 2018; Accepted: 13 December 2018; Published: 19 December 2018

Abstract: The synthesis of an alloy nanocluster that is atomically precise is the key to understanding the metal synergy effect at the atomic level. Using the $Ag_2Au_{25}(SR)_{18}$ nanocluster as a model, we reported a third approach for the metal exchange reaction, that is, intramolecular metal exchange. The surface adsorbed metal ions (i.e., Ag) can be exchanged with the kernel metal atoms (i.e., Au) that are promoted by thiol ligands. The exchanged gold atoms can be further stripped by the thiol ligands, and produce the $Ag_xAu_{25-x}(SR)_{18}{}^-$ nanocluster.

Keywords: alloy; metal exchange; atomically precise

---

1. Introduction

Precise alloy nanoclusters have recently attracted great interest due to their quantum size effect-induced unique optical and catalysis properties [1,2]. More importantly, with their well determined structure, the relationship between their structure and properties can be grasped at the atomic level. Unlike for the large-sized gold nanoparticles (>3 nm), the method of synthesizing alloy nanoclusters is relatively limited. For instance, galvanic replacement is widely used in the synthesis of alloy nanoclusters. For example, the Ag-Pd alloy nanoparticle could be synthesized by reacting the silver nanoparticle with Pd(II) salt [3]. However, it is a challenge to synthesize an alloy nanocluster with this method, since the metal nanocluster, which is comprised by active metals, are quite unstable and rarely reported [4].

Previously, the metal exchange method, which uses homo-gold nanoclusters as templates to synthesize alloy nanoclusters, has been reported [5–7]. More importantly, the metal exchange process can jump out of the metal activity sequence, which provides a new way to synthesize the gold nanocluster with a highly active metal (e.g., Cd) [8,9]. So far, two types of metal exchange reactions have been reported: (i) Metal exchange with metal complexes [10–15]. This reaction can be created between thiolated metal complexes (e.g., Ag(SR)) and thiolated gold nanoclusters. (ii) Metal exchange between nanoclusters [16–18]. Recently, metal exchange was found between two nanoclusters. Pradeep and co-workers reported the reaction between $Ag_{25}(SR)_{18}{}^-$ and $Au_{25}(SR)_{18}{}^-$ to produce the $Ag_xAu_{25-x}(SR)_{18}{}^-$ nanocluster [16]. Based on these two methods, a series of alloy nanoclusters with atomically precise structures have been synthesized; meanwhile, such alloy products have been widely used to study metal synergistic effects in chirality, [19] optics [20,21], and catalysis [22] at the atomic level. It is still necessary to develop new alloying methods, especially a controllable alloying method.

In this work, we reported a third type of metal exchange: self-alloying induced by intramolecular metal exchange. Specifically, the intramolecular metal exchange reaction between Ag atoms and Au atoms in the $Ag_2Au_{25}(SR)_{18}$ nanocluster can occur in the existence of thiol ligands and produce the

$Ag_xAu_{25-x}(SR)_{18}^-$ nanocluster. The UV-Vis, thin layer chromatography, and mass spectra indicate the efficient transformation of this self-alloying process.

2. Materials and Methods

2.1. Materials

Tetrachloroauric(III)acid ($HAuCl_4 \cdot 3H_2O$, >99.99% metals basis), silver nitrate ($AgNO_3$, 99%, metal basis), sodium borohydride ($NaBH_4$, 99.9%), phenylethyl mercaptan (PET, 99%), tetraoctylammonium bromide (TOABr, 98.0%), methylene chloride (CH_2Cl_2, HPLC grade), acetonitrile (MeCN, HPLC grade), methanol (MeOH, HPLC grade), and tetrahydrofuran (THF, HPLC grade) were purchased from Sigma-Aldrich (Shanghai, China). Pure water was purchased from Wahaha Co. Ltd., Hangzhou, Zhejiang, China.

2.2. Synthesis of the $Au_{25}(PET)_{18}^-$ Nanocluster

The synthesis steps were similar to a previously reported process but with minor modifications [23]. $HAuCl_4 \cdot 3H_2O$ (0.2 g/mL, 0.4 mL) and TOABr (0.254 g, 0.47 mmol) was dissolved in 5 mL of THF. Then, 400 μL (2.8 mmol) of PET was added to the flask under a slow stirring speed ~60 rpm. After the solution turned clear (~30 min), the stirring speed was increased to fast (~1200 rpm). At the same time, an aqueous solution of $NaBH_4$ (0.1550g, 4 mmol) was quickly added to the above solution to initiate the reaction. The reaction was allowed to proceed overnight. After that, the solution was washed with 5 mL pure water. The organic phase was dried via rotoevaporation. MeOH (~20 mL) was added to remove the byproducts and this centrifugation cycle was repeated at least 3 times. Then, 10 mL of MeCN was used to extract pure $Au_{25}(PET)_{18}^-TOA^+$ nanoclusters (NCs).

2.3. Synthesis of the $Ag_2Au_{25}(PET)_{18}^+$ Nanocluster

Ag_2Au_{25} was prepared by modifying a previous method [24]. Briefly, 10 mg of $Au_{25}(PET)_{18}^-TOA^+$ (1.3 μmol) NCs was dissolved in 10 mL of MeCN. After that, 0.67 mg of $AgNO_3$ (2.2 equivalents per mole of Au_{25}), dissolved in MeCN, was added into the solution. The color of the solution then quickly turned from brown to dark green. The Ag_2Au_{25} nanocluster was then precipitated out of the solution. MeCN (~10 mL) was added to remove the unreacted $AgNO_3$ and other byproducts (e.g., $TOA^+NO_3^-$).

2.4. Intramolecular Metal Exchange of the Ag_2Au_{25} Nanocluster

2 mg of the Ag_2Au_{25} nanocluster was dissolved in 10 mL of methylene chloride (DCM). After that, 0.1 mL PET was added into the solution. The color of the solution slowly turned from green to brown. This reaction was monitored by UV-Vis and MALDI-TOF-MS spectra. After ~70 min, all the Ag_2Au_{25} nanoclusters were converted into $Ag_xAu_{25-x}(SR)_{18}^-$ (x = 0–2).

2.5. Characterization

All UV-Vis absorption spectra of the nanoclusters dissolved in CH_2Cl_2 were recorded using an Agilent 8453 diode array spectrometer (Shanghai China), whose background correction was made using a CH_2Cl_2 blank. The X-ray photoelectron spectroscopy (XPS) measurement was performed on a Thermo ESCALAB 250 (Waltham, MA, USA), configured with a monochromated $AlK\alpha$ (1486.8 eV) 150 W X-ray source, with a 0.5 mm circular spot size, a flood gun to counter charge the effects, and with the analysis chamber base pressure lower than 1×10^{-9} mbar. Inductively coupled plasma-atomic emission spectrometry (ICP-AES) measurements were performed on an Atomscan Advantage instrument made by Thermo Fisher (Waltham, MA, USA). The nanoclusters were digested by concentrated nitric acid and the concentration of the nanoclusters were set to 0.5 mg L^{-1} approximately. MALDI-TOF-MS was recorded on a Bruker Autoflex III smart beam instrument (Karlsruhe, Germany), using trans-2-[3-(4-tert-butylphenyl)-2-methyl-2-propenylidene] malononitrile (DCTB) as the matrix.

3. Results and Discussion

The $Ag_2Au_{25}(SR)_{18}$ nanoparticle (Figure 1, abbreviated as Ag_2Au_{25} hereafter) was made by reaction with the $Au_{25}(SR)_{18}{}^-$ nanocluster, with two equivalents of $AgNO_3$ in acetonitrile. We previously reported that the $Ag(SR)$ complex reaction with $Au_{25}(SR)_{18}{}^-$ (dissolved in toluene or dichloromethane) will produce the $Ag_xAu_{25-x}(SR)_{18}{}^-$ nanocluster [6]. However, by using the inorganic $AgNO_3$ instead of $Ag(SR)$, the silver atoms do not replace the gold atoms in the $Au_{25}(SR)_{18}{}^-$ nanocluster but, instead, anchor on the surface of the Au_{25} nanocluster and produce the Ag_2Au_{25} nanocluster. Interestingly, different kinds of metal salt precursors led to the different alloying result, that is, the metal-exchange or metal adsorption.

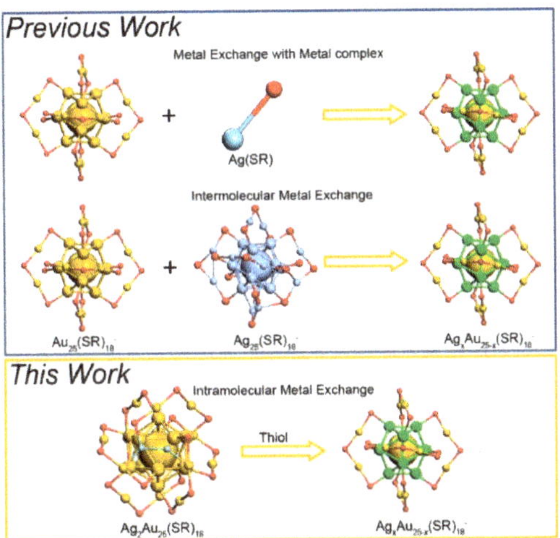

Figure 1. Synthesis of atomically precise alloy nanoclusters with different methods.

We further reacted the Ag_2Au_{25} with the thiol ligands. And applied the UV-Vis absorption spectra to monitor the reaction process. As shown in Figure 2, the HOMO-LUMO peak at ~700 nm of Ag_2Au_{25} gradually blue shifted to ~625 nm. Meanwhile, three absorption points, at ~625 nm, ~475 nm, and ~410 nm, were found that indicated the quantitative conversion. The final product was determined to be $Ag_xAu_{25-x}(SR)_{18}{}^-$ with the x ranging from 0 to 2 (Figure 3b). The metal ratio was further confirmed by the XPS and ICP tests (Table 1).

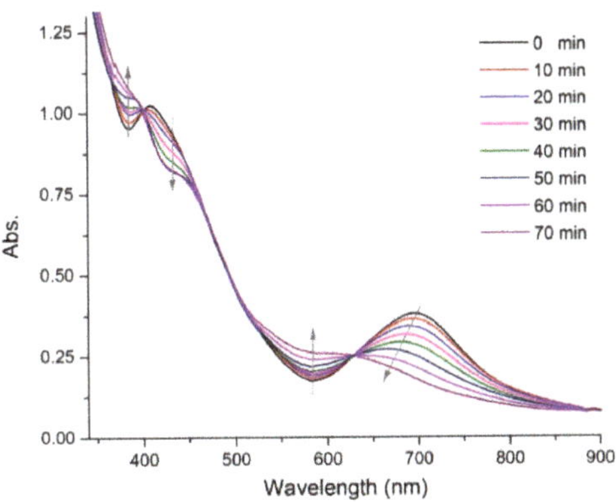

Figure 2. Time-dependent UV-Vis spectra of the $Ag_2Au_{25}(SR)_{18}$ nanocluster reacting with PET ligands in dichloromethane.

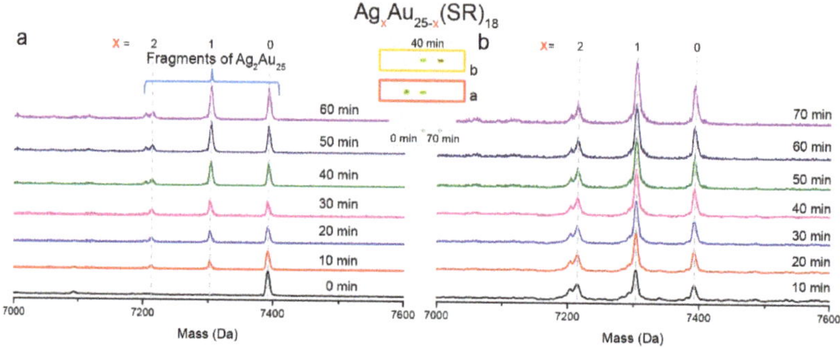

Figure 3. Time-dependent MALDI-TOF-MS spectra of the $Ag_2Au_{25}(SR)_{18}$ reaction with PET. The inset picture is the Thin-layer chromatography (TLC) of samples at 0 min (Ag_2Au_{25}), 40 min, and 70 min, respectively. The positive model MALDI-TOF-MS results of the two components separated by TLC are shown in (**a**) and (**b**), respectively. Note that, the peaks in (**a**) are the fragments of Ag_2Au_{25} according to [24]. The full spectrum of Ag_2Au_{25} is shown in Figure S1.

Table 1. The atomic ratio of Au:Ag in the $Ag_xAu_{25-x}(SR)_{18}^-$ nanocluster, calculated by Inductively coupled plasma-atomic emission spectrometry (ICP-AES) and X-ray photoelectron spectroscopy (XPS) measurements.

Results	Au Atom	Ag Atom
ICP experimental ratio	96.22%	3.78%
XPS experimental ratio	96.04%	3.96%
Theoretical	24/25 (96%)	1/25 (4%)

Time-dependent TLC and MALDI-TOF-MS spectra were applied to monitor the reaction process (Figure 3). The TLC results showed the high purity of the Ag_2Au_{25} nanocluster, after addition of the thiol ligands (PET) into the solution, the new compound was formed as demonstrated by the TLC. We further tested the mass spectrum of the two compounds, respectively. The mass spectrum suggested that the molecular ion peak of Ag_2Au_{25} was not changed, whereas with the different fragment peaks, the pure Ag_2Au_{25} nanocluster had one fragment peak at 7391 Da, which was caused by losing two Ag atoms at the surface. Interestingly, after the addition of the thiol ligands, we found three fragment peaks at 7391 Da, 7302 Da, and 7213 Da. The mass difference (89 Da) was equal to the mass difference between Au (197 Da) and Ag (108 Da). The different fragmentation after the addition of the thiol ligands suggested that the silver atoms at the surface were exchanged by gold atoms. The new compounds were determined to be $Ag_xAu_{25-x}(SR)_{18}$ nanoclusters, which indicated the occurrence of the metal exchange process.

There are two main approaches in which this metal exchange can occur: (i) In direct metal exchange. In this way, thiol ligands first strip the adsorbed silver ions from the Ag_2Au_{25} nanocluster. This process results in the Ag(SR) complex and the $Au_{25}(SR)_{18}^-$ nanocluster. Then, metal exchange occurrs between these two products and results in the $Ag_xAu_{25}(SR)_{18}^-$ nanocluster. (ii) In intramolecular metal exchange, that is, adsorbed silver ions are exchanged with the kernel gold atoms, which are further stripped by thiol ligands. In order to study this reaction, the mass analysis was applied on the new compounds, which were separated from the TLC. The mass spectra showed similar results during the reaction. The absence of $Ag_2Au_{25}(SR)_{18}$ indicated this composition had reacted. Three intense peaks, at 7391 Da, 7302 Da, and 7213 Da, were found, which were assigned to Ag_2Au_{23}, Ag_1Au_{24}, and Au_{25}, respectively. It is worth nothing that, the ratio of the three peaks was maintained at 1:2:1 during the reaction. For comparison, we applied the metal exchange reaction between the $Au_{25}(SR)_{18}^-$ nanocluster with the Ag(SR) complex; the time dependent mass spectra and UV-Vis absorption spectra are shown in Figure S2. Consequently, the intermolecular metal exchange reaction between Ag(SR) with $Au_{25}(SR)_{18}^-$ was quite different from the present work. The proportion of doped silver atoms gradually increased with time. These results ruled out the idea that the metal exchange in the Ag_2Au_{25} nanocluster resulted from the thiol ligand pulling out the adsorbed silver atoms, and the resultant intermolecular metal exchange reaction.

The intramolecular metal exchange in the Ag_2Au_{25} nanocluster contained two steps (Scheme 1): (i)self-metal-exchange; (ii) metal stripping. In the first step, surface silver atoms exchanged the gold atoms at the kernel, with the promotion of the thiol ligands, and produced the isomeric Ag_2Au_{25} nanocluster. In the second step, the surface atoms dissociated from the surface of the cluster and produced the Ag_xAu_{25-x} nanocluster. The 1:2:1 ratio, which was found from the mass spectra of the final product, revealed that the silver to gold ratio was 1:1.

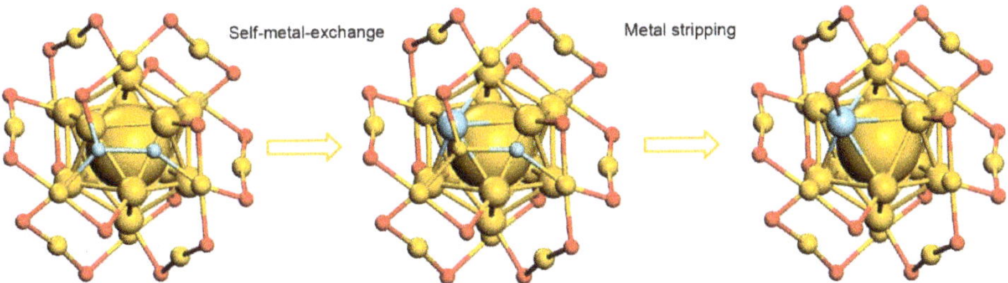

Scheme 1. Th proposed intramolecular metal exchange process.

4. Conclusions

In summary, we reported the intramolecular metal exchange in the Ag_2Au_{25} nanocluster. The intramolecular exchange involves two steps. First, silver atoms at the surface interchange with the gold atoms at the kernel, then the metal ions on the cluster surface are stripped from the cluster. With the intramolecular reaction, we obtained the $Ag_xAu_{25-x}(SR)_{18}^-$ ($x = 0-2$) nanocluster with a quantitative yield. Our work not only provides a new approach for the metal exchange reaction, but also provides a new perspective for understanding the metal behavior at the nanoscale.

Supplementary Materials: The following are available online at http://www.mdpi.com/2079-4991/8/12/1070/s1, Figure S1: The full spectra of Ag_2Au_{25} nanocluster with different laser energy during the MALDI-TOF-MS analysis. The intensity of Au_{25} decreases with the decrease of laser intensity, which indicate the Au_{25} is the fragment of Ag_2Au_{25}; Figure S2: Time-dependent MALDI-TOF-MS spectra of metal exchange between $Au_{25}(SR)_{18}^-$ and $Ag(SR)$ (4 equivalents) complex. (a) 0 min; (b) 5 min; and (c) 15 min.

Author Contributions: S.W. and M.Z. designed the study; Y.L. and M.C. performed the experiments and analyzed the data. All authors discussed the results and commented on the manuscript.

Funding: This research was funded by NSFC, grant number U1532141, 21631001, 21871001, and 21803001; PhD funding of Anhui University, grant number J10113190012; the Ministry of Education; and the Education Department of Anhui Province and 211 Project of Anhui University.

Conflicts of Interest: The authors declare no conflict of interest.

References

1. Jin, R.; Zeng, C.; Zhou, M.; Chen, Y. Atomically Precise Colloidal Metal Nanoclusters and Nanoparticles: Fundamentals and Opportunities. *Chem. Rev.* **2016**, *116*, 10346–10413. [CrossRef] [PubMed]
2. Chakraborty, I.; Pradeep, T. Atomically Precise Clusters of Noble Metals: Emerging Link between Atoms and Nanoparticles. *Chem. Rev.* **2017**, *117*, 8208–8271. [CrossRef] [PubMed]
3. Gilroy, K.D.; Ruditskiy, A.; Peng, H.-C.; Qin, D.; Xia, Y. Bimetallic Nanocrystals: Syntheses, Properties, and Applications. *Chem. Rev.* **2016**, *116*, 10414–10472. [CrossRef] [PubMed]
4. Nguyen, T.D.; Jones, Z.R.; Goldsmith, B.R.; Buratto, W.R.; Wu, G.; Scott, S.L.; Hayton, T.W. A Cu_{25} Nanocluster with Partial Cu(0) Character. *J. Am. Chem. Soc.* **2015**, *137*, 13319–13324. [CrossRef] [PubMed]

5. Wang, S.; Li, Q.; Kang, X.; Zhu, M. Customizing the Structure, Composition, and Properties of Alloy Nanoclusters by Metal Exchange. *Acc. Chem. Res.* **2018**, *51*, 2784–2792. [CrossRef] [PubMed]

6. Wang, S.; Song, Y.; Jin, S.; Liu, X.; Zhang, J.; Pei, Y.; Meng, X.; Chen, M.; Li, P.; Zhu, M. Metal Exchange Method Using Au_{25} Nanoclusters as Templates for Alloy Nanoclusters with Atomic Precision. *J. Am. Chem. Soc.* **2015**, *137*, 4018–4021. [CrossRef]

7. Wang, S.; Abroshan, H.; Liu, C.; Luo, T.-Y.; Zhu, M.; Kim, H.J.; Rosi, N.L.; Jin, R. Shuttling single metal atom into and out of a metal nanoparticle. *Nat. Commun.* **2017**, *8*, 848. [CrossRef]

8. Yang, S.; Wang, S.; Jin, S.; Chen, S.; Sheng, H.; Zhu, M. A metal exchange method for thiolate-protected tri-metal $M_1Ag_xAu_{24-x}(SR)_{18}$ (M = Cd/Hg) nanoclusters. *Nanoscale* **2015**, *7*, 10005–10007. [CrossRef]

9. Yang, S.; Chen, S.; Xiong, L.; Liu, C.; Yu, H.; Wang, S.; Rosi, N.L.; Pei, Y.; Zhu, M. Total Structure Determination of $Au_{16}(S\text{-}Adm)_{12}$ and $Cd_1Au_{14}(StBu)_{12}$ and Implications for the Structure of $Au_{15}(SR)_{13}$. *J. Am. Chem. Soc.* **2018**, *140*, 10988–10994. [CrossRef]

10. Jin, S.; Du, W.; Wang, S.; Kang, X.; Chen, M.; Hu, D.; Chen, S.; Zou, X.; Sun, G.; Zhu, M. Thiol-Induced Synthesis of PhosphineProtected Gold Nanoclusters with Atomic Precision and Controlling the Structure by Ligand/Metal Engineering. *Inorg. Chem.* **2017**, *56*, 11151–11159. [CrossRef]

11. Li, Q.; Lambright, K.J.; Taylor, M.G.; Kirschbaum, K.; Luo, T.-Y.; Zhao, J.; Mpourmpakis, G.; Mokashi-Punekar, S.; Rosi, N.L.; Jin, R. Reconstructing the Surface of Gold Nanoclusters by Cadmium Doping. *J. Am. Chem. Soc.* **2017**, *139*, 17779–17782. [CrossRef]

12. Li, Q.; Wang, S.; Kirschbaum, K.; Lambright, K.J.; Das, A.; Jin, R. Heavily doped $Au_{25-x}Ag_x(SC_6H_{11})_{18}$ nanoclusters: Silver goes from the core to the surface. *Chem. Commun.* **2016**, *52*, 5194–5197. [CrossRef]

13. Du, W.; Jin, S.; Xiong, L.; Chen, M.; Zhang, J.; Zou, X.; Pei, Y.; Wang, S.; Zhu, M. $Ag_{50}(Dppm)_6(SR)_{30}$ and Its Homologue $Au_xAg_{50-x}(Dppm)_6(SR)_{30}$ Alloy Nanocluster: Seeded Growth, Structure Determination, and Differences in Properties. *J. Am. Chem. Soc.* **2017**, *139*, 1618–1624. [CrossRef]

14. Kang, X.; Xiong, L.; Wang, S.; Yu, H.; Jin, S.; Song, Y.; Chen, T.; Zheng, L.; Pan, C.; Pei, Y.; et al. Shape-Controlled Synthesis of Trimetallic Nanoclusters: Structure Elucidation and Properties Investigation. *Chem. Eur. J.* **2016**, *22*, 17145–17150. [CrossRef] [PubMed]

15. Hossain, S.; Ono, T.; Yoshioka, M.; Hu, G.; Hosoi, M.; Chen, Z.; Nair, L.V.; Niihori, Y.; Kurashige, W.; Jiang, D.-E.; et al. Thiolate-Protected Trimetallic $Au_{\sim 20}Ag_{\sim 4}Pd$ and $Au_{\sim 20}Ag_{\sim 4}Pt$ Alloy Clusters with Controlled Chemical Composition and Metal Positions. *J. Phys. Chem. Lett.* **2018**, *9*, 2590–2594. [CrossRef] [PubMed]

16. Krishnadas, K.R.; Baksi, A.; Ghosh, A.; Natarajan, G.; Pradeep, T. Structure-conserving spontaneous transformations between nanoparticles. *Nat. Commun.* **2016**, *7*, 13447. [CrossRef]

17. Zhang, B.; Salassa, G.; Bürgi, T. Silver migration between $Au_{38}(SC_2H_4Ph)_{24}$ and doped $Ag_xAu_{38-x}(SC_2H_4Ph)_{24}$ nanoclusters. *Chem. Commun.* **2016**, *52*, 9205–9207. [CrossRef]

18. Zhang, B.; Safonova, O.V.; Pollitt, S.; Salassa, G.; Sels, A.; Kazan, R.; Wang, Y.; Rupprechter, G.; Barrabes, N.; Bürgi, T. On the mechanism of rapid metal exchange between thiolate-protected gold and gold/silver clusters: A time-resolved in situ XAFS study. *Phys. Chem. Chem. Phys.* **2018**, *20*, 5312–5318. [CrossRef] [PubMed]

19. Zhang, B.; Bürgi, T. Doping Silver Increases the $Au_{38}(SR)_{24}$ Cluster Surface Flexibility. *J. Phys. Chem. C* **2016**, *120*, 4660–4666. [CrossRef]

20. Wang, S.; Meng, X.; Das, A.; Li, T.; Song, Y.; Cao, T.; Zhu, X.; Zhu, M.; Jin, R. A 200-fold Quantum Yield Boost in the Photoluminescence of Silver Doped Ag_xAu_{25-x} Nanoclusters: The 13th Silver Atom Matters. *Angew. Chem. Int. Ed.* **2014**, *53*, 2376–2380. [CrossRef]

21. Bootharaju, M.S.; Joshi, C.P.; Parida, M.R.; Mohammed, O.F.; Bakr, O.M. Templated Atom-Precise Galvanic Synthesis and Structure Elucidation of a $[Ag_{24}Au(SR)_{18}]^-$ Nanocluster. *Angew. Chem.* **2016**, *128*, 934–938. [CrossRef]

22. Deng, H.; Wang, S.; Jin, S.; Yang, S.; Xu, Y.; Liu, L.; Xiang, J.; Hu, D.; Zhu, M. Active metal (cadmium) doping enhanced the stability of inert metal (gold) nanocluster under O_2 atmosphere and the catalysis activity of benzyl alcohol oxidation. *Gold Bull.* **2015**, *48*, 161–167. [CrossRef]

23. Parker, J.F.; Weaver, J.E.F.; McCallum, F.; Fields-Zinna, C.A.; Murray, R.W. Synthesis of Monodisperse $[Oct_4N]^+$ $[Au_{25}(SR)_{18}]^-$ Nanoparticles, with Some Mechanistic Observations. *Langmuir* **2010**, *26*, 13650–13654. [CrossRef]
24. Yao, C.; Chen, J.; Li, M.; Liu, L.; Yang, J.; Wu, Z. Adding Two Active Silver Atoms on Au_{25} Nanoparticle. *Nano Lett.* **2015**, *15*, 1281–1287. [CrossRef] [PubMed]

 *nanomaterials*

Article

In Situ Decoration of Gold Nanoparticles on Graphene Oxide via Nanosecond Laser Ablation for Remarkable Chemical Sensing and Catalysis

Parvathy Nancy [1], Anju K Nair [2], Rodolphe Antoine [3], Sabu Thomas [4,5,*] and Nandakumar Kalarikkal [1,4,*]

1   School of Pure and Applied Physics, Mahatma Gandhi University, Kottayam 686560, India
2   Department of Physics, St. Teresas's College, Ernamkulam 682011, India
3   Institut Lumière Matière, UMR 5306 CNRS, Université Claude Bernard Lyon 1, Domaine Scientifique de La Doua, Batiment Kastler, 10 rue Ada Byron, 69622 Villeurbanne CEDEX, France
4   International and Inter University Centre for Nanoscience and Nanotechnology, Mahatma Gandhi University, Kottayam 686560, India
5   School of Chemical Sciences, Mahatma Gandhi University, Kottayam 686560, India
*   Correspondence: nkkalarikkal@mgu.ac.in (N.K.); sabuthomas@mgu.ac.in (S.T.)

Received: 5 August 2019; Accepted: 22 August 2019; Published: 26 August 2019

Abstract: Gold decorated graphene-based nano-hybrids find extensive research interest due to their enhanced chemical catalytic performance and biochemical sensing. The unique physicochemical properties and the very large surface area makes them propitious platform for the rapid buildouts of science and technology. Graphene serves as an outstanding matrix for anchoring numerous nanomaterials because of its atomically thin 2D morphological features. Herein, we have designed a metal-graphene nano-hybrid through pulsed laser ablation. Commercially available graphite powder was employed for the preparation of graphene oxide (GO) using modified Hummers' method. A solid, thin gold (Au) foil was ablated in an aqueous suspension of GO using second harmonic wavelength (532 nm) of the Nd:YAG laser for immediate generation of the Au-GO nano-hybrid. The synthesis strategy employed here does not entail any detrimental chemical reagents and hence avoids the inclusion of reagent byproducts to the reaction mixture, toxicity, and environmental or chemical contamination. Optical and morphological characterizations were performed to substantiate the successful anchoring of Au nanoparticles (Au NPs) on the GO sheets. Remarkably, these photon-generated nano-hybrids can act as an excellent surface enhanced Raman spectroscopy (SERS) platform for the sensing/detection of the 4-mercaptobenzoic acid (4-MBA) with a very low detection limit of 1×10^{-12} M and preserves better reproducibility also. In addition, these hybrid materials were found to act as an effective catalyst for the reduction of 4-nitrophenol (4-NP). Thus, this is a rapid, mild, efficient and green synthesis approach for the fabrication of active organometallic sensors and catalysts.

Keywords: gold nanoparticles; graphene oxide; laser ablation; Au-GO nano-hybrid

1. Introduction

Nano sized particles of noble metals, particularly gold NPs (Au NPs), have ample scope in research due to their unique optical, electrical and catalytic properties and their potential applications in material science, physics, chemistry and in many interdisciplinary fields [1–3]. The Au NPs have the tendency of aggregation when they are in a solution on account of their superior surface energy, which results in a reduced catalytic activity and stability [4]. This disadvantage can be solved by anchoring Au NPs on various matrices such as SiO_2, polymer and carbon microspheres, etc. [5–7]. Among these matrices, graphene derivatives like graphene oxide (GO), and reduced graphene oxide (rGO) have been subjected to vast study owing to an excessive surface area, planar structure, and exceptional physicochemical properties [8]. GO is a better choice when compared with graphene in terms of a low production cost, high yield manner and more application suitability. Having a rich O_2 containing functional groups on its edges and basal planes make GO hydrophilic and a good support for the surface functionalization and nanoparticle immobilization. Until now, extensive research has been done on the development of GO or rGO supported Au NPs for applications in the field of catalysts, sensors and medicine. Au NPs and GO nano-hybrids shows novel and increased physicochemical properties, which is considered to be accomplished from the strong interaction between these two constituents [9–11].

Graphene sheets embellished with metal nanoparticles are brilliant materials having potential use in energy storage, optoelectronics, bio sensors or catalysts, because of their unique physical and chemical characteristics [12,13]. The surface plasmon resonance of the metallic nanoparticles, interaction of graphene sheets with light, formation of bio sensors by means of bio conjugation, inflation of catalytic activity of nanoparticles [14,15], etc. are controlled by the unique conductivity and volatile features of graphene and metal nanoparticles, respectively. Usually, the graphene- metal nano-hybrids are fabricated by the synchronized reduction of metal salts and graphene derivatives. In such methods, surfactants or co-ligands on the synthesized nanoparticles are utilized to augment stability and to reduce aggregation behavior. At the same time, impurities are also produced along with the entire procedure and final products [16,17]. Generally, the metal nanoparticles are capped with ligands in the surface, which foster a cross-chemical effect. As a solution to this dilemma, ligand free metal nanoparticles have been synthesized through alternative top-end and easy pathways, functioning on the basis of principles of green protocols [18,19]. Pulsed laser ablation in liquids (PLAL) is such a method, in which a solid target immersed in a colloid or liquid is irradiated with a suitable pulsed laser beam [20–23]. Here in PLAL, the possible synergy between the electromagnetic field of the laser pulses and the atoms on the material surface leads to the removal of electrons from the surface, culminating in creation of an electronic cloud and an electron deficiency localized area on the surface of the target. The electronic cloud in turn will attract the surface ions by means of electromagnetic force, resulting in the formation of clusters of atoms and ionized atoms. A cavitation bubble formation is the next process in line, which cages the material that forms crystalline nanoparticles and finally discharges the nanoparticles in to the liquid environment as the bubble blots out [24,25]. The ligand free Au NPs shows better surface activity due to the lack of ligands on its surface so that it can retain on the GO surface by the combined effect of three distinct reasons (a) interplay among the d-orbitals of metal nanoparticles and the pi-electron aromatic network of graphene (b) the strong electrostatic interaction between metal nanoparticles and graphene promotes the effective anchoring of AuNPs on GO (c) desirable distribution of nanoparticles at the defects/wrinkles found in the graphene surface [26,27]. There are reports of synthesis of ligand free Au NPs by PLAL and subsequent blending with graphene with the aid of a variety of lasers [28–30]. The prospect of direct synthesis of particles from graphene sheets in liquid medium signifies that the electromagnetic effects from the laser radiation causing some reactive process in GO may govern the reduction process of graphene oxide. Thus, the laser irradiation eliminates the O_2 functional groups from the graphene sheet, whilst possessing the large inter layer spacing, that characterizes the electronic decoupling of the individual layer in the material and this decoupling effect may give rise to superconductivity effects sometimes [31,32]. In this work, we focus on the decoration of graphene oxide sheets with Au NPs in a novel single step approach. The impact given by a nanosecond radiation-based technique upon the control of synthesis of highly pure Au NPs in a liquid ambience containing graphene oxide layers has been studied. Apart from the conventional synthesis methods and the previously reported synthesis via PLAL, we present the fabrication of the Au-GO hybrid material, especially in the nanosecond regime for the first time. The in situ generation and decoration of the Au NPs on the GO layers are established by photons and the purity, structure, size and stability of the AuNPs were fine-tuned by different laser parameters like laser fluence, repletion rate, laser wavelength, irradiation time, etc. To establish the hypothesis, Au NPs were synthesized in a homogenous aqueous suspension of graphene oxide layers. Moreover, this methodology has proved that it is an impressive protocol to derive graphene oxide-metal nanoparticle assemblies with a potential to be used in a verity of applications where undesirable effects of harmful chemicals can be avoided successfully.

2. Experimental Procedure

In order to generate the Au-GO nano-hybrid, a thin gold foil of thickness of 0.1 mm (SIGMA ALDRICH, Missouri, MO, USA, pure trace metal, 99.99%) was properly fixed in the inner side of a glass cuvette, which is filled with 30 mL of an aqueous suspension of the GO solution. The second harmonic beam of an Nd:YAG laser (Litron LPY 674G-10) having a pulse width of 8 ns was focused on to that gold foil with a bi-convex lens having focal length 15cm at a 10 Hz repletion rate. The spot diameter of the beam was found to be 0.01 mm. Thus, the Au NPs were born and anchored in situ on to the GO single layers through the nanosecond laser radiation strategy technically known as PLAL. GO was prepared previously from the standard graphite powder using a modified Hummers' method [33]. The whole experiment was performed completely in room temperature.

Figure 1 depicts the experimental arrangement of laser ablation. The Au target is ablated for fifteen minutes (15 min) at four distinct laser energy densities such as 5.3 Jcm^{-2}, 7.9 Jcm^{-2}, 10.5 Jcm^{-2}, and 13.2 Jcm^{-2}. The GO solution was kept under stirring with the aid of a magnetic stirrer for uniform distribution/decoration of the Au NPs on the GO matrix.

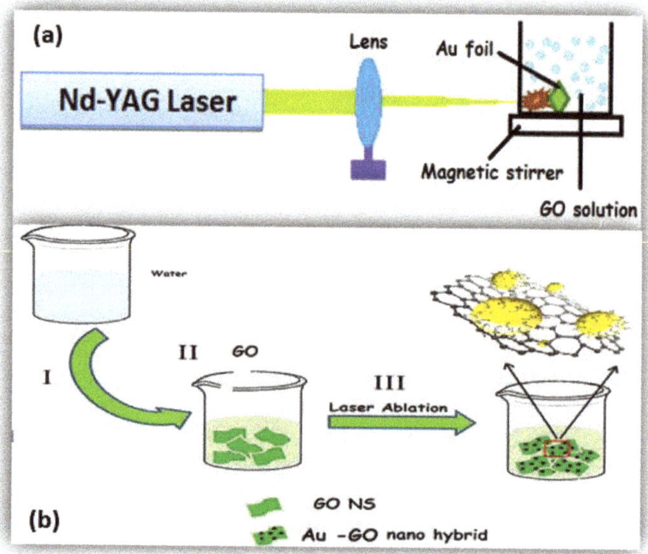

Figure 1. (a) The experimental framework of fabrication of Au-GO nano-hybrid; (b) synthesis procedure for the Au-GO nano-hybrid.

3. Results and Discussion

Here, we focus on the facile generation of the Au-GO nano-hybrid using a laser ablation in nanoseconds. The influence of the nanosecond radiation on the controlled synthesis of uniformly dispensed Au NPs in the aqueous suspension of GO sheets and its applications in chemical sensing and catalysis has been studied. With this focus, Au NPs were produced in situ and were anchored on to single layered graphene-oxide sheets. Our results demonstrate that the growth of ligand-free metal nanoparticles on graphene oxide sheets in a single reaction procedure using a laser ablation is a highly beneficial technique for fast, simple and environmentally friendly synthesis of nanoparticles, hybrids and composites.

3.1. UV-Vis Absorption Spectroscopy

Figure 2a represents the absorption spectrum of a pure GO sheet. The GO sheets display an absorption at 234 nm, which is attributed to the C–C bond's $\pi \longrightarrow \pi^*$ interaction; the tiny shoulder peak at 304 nm is attributed to the C=O bond's n $\longrightarrow \pi^*$ interaction. Figure 2b represents the absorption spectra of the Au-GO nano-hybrid. After the ablation process at four distinct laser fluences (5.3 Jcm^{-2}, 7.9 Jcm^{-2}, 10.5 Jcm^{-2}, and 13.2 Jcm^{-2}), the absorption peak of the GO in all prepared samples slowly shifted to 255 nm, which implies that a few GO sheets have undergone a reduction due to the interactivity of the GO with a high energy laser beam (Figure 2b). Together with this, the Au-GO nano-hybrid material evince a wide absorption peak at ~526 nm. This is the contribution by the surface plasmon resonance (SPR) of the Au NPs [34,35]. These results indicate that the Au NPs were generated and successfully anchored on the GO matrix. The results also elucidate that the laser fluence is a key-regulating parameter and has a vital role in the size and concentration of Au NPs on the GO matrix.

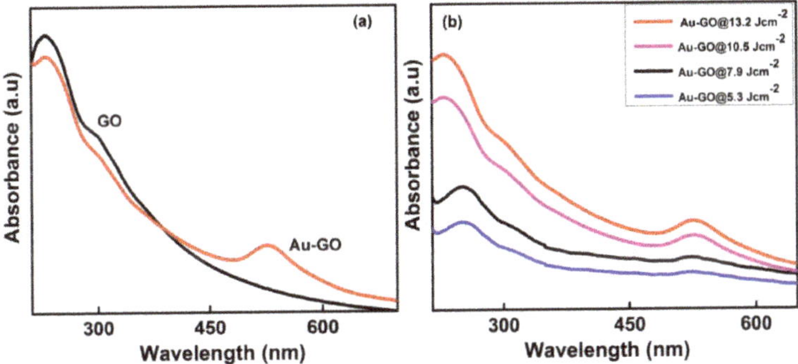

Figure 2. UV-Vis absorption spectrum. (**a**) GO and Au-GO nano-hybrid; (**b**) Au-GO at different laser fluences.

3.2. XPS and Raman Analysis

The wide-scan XPS spectra of the Au-GO nano-hybrid shown in Figure 3a upholds the existence elements of Au, C and O. In order to examine the oxidation states of Au and GO, the X-ray photoelectron spectroscopy (XPS) was employed. Figure 3a depicts the wide scan spectra of the Au-GO hybrid material. In fact, the high-resolution XPS peaks of the deconvoluted C1s spectrum of the Au-GO nano-hybrid shows three different peaks due to the existence of the C=O (288.2 eV), C–O (286.6 eV) and C–C (286.4 eV) functional groups (Figure 3c), while the binding energy at 84 eV and 87 eV were attributed to the Au4f$_{7/2}$ and Au4f$_{5/2}$ peaks (Figure 3b). This result confirms the effective formation of the Au layer over GO.

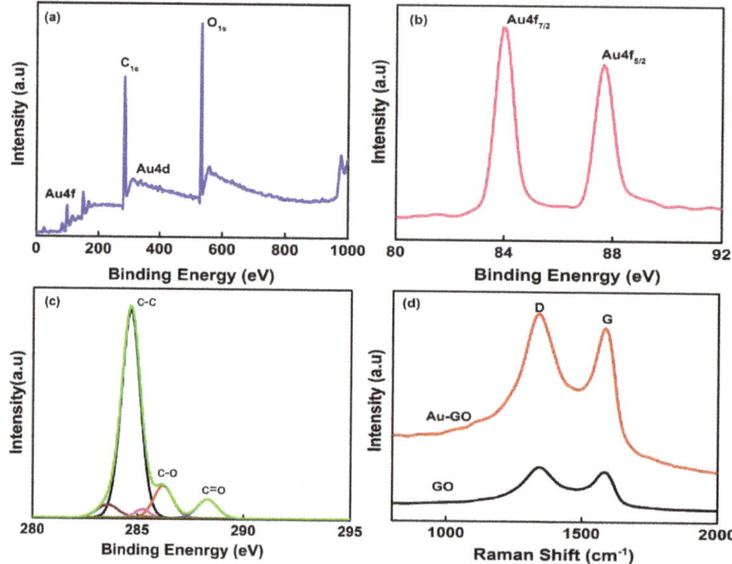

Figure 3. X-ray photoelectron spectroscopy (XPS) and Raman spectra of the Au-GO nano-hybrid. (**a**) Wide scan spectra; (**b**) Au4f spectrum; (**c**) C1s spectrum; (**d**) Raman spectra of the Au-GO and GO.

The Raman spectroscopy was employed additionally to explore the structure of the nano-hybrid and the interactivity between the Au nanoparticles and GO sheets. Here, in the Raman spectra Figure 3d reveals an intense D peak at 1349 cm^{-1} and a G peak at 1580 cm^{-1}, which affirms the genesis of a few-layered GO with its graphitic structure. The occurrence of an intense D band at 1349 cm^{-1} in the GO sample clearly indicates the existence of defect sites in the graphene layers. These defects are often assigned to the oxidation of graphite and doping effects in the hexagonal carbon lattice. Furthermore, the I_D/I_G ration (I_D/I_G = 1.09) for the Au-GO hybrid is large compared with the bare GO (I_D/I_G = 0.89); it substantiates the existence of several defects in the Au-GO created during the laser ablation. The evident spike in the I_D/I_G value from GO to Au-GO indicates the successful anchoring of Au on the GO network. Moreover, the nano-hybrids exhibit increased Raman intensity, which was because of the effect of the localized electromagnetic field of the Au nanoparticles and thus it can be a competent substrate for SERS detection.

3.3. Morphological Analysis

The surface morphology and size of the Au-GO nano-hybrids were studied using a transmission electron microscopy (TEM) and field emission scanning electron microscopy (FESEM).

The TEM image (Figure 4) of the Au-GO nano-hybrid shows that the small Au nanoparticles of a diameter around ~30 nm are anchored on to the surface of the GO. The HRTEM analysis shows that the synthesized Au-GO material is unbound of any other chemical impurities and elements. The high crystalline nature of the material was clear from the lattice fringes. It also provides clear details regarding the d spacing of the Au nanoparticles. The Au nanoparticles exhibit a darker contrast with an interlayer spacing of 0.202 nm that corresponds to the d-spacing of a (200) plane of Au. In addition, the transmission electron microscope-energy dispersive spectra (TEM-EDS) confirms the presence of highly-pure gold nanoparticles decorated over the GO matrix; it is shown in Figure 5. The EDS reveals that, the most evident intensity peaks are Au, C and O corresponding to the Au NPs and GO sheets.

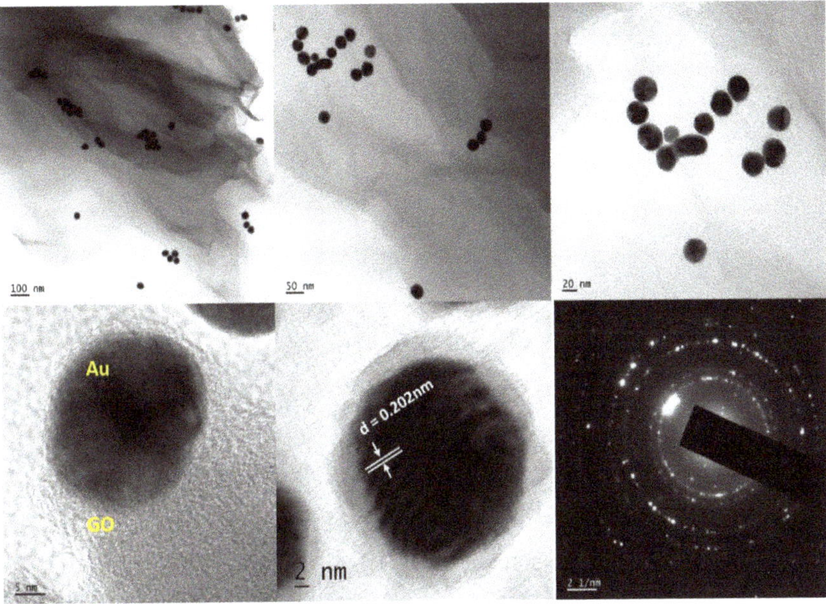

Figure 4. TEM and HRTEM images of the Au-GO nano-hybrid.

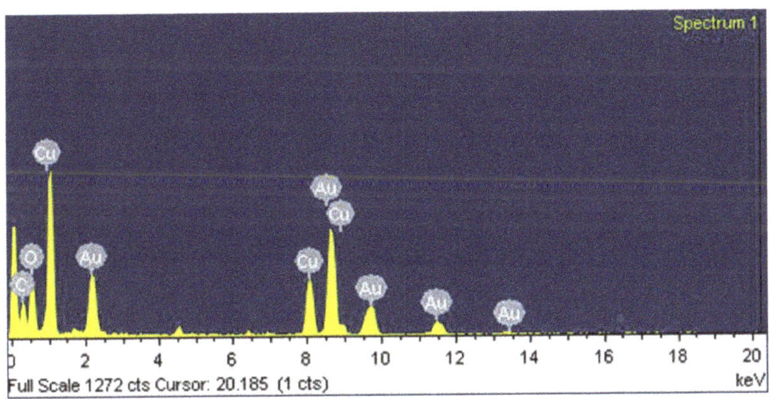

Figure 5. TEM-EDS of the Au-GO nano-hybrid.

Also, Figure 6 shows the FESEM image and the elemental mapping of the Au-GO nano-hybrid in which the elements Au, C and O are mapped in blue, pink and yellow colour, respectively; the mapping clearly suggests the co-existence of Au and GO in the laser generated nano-hybrid. Figure 6a clearly depicts that the number density of the Au NPs are very high in the vicinity of wrinkles in the GO sheets. These are one of the main reasons for the adsorption of the AuNPs on the GO. The wrinkles present in the GO corresponds to the defect sites and it facilitates the high rate of accumulation and immobilization of the Au NPs on such areas.

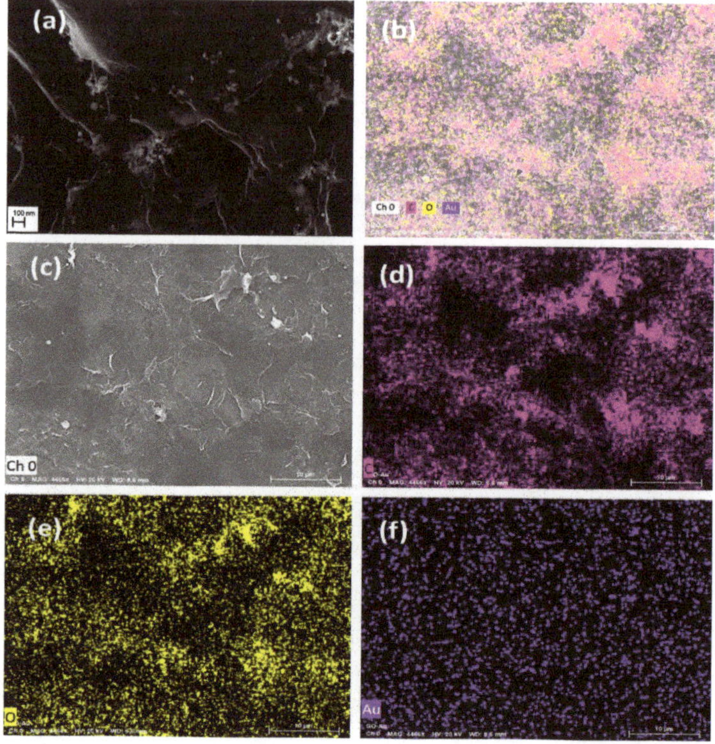

Figure 6. (**a**) Field emission (FESEM) image of the Au-GO nano-hybrid (**b**–**f**) elemental mapping of Au-GO nano-hybrid, where C, O and Au are mapped in pink, yellow and blue colours respectively.

3.4. SERS Activity of 4-Mercaptobenzoic Acid (4-MBA)

For the SERS measurement, 10 µL of the samples were coated on a silicon wafer (~1 cm^2). The 4-mercaptobenzoic acid solution (4-MBA solution, 10 µM concentration, and 10 µL quantity) was prepared and drop casted on a silicon wafer and then dried in room temperature. These stock solutions were gradually diluted with de-ionized water to obtain various concentrations of analytes. A confocal Raman spectrometer (WITec alpha 300RA) with a 532 nm excitation wavelength was used for SERS measurements of hence prepared samples. The SERS data were collected over the span of 200–2000 cm^{-1} with 10 s as an integration time.

The relative SERS enhancement capabilities of the Au-GO nano-hybrids were measured using 4-mercaptobenzoic acid (4-MBA) as the probe molecule, which is highly Raman active and adsorbed on the surface of these hybrid nanostructures. For comparison, Au NPs are also prepared to confirm the SERS activity of the 4-MBA molecules. The 4-MBA is an organic molecule with a thiol group and a carboxylic acid group on the two ends that usually exhibits a strong chemical coupling with metallic surfaces [36]. It is well known that metal nanoparticles incorporated nanostructures give higher SERS enhancement than the bare metal nanoparticles. Metal nanoparticles usually contribute an electromagnetic field enhancement to the SERS signals. Figure 7a represents the SERS spectra of the Au-GO synthesized at different laser energies, capped with 4-MBA (10 µM). In Figure 7a, the SERS spectra exhibit two prominent Raman peaks at 1076 cm^{-1} and 1586 cm^{-1} that is attributed to the ν(C–C) breathing modes of a benzene ring. The other weak bands, 1412 cm^{-1} emerged from the stretching mode of ν_s (COO–) and the band at 1182 cm^{-1} from the C–H deformation modes; they were found to be consistent with the previous report [37]. The ν(C–C) mode depicts a dominant peak intensity than the other modes of vibration due to the enhanced coupling between the transition dipole moment and local electric field [38,39]. The samples at 13.2 Jcm^{-2} laser energy displayed more SERS intensity. These observations confirm that the use of Au resulted in a higher surface area, providing amplification of the hot spots. Therefore, such a simple approach of the controlled decoration of Au over GO could successfully enhance the SERS activity. The peak intensities at 1076 cm^{-1} and 1586 cm^{-1} were chosen to calculate the SERS enhancement factor for the Au decorated GO sheet. The SERS enhancement factor is evaluated by comparing the SERS signals obtained from the 4-MBA capped structures to the Raman intensities acquired from the bulk 4-MBA molecule using the formula:

$$\text{Enhancement factor (EF)} = I_{SERS}/I_{NR} \times C_{NR}/C_{SERS} \tag{1}$$

where, I_{SERS} and I_{NR} correspond to the calculated intensity of the SERS spectra of the molecule which is adsorbed on the nano-hybrid surface and the normal Raman spectrum of the bulk molecule, respectively. C_{NR} and C_{SERS} are the corresponding 4-MBA concentrations for the normal Raman and SERS substrates [40]. We have expected that the 4-MBA molecules are evenly distributed on the SERS substrate. So, 10 µL of the 4-MBA solution having a 10^{-7} M concentration was pipetted and casted on the SERS substrate and the droplet evaporated in air for the experiment.

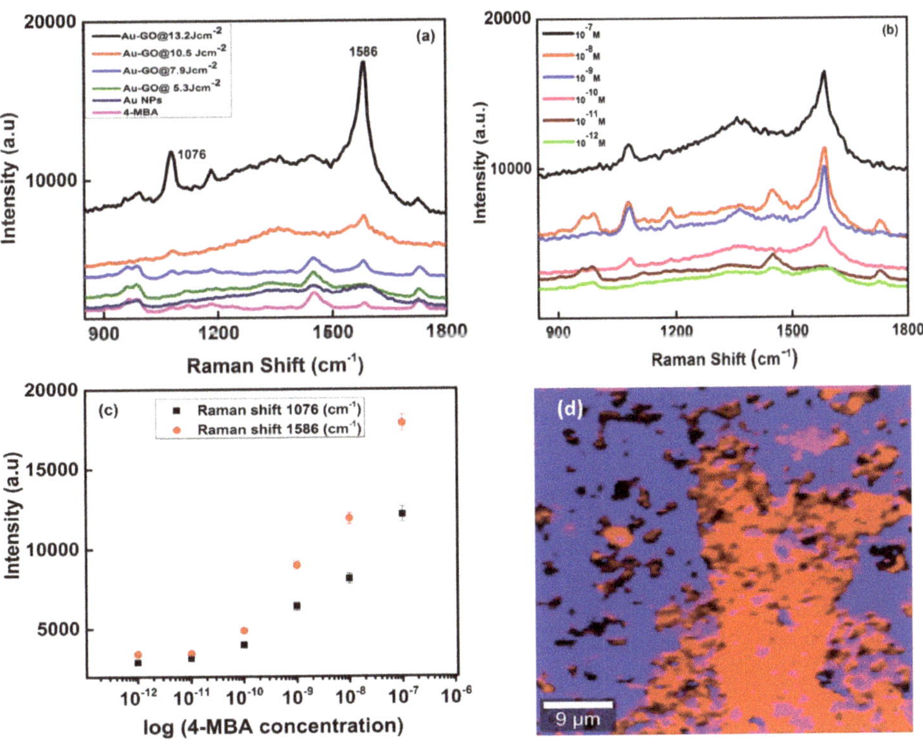

Figure 7. (a) The surface enhanced Raman spectroscopy (SERS) spectra of the samples at 10 µM 4-MBA; (b) the SERS spectra of Au-GO at different concentrations of 4-MBA; (c) calibration data with respect to the average SERS intensities of the peaks of 4-MBA at 1076 cm^{-1} and 1586 cm^{-1} with the 4-MBA concentrations; (d) the Raman image of the Au-GO, the red region corresponds to the 'hot spots'.

The calculated SERS enhancement factor (EF) for the Au-GO@13.2 Jcm^{-2} at 1076 cm^{-1} is 7.9 × 10^6, respectively. Therefore, such nano-hybrids demonstrate higher SERS efficiency than the bare Au NPs and their excellent stability is in very good agreement with the literature [41].

Moreover, it should be noted that the Au nanoparticles with small dimensions are distributed throughout the GO surface thereby creating 'hot spots'. Figure 7b represents the SERS spectra at different concentrations of 4-MBA, i.e., peaks from 10^{-7} M to 10^{-12} M were adsorbed on the surface of Au-GO@13.2 Jcm^{-2} sample. From Figure 7b, it is evident that the peaks corresponding to 4-MBA can be sensed even at a very low concentration of 4-MBA at 10^{-12} M thereby indicating a very high sensitivity of SERS substrate. Figure 7c represents the SERS Raman mapping image of the Au-GO@13.2 Jcm^{-2} sample, in which the red region corresponds to the 'hot spots'. The Au-GO nano-hybrids are believed to be a favourable candidate for the SERS applications due to their surface roughness, which in turn provides a higher surface area for the more adsorption of the SERS-active molecules. The huge enhancement of the SERS signals from the Au-GO nano-hybrids is attributed to the presence of ultra-pure Au NPs, which serves as 'hot spots' during the SERS analysis and results in the strong coupling between the surface plasmons of Au NPs thereby enhancing the local electromagnetic field [42]. In the case of our samples, Au nanostructures are randomly distributed over the surface of the GO sheets. With improved structural design and enhanced SERS activity, these nano-hybrids could act as a new type of plasmonic nanomaterial with great potential for various applications including detection of trace chemicals, biomolecules, and in the detection of pathogens for food safety. Hence, the Au decorated GO could be excellent substrates for SERS due to their tunable plasmonic properties, more accessible surface area and chemical stability.

3.5. Au-GO Nano-Hybrids for the Catalytic Reduction of 4-nitrophenol

The catalytic reduction of 4-nitrophenol (4-NP) to 4-aminophenol (4-AP) by NaBH$_4$ was performed as a time-controlled reaction in order to establish the catalytic activity of the Au-GO nano-hybrids. For that, 2.2 mL of water was mixed with 30 µL of 4-NP at a 10 mM concentration in a quartz cuvette at room temperature. To this solution, when 200 µL of 0.1 M, the NaBH$_4$ solution was added. There was an instant change in the solution colour from light yellow to dark yellow. To this mixture, 0.5 mL of bare Au NPs and Au-GO nano-hybrid samples were added and the UV-Vis absorption spectra of the samples were recorded at certain specific intervals in the range of 200–800 nm at room temperature.

For a quantitative investigation of the catalytic activity of the Au NPs and Au decorated GO sheets (Au-GO nano-hybrids), the reduction of 4-nitrophenol (4-NP) to 4-aminophenol (4-AP) by NaBH$_4$ was chosen as a standard at room temperature [43]. The catalytic reduction reaction can easily be demonstrated by the UV-Vis absorption spectroscopy. The absorption peak of 4-NP is shifted to a higher wavelength (400 nm), upon the addition of NaBH$_4$ due to the formation of an intermediate compound 4-nitrophenolate. In the absence of the as-prepared catalyst (Au-GO nano-hybrids), the absorption peak of 4-nitrophenolate at 400 nm remains unaltered over several times thereby confirming that the reduction reaction did not take place. Figure 8a–d represents the time-dependent UV-Vis absorption spectra corresponding to the catalytic reduction of 4-nitrophenol to 4-aminophenol by the Au and Au-GO nano-hybrids at different laser energies, respectively. In the absence of the catalyst (Au-GO nano-hybrid), the 400 nm absorption peak (due to 4-nitrophenolate) remains constant. However, after the small addition of the Au-GO samples, the absorption peak intensity at 400 nm slowly decreased and a new peak centered at 300 nm emerged corresponding to the 4-aminophenol and the colour of the solution turned transparent from the previous dark yellow. Herein, the nano-hybrid acted as a catalyst promoting the electrons transfer from BH$_4{}^-$ to 4-NP, facilitating the reduction process.

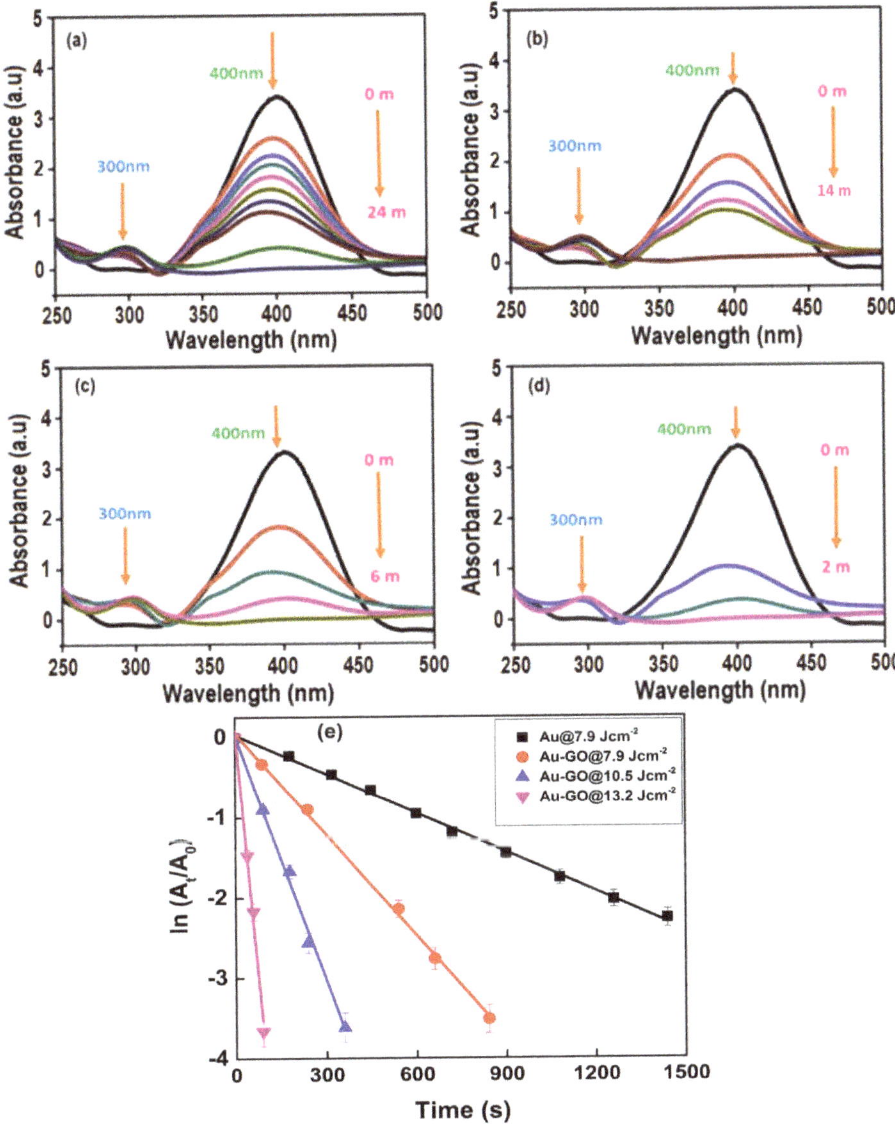

Figure 8. Time-resolved UV-Vis absorption spectra attributed to the catalytic reduction of 4-nitrophenol (4-NP) to 4-aminophenol (4-AP) with (**a**) Au NPs)@7.9 Jcm^{-2}; (**b**) Au-GO@7.9 Jcm^{-2}; (**c**) Au-GO@10.5 Jcm^{-2}; (**d**) Au-GO@13.2 Jcm^{-2}; (**e**) plot of ln(A_t/A_0) against the reaction time of various proportions of the Au-GO samples.

The time required for the completion of the reaction varied with the different Au-GO hybrids and Au itself. With the use of the Au-GO@13.2 Jcm^{-2} laser fluence, the reduction reaction was completed within 2 min as shown in Figure 8d and was found quicker than earlier reports. It should be noted that upon increasing the concentration of Au, the reaction time essential for the reduction was shortened. Additionally, the concentration of NaBH$_4$ extensively goes beyond that of 4-NP, the reduction rate was found to be independent of the NaBH$_4$ concentration. Therefore, the first order kinetics approximation may be suitable for calculating the apparent rate constants (k_{app}) [44,45]. The reduction rate constant can be calculated according to pseudo first-order kinetics,

$$\ln(C_t/C_o) = \ln(A_t/A_o) = k_{app}\, t \tag{2}$$

where k_{app} corresponds to the apparent rate constant, and A_t and A_o represents the absorption of 4-NP at time t and its initial time, respectively. Figure 8e represents the relationship between ln (A_t/A_o) and the reaction time t, which exactly match with the first order reaction kinetics. Each sample exhibits a good linear fit of $\ln(A_t/A_o)$ with the reaction time. The k_{app} values were obtained from the slope of the plot and the details are listed in Table 1. For Au-GO@13.2 Jcm^{-2} [k_{app} = 40.2 × 10^{-3} s^{-1}], Au-GO@10.5 Jcm^{-2} [k_{app} = 10.2 × 10^{-3} s^{-1}), Au-GO@7.9 Jcm^{-2} [k_{app} = 4.2 × 10^{-3} s^{-1}] and for Au@ 7.9 Jcm^{-2} [k_{app} = 1.6 × 10^{-3} s^{-1})]. From the results, it is evident that there is a significant increase in the k_{app} value of the nano-hybrids synthesised at higher laser fluences. The k_{app} value for the sample Au-GO@13.2 Jcm^{-2} is found to be 40.2 × 10^{-3} s^{-1}, which is significantly higher than the other two samples and AuNPs itself. Hence, the catalytic activity of the Au-GO nano-hybrids obtained at higher laser fluences was found to be greater than that of the bare Au NPs.

Table 1. Catalytic reduction time and k_{app} of AuNPs and Au-GO nano-hybrids.

Product	Au@7.9 Jcm^{-2}	Au-GO@7.9 Jcm^{-2}	Au-GO@10.5 Jcm^{-2}	Au-GO@13.2 Jcm^{-2}
Reduction time (in seconds)	1440	840	360	120
k_{app} (10^{-3} s^{-1})	1.6	4.2	10.2	40.2

Moreover, the catalytic reduction activity of Au-GO nano-hybrid with other materials has also been compared and the details are summarized in Table 2.

Table 2. Comparison of the apparent rate constants k_{app} of different catalysts for the reduction of 4-nitrophenol.

Samples	Apparent Rate Constant k_{app}, (10^{-3} s^{-1})	Reference
Au-GO@13.2 Jcm^{-2} via PLA	40.2	This work
Au-GO@10.5 Jcm^{-2} via PLA	10.2	This work
Au-GO@7.9 Jcm^{-2} via PLA	4.2	This work
Au@7.9 Jcm^{-2} via PLA	1.6	This work
rGO-AgAu bimetallic nanocomposite via green synthesis	1.4	[46]
Au/graphene hydrogel via chemical reduction	3.1	[47]
Ag-Pt NWs via chemical reduction	6.9	[48]
Ag/Au bimetallic nanostructures via chemical reduction	6.1	[49]
Ag–Au–C composite via chemical reduction	1.6	[50]

4. Conclusions

In this work, we establish the benefits of using a laser-assisted synthesis protocol for the production of highly pure, ligand-free gold nanoparticles that are directly anchored on to the surface of the graphene oxide matrix. The tuning of laser parameters influences the size and concentration of the AuNPs. There is not much considerable alteration in the properties of graphene oxide after laser ablation except for a modest increase in the GO reduction. Throughout the laser ablation process for the in situ generation of gold nanoparticles on the GO matrix, the inherent properties of GO were observed to be unaltered. Here, the graphene oxide is functioning as a capping agent that controls the size of the Au nanoparticles. We further demonstrated that the Au-GO hybrid material exhibits outstanding catalytic and SERS performance when compared with the metal nanoparticles alone. Both the synthesis protocol by laser ablation and the applications of the designed nano-hybrid provides motivations for future advancement in the materials science platform, especially in the catalysis and chemical sensing.

Author Contributions: N.K. and S.T. designed the study; P.N. performed the experiments, analyzed the data and wrote the article. A.K.N. and R.A contributed to the manuscript discussion and result analysis. All authors discussed the results and commented on the manuscript.

Funding: This research was funded by BRNS-DAE, Govt. of India, grant number 39/29/2015-BRNS/39009 and DST-FIST (Sr.No-417 dated 27-02-2017), Govt. of India.

Acknowledgments: The authors acknowledge the financial support from DST for funding through PURSE PII, FIST and Nano Mission programs and BRNS-DAE, Govt. of India, TEM-IIUCNN-MGU, SAIF-MGU and FESEM-CUSAT.

Conflicts of Interest: The authors proclaim no conflict of interest.

References

1. Zhou, W.; Gao, X.; Liu, D.; Chen, X. Gold Nanoparticles for In Vitro Diagnostics. *Chem. Rev.* **2015**, *115*, 10575–10636. [CrossRef] [PubMed]

2. Hu, L.; Liu, Y.J.; Xu, S.; Li, Z.; Guo, J.; Gao, S.; Lu, Z.; Si, H.; Jiang, S.; Wang, S. Facile and low-cost fabrication of Ag-Cu substrates via replacement reaction for highly sensitive SERS applications. *Chem. Phys. Lett.* **2017**, *667*, 351–356. [CrossRef]

3. Yue, G.; Li, S.; Li, D.; Liu, J.; Wang, Y.; Zhao, Y.; Wang, N.; Cui, Z.; Zhao, Y. Coral-like Au/TiO$_2$ Hollow Nanofibers with Through-Holes as a High-Efficient Catalyst through Mass Transfer Enhancement. *Langmuir* **2019**, *35*, 4843–4848. [CrossRef] [PubMed]

4. Lee, J.E.; Bera, S.; Choi, Y.S.; Lee, W.I. Size-dependent plasmonic effects of M and M@SiO 2 (M = Au or Ag) deposited on TiO 2 in photocatalytic oxidation reactions. *Appl. Catal. B Environ.* **2017**, *214*, 15–22. [CrossRef]

5. Huang, Z.; Tang, Y.; Li, J.; Yang, M.; Tan, L.; Zhang, X.; Gao, H.; Ma, Q.; Wang, G. Oriented immobilization of Au nanoparticles on C@P4VP core–shell microspheres and their catalytic performance. *New J. Chem.* **2015**, *39*, 2949–2955.

6. Bai, S.; Shen, X. Graphene–inorganic nanocomposites. *RSC Adv.* **2012**, *2*, 64–98. [CrossRef]

7. Wang, J.; Lu, X.; Huang, N.; Zhang, H.; Li, R.; Li, W. Temperature-responsive multifunctional switchable nanoreactors of poly (N-isopropylacrylamide)/SiO$_2$/lanthanide-polyoxometalates/Au: Controlled on/off catalytic and luminescent system. *Mater. Sci. Eng. B* **2017**, *224*, 1–8. [CrossRef]

8. Zhang, Y.; Liu, S.; Lu, W.; Wang, L.; Tian, J.; Sun, X. In situ green synthesis of Au nanostructures on graphene oxide and their application for catalytic reduction of 4-nitrophenol. *Catal. Sci. Technol.* **2011**, *1*, 1142. [CrossRef]

9. Fujigaya, T.; Kim, C.; Hamasaki, Y.; Nakashima, N. Growth and Deposition of Au Nanoclusters on Polymer-wrapped Graphene and Their Oxygen Reduction Activity. *Sci. Rep.* **2016**, *6*, 21314. [CrossRef]

10. Parlak, O.; Turner, A.P.F.; Tiwari, A. pH-induced on/off-switchable graphene bioelectronics. *J. Mater. Chem. B* **2015**, *3*, 7434–7439. [CrossRef]

11. Cao, A.; Liu, Z.; Chu, S.; Wu, M.; Ye, Z.; Cai, Z.; Chang, Y.; Wang, S.; Gong, Q.; Liu, Y. A facile one-step method to produce graphene–CdS quantum dot nanocomposites as promising optoelectronic materials. *Adv. Mater.* **2010**, *22*, 103–106. [CrossRef]

12. Cai, B.; Wang, S.; Huang, L.; Ning, Y.; Zhang, Z.; Zhang, G.J. Ultrasensitive Label-Free Detection of PNA–DNA Hybridization by Reduced Graphene Oxide Field-Effect Transistor Biosensor. *ACS Nano* **2014**, *8*, 2632–2638. [CrossRef]

13. Hughes, M.D.; Xu, Y.J.; Jenkins, P.; McMorn, P.; Landon, P.; Enache, D.I.; Carley, A.F.; Attard, G.A.; Hutchings, G.J.; King, F.; et al. Tunable gold catalysts for selective hydrocarbon oxidation under mild conditions. *Nature* **2005**, *437*, 1132–1135. [CrossRef]

14. Huang, J.; Akita, T.; Faye, J.; Fujitani, T.; Takei, T.; Haruta, M. Propene Epoxidation with Dioxygen Catalyzed by Gold Clusters. *Angew. Chem.* **2009**, *121*, 8002–8006. [CrossRef]

15. Turner, M.; Golovko, V.B.; Vaughan, O.P.H.; Abdulkin, P.; Berenguer-Murcia, A.; Tikhov, M.S.; Johnson, B.F.G.; Lambert, R.M. Selective oxidation with dioxygen by gold nanoparticle catalysts derived from 55-atom clusters. *Nature* **2008**, *454*, 981–983. [CrossRef]

16. Zhao, Y.; Zhu, Y. Graphene-based hybrid films for plasmonic sensing. *Nanoscale* **2015**, *7*, 14561–14576. [CrossRef]

17. Primo, A.; Esteve-Adell, I.; Blandez, J.F.; Dhakshinamoorthy, A.; Álvaro, M.; Candu, N.; Coman, S.M.; Pârvulescu, V.I.; Garcia, H. High catalytic activity of oriented 2.0.0 copper(I) oxide grown on graphene film. *Nat. Commun.* **2015**, *6*, 8561. [CrossRef]

18. Gao, F.; Wang, Q.; Gao, N.; Yang, Y.; Cai, F.; Yamane, M.; Gao, F.; Tanaka, H. Hydroxyapatite/chemically reduced graphene oxide composite: Environment-friendly synthesis and high-performance electrochemical sensing for hydrazine. *Biosens. Bioelectron.* **2017**, *97*, 238–245. [CrossRef]

19. Wen, F.; Dong, Y.; Feng, L.; Wang, S.; Zhang, S.; Zhang, X. Horseradish Peroxidase Functionalized Fluorescent Gold Nanoclusters for Hydrogen Peroxide Sensing. *Anal. Chem.* **2011**, *83*, 1193–1196. [CrossRef]

20. Ziefuß, A.R.; Reichenberger, S.; Rehbock, C.; Chakraborty, I.; Gharib, M.; Parak, W.J.; Barcikowski, S. Laser Fragmentation of Colloidal Gold Nanoparticles with High-Intensity Nanosecond Pulses is Driven by a Single-Step Fragmentation Mechanism with a Defined Educt Particle-Size Threshold. *J. Phys. Chem. C* **2018**, *122*, 22125–22136. [CrossRef]

21. Nancy, P.; James, J.; Valluvadasan, S.; Kumar, R.A.; Kalarikkal, N. Laser–plasma driven green synthesis of size controlled silver nanoparticles in ambient liquid. *Nano-Struct. Nano-Objects* **2018**, *16*, 337–346. [CrossRef]

22. James, J.; Nancy, P.; Vignaud, G.; Grohens, Y.; Thomas, S.; Kalarikkal, N. Laser assisted synthesis of graphene quantum dots for multifunctional applications. *AIP Conf. Proc.* **2019**, *2100*, 020131.

23. Nancy, P.; Nair, A.K.; James, J.; Kalarikkal, N. Green synthesis of graphene oxide/Ag nanocomposites via laser ablation in water for SERS applications. *AIP Conf. Proc.* **2019**, *2100*, 020025.

24. Haruta, M. Size- and support-dependency in the catalysis of gold. *Catal. Today* **1997**, *36*, 153–166. [CrossRef]

25. Tang, X.Z.; Cao, Z.; Zhang, H.B.; Liu, J.; Yu, Z.Z. Growth of silver nanocrystals on graphene by simultaneous reduction of graphene oxide and silver ions with a rapid and efficient one-step approach. *Chem. Commun.* **2011**, *47*, 3084. [CrossRef]

26. Lai, J.; Niu, W.; Luque, R.; Xu, G. Solvothermal synthesis of metal nanocrystals and their applications. *Nano Today* **2015**, *10*, 240–267. [CrossRef]

27. Wu, Z.S.; Yang, S.; Sun, Y.; Parvez, K.; Feng, X.; Müllen, K. 3D Nitrogen-Doped Graphene Aerogel-Supported Fe_3O_4 Nanoparticles as Efficient Electrocatalysts for the Oxygen Reduction Reaction. *J. Am. Chem. Soc.* **2012**, *134*, 9082–9085. [CrossRef]

28. Torres-Mendieta, R.; Ventura-Espinosa, D.; Sabater, S.; Lancis, J.; Mínguez-Vega, G.; Mata, J.A. In situ decoration of graphene sheets with gold nanoparticles synthetized by pulsed laser ablation in liquids. *Sci. Rep.* **2016**, *6*, 30478. [CrossRef]

29. Lau, M.; Haxhiaj, I.; Wagener, P.; Intartaglia, R.; Brandi, F.; Nakamura, J.; Barcikowski, S. Ligand-free gold atom clusters adsorbed on graphene nano sheets generated by oxidative laser fragmentation in water. *Chem. Phys. Lett.* **2014**, *610*, 256–260. [CrossRef]

30. Senyuk, B.; Behabtu, N.; Martinez, A.; Lee, T.; Tsentalovich, D.E.; Ceriotti, G.; Tour, J.M.; Pasquali, M.; Smalyukh, I.I. Three-dimensional patterning of solid microstructures through laser reduction of colloidal graphene oxide in liquid-crystalline dispersions. *Nat. Commun.* **2015**, *6*, 7157. [CrossRef]

31. Mafuné, F.; Kohno, J.Y.; Takeda, Y.; Kondow, T. Growth of Gold Clusters into Nanoparticles in a Solution Following Laser-Induced Fragmentation. *J. Phys. Chem. B* **2002**, *106*, 8555–8561. [CrossRef]

32. Giammanco, F.; Giorgetti, E.; Marsili, P.; Giusti, A. Experimental and Theoretical Analysis of Photofragmentation of Au Nanoparticles by Picosecond Laser Radiation. *J. Phys. Chem. C* **2010**, *114*, 3354–3363. [CrossRef]

33. Liu, G.; Jin, W.; Xu, N. Graphene-based membranes. *Chem. Soc. Rev.* **2015**, *44*, 5016–5030. [CrossRef]

34. Jain, P.K.; Huang, X.; El-Sayed, I.H.; El-Sayed, M.A. Review of Some Interesting Surface Plasmon Resonance-enhanced Properties of Noble Metal Nanoparticles and Their Applications to Biosystems. *Plasmonics* **2007**, *2*, 107–118. [CrossRef]

35. Huang, X.; El-Sayed, M.A. Gold nanoparticles: Optical properties and implementations in cancer diagnosis and photothermal therapy. *J. Adv. Res.* **2010**, *1*, 13–28. [CrossRef]

36. Yan, J.; Han, X.; He, J.; Kang, L.; Zhang, B.; Du, Y.; Zhao, H.; Dong, C.; Wang, H.L.; Xu, P. Highly Sensitive Surface-Enhanced Raman Spectroscopy (SERS) Platforms Based on Silver Nanostructures Fabricated on Polyaniline Membrane Surfaces. *ACS Appl. Mater. Interfaces* **2012**, *4*, 2752–2756. [CrossRef]

37. Michota, A.; Bukowska, J. Surface-enhanced Raman scattering (SERS) of 4-mercaptobenzoic acid on silver and gold substrates. *J. Raman Spectrosc.* **2003**, *34*, 21–25. [CrossRef]

38. Xu, X.; Ma, Y.; Du, Y.; Jiang, T.; Zhou, J.; Zhao, Z. Sensitive surface-enhanced Raman scattering activity of triple gold/silver/graphene oxide nanostructures decorated on gold nanowire arrays. *Mater. Res. Express* **2018**, *5*, 015013. [CrossRef]

39. Dalla Marta, S.; Novara, C.; Giorgis, F.; Bonifacio, A.; Sergo, V. Optimization and characterization of paper-made Surface Enhanced Raman Scattering (SERS) substrates with Au and Ag NPs for quantitative analysis. *Materials* **2017**, *10*, 1365. [CrossRef]

40. Cheng, C.; Yan, B.; Wong, S.M.; Li, X.; Zhou, W.; Yu, T.; Shen, Z.; Yu, H.; Fan, H.J. Fabrication and SERS Performance of Silver-Nanoparticle-Decorated Si/ZnO Nanotrees in Ordered Arrays. *ACS Appl. Mater. Interfaces* **2010**, *2*, 1824–1828. [CrossRef]

41. Wang, Z.; Wu, S.; Ciacchi, L.C.; Wei, G. Graphene-based nanoplatforms for surface-enhanced Raman scattering sensing. *Analyst* **2018**, *143*, 5074–5089. [CrossRef]

42. Jiang, X.; Sun, X.; Yin, D.; Li, X.; Yang, M.; Han, X.; Yang, L.; Zhao, B. Recyclable Au–TiO$_2$ nanocomposite SERS-active substrates contributed by synergistic charge-transfer effect. *Phys. Chem. Chem. Phys.* **2017**, *19*, 11212–11219. [CrossRef]

43. Zeng, J.; Zhang, Q.; Chen, J.; Xia, Y. A comparison study of the catalytic properties of Au-based nanocages, nanoboxes, and nanoparticles. *Nano Lett.* **2009**, *10*, 30–35. [CrossRef]

44. Sarkar, C.; Dolui, S.K. Synthesis of copper oxide/reduced graphene oxide nanocomposite and its enhanced catalytic activity towards reduction of 4-nitrophenol. *RSC Adv.* **2015**, *5*, 60763–60769. [CrossRef]

45. Wu, T.; Ma, J.; Wang, X.; Liu, Y.; Xu, H.; Gao, J.; Wang, W.; Liu, Y.; Yan, J. Graphene oxide supported Au–Ag alloy nanoparticles with different shapes and their high catalytic activities. *Nanotechnology* **2013**, *24*, 125301. [CrossRef]

46. Çıplak, Z.; Getiren, B.; Gökalp, C.; Yıldız, A.; Yıldız, N. Green synthesis of reduced graphene oxide-AgAu bimetallic nanocomposite: Catalytic performance. *Chem. Eng. Commun.* **2019**, 1–15. [CrossRef]

47. Liu, Y.; Li, J.; Liu, C.Y. Au/graphene hydrogel: Synthesis, characterization and its use for catalytic reduction of 4-nitrophenol. *J. Mater. Chem.* **2012**, *22*, 8426. [CrossRef]

48. Liu, R.; Guo, J.; Ma, G.; Jiang, P.; Zhang, D.; Li, D.; Chen, L.; Guo, Y.; Ge, G.L. Alloyed Crystalline Au-Ag Hollow Nanostructures with High Chemical Stability and Catalytic Performance. *ACS Appl. Mater. Interfaces* **2016**, *8*, 16833–16844. [CrossRef]

49. Huang, J.; Vongehr, S.; Tang, S.; Lu, H.; Shen, J.; Meng, X. Ag Dendrite-Based Au/Ag Bimetallic Nanostructures with Strongly Enhanced Catalytic Activity. *Langmuir* **2009**, *25*, 11890–11896. [CrossRef]

50. Tang, S.; Vongehr, S.; Meng, X. Controllable Incorporation of Ag and Ag–Au nanoparticles in carbon spheres for tunable optical and catalytic properties. *J. Mater. Chem.* **2010**, *20*, 5436. [CrossRef]

MDPI

St. Alban-Anlage 66

4052 Basel

Switzerland

Tel. +41 61 683 77 34

Fax +41 61 302 89 18

www.mdpi.com

Nanomaterials Editorial Office

E-mail: nanomaterials@mdpi.com

www.mdpi.com/journal/nanomaterials